World-Systems Evolution and Global Futures

Series editor

Christopher Chase-Dunn, University of California, Riverside, CA, USA

Barry K. Gills, Political and Economic Studies, University of Helsinki, Helsinki, Finland

Leonid E. Grinin, Higher School of Economics, National Research University, Moscow, Russia

Andrey V. Korotayev, Higher School of Economics, National Research University, Moscow, Russia

This series seeks to promote understanding of large-scale and long-term processes of social change, in particular the many facets and implications of globalization. It critically explores the factors that affect the historical formation and current evolution of social systems, on both the regional and global level. Processes and factors that are examined include economies, technologies, geopolitics, institutions, conflicts, demographic trends, climate change, global culture, social movements, global inequalities, etc.

Building on world-systems analysis, the series addresses topics such as globalization from historical and comparative perspectives, trends in global inequalities, core-periphery relations and the rise and fall of hegemonic core states, transnational institutions, and the long-term energy transition. This ambitious interdisciplinary and international series presents cutting-edge research by social scientists who study whole human systems and is relevant for all readers interested in systems approaches to the emerging world society, especially historians, political scientists, economists, sociologists, geographers and anthropologists.

More information about this series at http://www.springer.com/series/15714

Eugene N. Anderson

The East Asian World-System

Climate and Dynastic Change

Eugene N. Anderson
University of California
Riverside, CA, USA

ISSN 2522-0985　　　　　　　ISSN 2522-0993　(electronic)
World-Systems Evolution and Global Futures
ISBN 978-3-030-16869-8　　　ISBN 978-3-030-16870-4　(eBook)
https://doi.org/10.1007/978-3-030-16870-4

© Springer Nature Switzerland AG 2019
This work is subject to copyright. All rights are reserved by the Publisher, whether the whole or part of the material is concerned, specifically the rights of translation, reprinting, reuse of illustrations, recitation, broadcasting, reproduction on microfilms or in any other physical way, and transmission or information storage and retrieval, electronic adaptation, computer software, or by similar or dissimilar methodology now known or hereafter developed.
The use of general descriptive names, registered names, trademarks, service marks, etc. in this publication does not imply, even in the absence of a specific statement, that such names are exempt from the relevant protective laws and regulations and therefore free for general use.
The publisher, the authors and the editors are safe to assume that the advice and information in this book are believed to be true and accurate at the date of publication. Neither the publisher nor the authors or the editors give a warranty, express or implied, with respect to the material contained herein or for any errors or omissions that may have been made. The publisher remains neutral with regard to jurisdictional claims in published maps and institutional affiliations.

This Springer imprint is published by the registered company Springer Nature Switzerland AG.
The registered company address is: Gewerbestrasse 11, 6330 Cham, Switzerland

To the Silent Gardeners
"The tree of humanity forgets the labour of the silent gardeners who sheltered it from the cold, watered it in time of drought, shielded it against wild animals; but it preserves faithfully the names mercilessly cut into its bark." Heinrich Heine, 1833 (as quoted in Gross 1983:323)

Acknowledgements

All thanks to Paul Buell, Christopher Chase-Dunn, Peter Grimes, Hiroko Inoue, and Andrey Korotayev for their help at all stages and to Peter Turchin for invaluable discussion. Exchanges with James Beattie, Philippe Beaujard, Sing Chew, Nicola Di Cosmo, Patrick Manning, William Thompson, Ling Zhang, and many other friends and colleagues who provided many helpful suggestions are deeply acknowledged. At an earlier stage of my career, I worked with Philip Huang and sat at the feet of James Lee, Kenneth Pomeranz, Richard von Glahn, Bin Wong, Yan Yunxiang, and others in our California Sinological world, learning from them what I know of Chinese political economy in later Imperial centuries. To all these and many more, my deepest gratitude. Thanks also to my family, including Kit the black Lab, Kangal the cattle dog, and Sage the border collie, who did everything they possibly could to prevent this book from being written ("aw, c'mon, just one more ball toss...one more walk...one more garden tour...").

Contents

1 Theoretical Overview 1
   1.1   On History 1
   1.2   Climate and Culture 2
   1.3   The Mandate of Heaven 5
   1.4   East Asia's Climate Record 9
   1.5   Geography as Destiny 22
   1.6   Cycles 28
   1.7   Cause and Explanation 29
   1.8   Human Nature: Basics Needed for Understanding History 32
   1.9   Rationality 33
   1.10 Human Good and Evil 35
   1.11 Population Growth 36
   1.12 Growth and Non-growth 40
   1.13 Power 41
   1.14 States 43
   1.15 World-Systems 46
   1.16 Conclusion 50
   References 50

2 Cycles and Cycling 61
   2.1   Ibn Khaldun's Theory of Dynastic Cycles 61
   2.2   State and Family 65
   2.3   Ibn Khaldun on Timing and Dynamics 66
   2.4   Ibn Khaldun on Decline 67
   2.5   Generalizing the Features of the Theory 69
   2.6   Barbarians 70
   2.7   Collapse and War 72
   2.8   A Wider Theory of Cycles: Resilience 73
   2.9   Coupled Cycles? 75
   2.10 Turchin, Nefedov, and Longer Cycles 76
   References 81

3 Before Empire: State Formation in China and Proto-states Elsewhere ... 83
3.1 Language-Systems Before World-Systems ... 83
3.2 East Asia: General Introduction ... 86
3.3 Early China ... 87
3.4 The Rise of Philosophy ... 92
3.5 Summary ... 94
References ... 95

4 The Creation of Stable Dynastic Empires in East and Southeast Asia ... 99
4.1 Qin and Its Fall ... 99
4.2 The Rise of Han and the Formation of Chinese Society ... 101
4.3 The Fall of Former Han ... 105
4.4 China Disunited ... 108
4.5 Southeast Asia Enters History ... 111
4.6 Korean Civilization ... 113
4.7 Japan Appears in the Records ... 114
4.8 Verdict ... 115
References ... 116

5 High Empire: The Glory Days of Early Medieval Eastern Asia ... 119
5.1 The Development of the East Asian World-system ... 119
5.2 Events in China After the Mid-Sixth Century ... 122
5.3 High Tang ... 125
5.4 Tang's Decline and Fall ... 128
5.5 Meanwhile in the South ... 130
5.6 Japan, 550–950 ... 130
5.7 China Divided After Tang ... 133
5.8 Verdict ... 134
References ... 135

6 The Rise of Central Asia: Coastal Golden Ages Increasingly Threatened by Conquest Dynasties from the Deep Interior ... 139
6.1 Medieval Warmth ... 139
6.2 World-system Dynamics ... 140
6.3 Japanese Crisis ... 142
6.4 Southern Approaches ... 144
6.5 China's Warm Period ... 144
6.6 A Climatic Interruption ... 146
6.7 Social Changes ... 147
6.8 Northern Invaders and Neighbors ... 151
6.9 The Fall of Song ... 155
References ... 156

7	**The Mongol Conquests of China and Korea and Invasion of Japan**	**159**
	7.1 Climate: From Warm to Frigid	159
	7.2 The Mongol Rise	162
	7.3 Bubonic Plague Goes West, Not East	165
	7.4 Mongol Decline and Fall	167
	References	169
8	**Long-Lived Dynasties: Ming and Its Contemporaries**	**171**
	8.1 Climate and Environment	171
	8.2 The Rise of Ming	173
	8.3 The Final Stroke	178
	8.4 Loyalty After the Fall	180
	8.5 Disasters in and After Ming	180
	8.6 The East Asian World-System in Ming Times	181
	8.7 Korea Changes Dynasty	182
	8.8 Japan Collapses into War	184
	8.9 The Ryukyus	187
	8.10 Sidebar: Mazu the Empress of Heaven	188
	References	188
9	**The Early Modern Period in the East Asian World-System**	**193**
	9.1 Climate in the Early Modern Period	193
	9.2 The Rise of Qing	193
	9.3 Qing's Malthusian Stress	197
	9.4 Waning and Decline	198
	9.5 Vietnam Declines	200
	9.6 In the Northeast	201
	9.7 East Asia Fails to Modernize	206
	References	208
10	**Lessons: Factors Driving the Rise and Fall of Dynasties**	**211**
	10.1 Climate and Chinese Dynasties	211
	10.2 Korea and Japan	220
	10.3 Vietnam	221
	10.4 Central Asia	221
	10.5 Putting It Together: World-System Dynamics	222
	References	222
11	**Comparisons: Cycles and Empires in Agrarian Worlds**	**225**
	11.1 Cycles Rising	225
	11.2 Models and Traditions	226
	11.3 East Asian Agriculture as Relief	227
	11.4 Innovation and Progress: Golden Ages and Leaden Ages	230

	11.5 Cycles' End	231
	11.6 Fall and Collapse of Entire Regions or Civilizations	234
	References	239
12	**What East Asia's Dynamics Teach Us about Climate, Society, and Change in the Modern and Future World**	**243**
	References	248

Theoretical Overview

> *I read the history of man, age after age,*
> *And little find therein but treachery and slaughter.*
> *No pestilence, no fiend could inflict half the evil*
> *Or half the desolation that man brings on man.*
> Anonymous Arab, tr. Wendell 1967: 24

1.1 On History

Cleopatra's nose: Had it been shorter, the whole face of the earth would have been changed.
   Pascal (tr. Roger Ariew 2005: 6)

My father, a historian, frequently quoted Blaise Pascal's line to show how trivial the contingencies that affect history can be. Short noses were considered unlovely in Pascal's time. The philosopher was implying that if Cleopatra had been so endowed, Caesar and then Mark Antony might not have romanced her, the Roman Republic might have been saved from empire and corruption, and we might all be wearing togas today. Perhaps Pascal knew that the nose is the most important of the body for facial recognition and assessment, as recently established by psychologists (Kleiner 2009).

Autocratic regimes are particularly prone to such contingencies, because the lone individual at the top is so important. If, as often happened in Asian history, the emperor is a child, or dies without issue, or goes mad, the country is in danger. China had its own Cleopatras. If, in the eighth century, Imperial Courtesan Yang had been less attractive, the Tang Dynasty might have had less trouble; the romantic story goes that she was having an affair with An Lushan, the general who rebelled and almost destroyed the dynasty.

Indeed, explaining and predicting history is not easy. Not only crazy chance, but also the stubborn irrationality of human actors, the impersonal forces of climate and weather, and many other imponderables affect the dynamics of societies. This book explores some factors currently alleged to "explain" history. It is written with a

© Springer Nature Switzerland AG 2019
E. N. Anderson, *The East Asian World-System*, World-Systems Evolution and Global Futures, https://doi.org/10.1007/978-3-030-16870-4_1

skeptical mind. I neither believe on faith nor write off without examination such factors as climate change, military actions, economic decisions, or short noses. On the other hand, recent advances in databasing, modeling, and the philosophy of explanation make it much more reasonable to predict and explain—two *different* enterprises (Hofman et al. 2017).

I am not a historian nor a Sinologist or Asianist. I do not have the knowledge of Asian languages or history that would be necessary for an exhaustive work. I am a human ecologist, concerned with ways that environment and human society create themselves through interaction. Therefore, this is an extremely modest exploratory work—a long essay, not a finished piece. I am using a broad brush to characterize a great deal of history that should be more closely examined. I have few answers and many questions.

1.2 Climate and Culture

In this book, I am concerned with the great dynastic cycles that dominate Chinese, Korean, Japanese, and Vietnamese history. I compare (nonformalized) agent-based models of the rise and fall of dynasties—models that attribute such things to individual human decisions and actions—with the effects of climatic change.

Several recent historians, such as John Brooke (2014) and Bruce Campbell (2016), hold that climate change determines history, including East Asia's recurrent dynastic changes and related cultural dynamics (see Fig. 1.1). Historians of China such as Timothy Brook (2010) and Bret Hinsch (1988) also stress the role of climate. They point out that cold, dry periods stress China's agricultural system, often to the breaking point, while warm, moist periods help agriculture but may flood the south (see Fig. 1.2). Hinsch thus predicts that dynasties will rise with warmth and fall with dry cold. Japan and Korea are more buffered against climate, but Vietnam displays some of the same vulnerability as China, and historians have been quick to note the relationship of climatic change to Vietnamese history (Kiernan 2017) (Fig. 1.1).

My reading suggests a more nuanced picture, in which climate is one of the factors that make up the background of human decisions. Climate may make or break, but usually (if not always) it is part of a more complex story. Agricultural production in general, geography of empire, military preparedness and lack of it, epidemics (Lee 2018), and government policy debates and changes all affect the rise and fall of empires. Ideology and philosophy can be important. The spread of Confucian and Legalist statecraft from China to Korea, Japan, and Vietnam tracks the consolidation of government in those realms, and led to a stable, bureaucratic, conservative government that was hard to bring down but often hard to change. (Among similar works looking at a range of factors is Sunil Amrith's important new book *Unruly Waters,* 2018.) (Fig. 1.2)

Political crises within dynastic cycles—notably court intrigues that lead to blow-ups—are particularly important. At the beginning of a dynasty, they tend to make it stronger. At the middle, they weaken it. At the end, they bring it down.

1.2 Climate and Culture

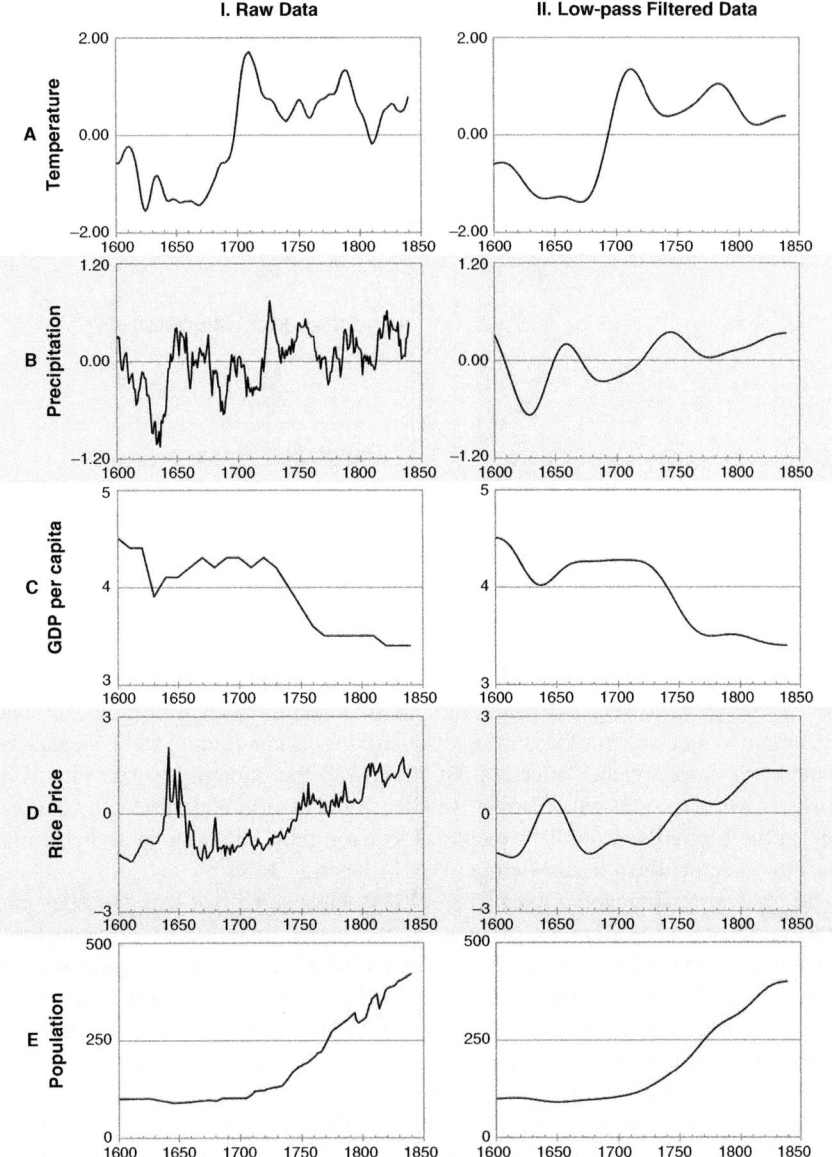

Fig. 1.1 Temperature, precipitation, and social measures in late imperial China. From Pei et al. 2016, *Environmental Research Letters* 11–064008, open access

All these various events work not by acting on their own, but by affecting human responses. History comes down in the final analysis to human decisions.

However, climate has a privileged place, being hard for anyone to ignore. Since all East Asian societies were based on intensive agriculture, or, in a few cases,

Fig. 1.2 Relations of climate change and agricultural migration. From Pei et al., *The Holocene*, 2018, by kind permission of Sage Publishers

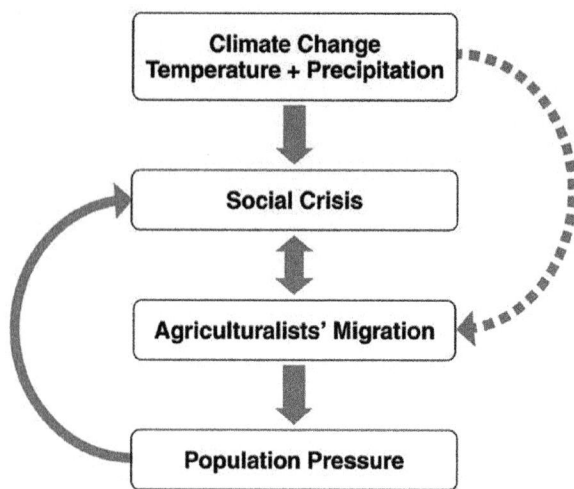

intensive stock rearing, climate is all-important. Tiny differences in rainfall or temperature could produce incalculable misery or stunning prosperity.

The natural tendency is to assume that the fall of a dynasty and the rise of a new one should come at a point when drastic climatic events have broken the power of the falling dynasty. One looks for extensive climatic traumas to cause rebellions and insurrections, while very extensive ones cause actual dynastic transitions. Climate can empower one side (perhaps through improving its conditions) while weakening another. Warmer weather after 200 BCE allowed the Xiongnu to rise rapidly to power in the far northwest. Warmer weather in the twelfth and thirteenth centuries helped the Mongols create their empire. Floods in the Taira stronghold helped the Minamoto defeat them in the Gempei War in Japan in 1185.

Several writers, notably Qiang Chen (2015), have suggested that dynasties (and other regimes worldwide) will rise in warm, wet periods and decline—or sometimes consolidate—in cold, dry ones. This is the hypothesis that will be evaluated in the present book. Warmer, wetter times should lead to more agricultural production and more stable weather regimes, enriching and stabilizing the collectivity. Colder, drier times devastate agriculture, especially in the north, the traditional centers of imperial power in China and Korea.

Wetter times could bring more flooding, and thus harm the state, as happened in the twelfth century (Zhang 2016). Hard times might "challenge" a dynasty, leading to successful or less successful responses. This idea does not predict the level of hard times or the nature of the response. Flooding in the twelfth and fourteenth centuries certainly qualifies; the state tried to respond, the first time unsuccessfully, the second time with temporary success. Climate Change X may or may not have Effect Y; we must look at the actual chain of events.

The counter-hypothesis is that dynastic changes happen because of internal dynamics endemic to dynastic systems. This counter-hypothesis can take many forms. The present book explores the specific theory of the fourteenth-century

Arab thinker Ibn Khaldun, especially as recently revised and updated by Peter Turchin. Other theories will be considered to a lesser extent.

Climatic models [notable recent attempts include Brooke (2014) and Chen (2015); see also Bello (2016)] have had some success at correlating rises and falls, but their actual *explanatory* power is limited by the fact that they correlate vast, worldwide, slow changes with sudden, highly specific events. They restore the importance of "Heaven" in the models. If indeed cold dry periods brought the great dynasties down (Chen 2015), Heaven, in the Chinese sense, was indeed a major actor (see also Li et al. 2017a). However, this is not clear. What is clear is that combining models of climate change with models of human decision-making are necessary to explain dynastic events. This book provides a skeleton outline of how this combination might explain changes in the East Asian world-system.

I test climate change against other models of dynastic circulation, notably the comprehensive theory developed by the Tunisian historian Ibn Khaldun in the fourteenth century and elaborated by Peter Turchin and collaborators in recent years (Turchin 2003, 2006, 2015, 2016; Turchin et al. 2013; Turchin and Nefedov 2009). I will suggest that climate is best understood as one of the influences—often an important one—on dynastic cycles, and can be included in Ibn Khaldun's theory. This book is not simply a test of the influence of climate; it is a humble and tentative step toward an actual understanding of dynastic cycles. Such an understanding must begin with Ibn Khaldun's theories.

1.3 The Mandate of Heaven

The Chinese and their neighbors in East Asia always wondered at the ways Earth and Humanity responded to Heaven. Those classic three realms of being interacted constantly. All too often, drought, heat, cold, floods, and pestilences were the result. Historians wondered how and why Heaven, *tian,* gave the "mandate", *ming,* that allowed a conqueror to establish a dynasty and lead it to greatness. The loss of this *tianming* meant dynastic collapse. Earth, *di,* ever passive, somehow kept producing enough food to support most people most of the time. Humans, *ren,* stood in the middle between heaven and earth, trying to balance everything.

Other East Asian societies adopted versions of this idea. Japan borrowed the idea (pronounced *temmei*): "Heaven, it was held, would grant its mandate…to rule the people to one whose virtue showed him to be worthy of Heaven's trust" (Masahide 1991: 405). Mongolia fused the concept with Persian and Turkic notions of heavenly selection of leaders.

In Chinese thought, humans are creatures of love and hate, meaning and alienation, ambition and laziness, overconfidence and miscalculation. Traditional Chinese historians were quite aware of these moods—perhaps more aware of them than are current Western theorists, who are sometimes blinded by the fiction of "rational choice." One key variable is perceived control or lack of it (Bandura 1982). Another is shifting loyalty: to self, group, empire. Another is group hate, less a force in China than in Europe (with its religious wars), but still not to be neglected.

The classic Chinese explanation for this cycling was that regimes had, or lost, the *tianming,* the "mandate of Heaven" (or "Heavenly fate"). *Ming* is divine favor or ruling, not a mere human "mandate." Heaven, in ancient China, was an active deity, and *ming* was the decision of Heaven about destinies. Arthur Waley (1996: 293–294) held that it was formed on the model of the king mandating feudal positions for his followers.

In the very earliest Chinese texts, Heaven directly intervened to punish people, often for neglecting sacrifices to the spirits of their ancestors. In the *Book of Songs,* Song 258 begins:

"Vast is that River of Stars,
   Shining and turning in the sky.
   The king cries out, 'But alas,
   What blame do you find with us?'" (translated by Joseph Allen, in Waley 1996: 270).

The poem continues, reporting that floods and droughts have ruined the food production. Hou Ji—Lord Millet, the deified grain-staple, the ultimate ancestor of the Zhou people—is not helping. The king must get divination, probably by heat-cracking shoulder blades and tortoise shells and interpreting the cracks, to find out what ancestor has been neglected. Providing too little ritual and sacrifice to a royal ancestor caused Heaven to punish the state.

This text shows how deeply the concept of heavenly punishment was embedded in Zhou Dynasty religion (Shun 1997: 15–21). In later thought, it became less literal, more understood as reward and punishment for royal performance of duties to the people. By Han, the founder of Chinese historiography, Sima Qian, wrote: "...we may see that the rulers at the beginning of each new dynasty never failed to conduct themselves with awe and reverence, but that their descendants little by little sank into indolence and vain pride" (Sima 1993, vol. 2: 6). Shortly after, a more detailed account of the process was provided in a famous essay by Han Dynasty historian Ban Biao (3 CE–54 CE). Many other writers followed (Loewe 1986: 735–736, 779–780).

Later, the term was taken as divine judgment on the effects of competent, assertive rule, gaining the mandate, vs. incompetent and failing rule, losing the mandate. More and more skeptics saw Heaven as simply a metaphor. Wang Chung (27–100 CE), a mercilessly skeptical Han philosopher who was a near-contemporary of Ban Biao, held that heaven was remote and indifferent and did not interfere with humans (Loewe 1986: 731; Wang 1907). Wang's skepticism, which comes close to the universal acid solvent, is extremely refreshing, and can be read with enormous profit today, but it made him so unpopular in his own time that his views had little effect. Much later, Sima Guang (1019–1086), the great Song Dynasty historian, "held that dynasties were merely the result of power struggles. Their survival and failure did not depend on Heaven's favor but on adherence to universal principles of administrative organization" (Bol 2003: 248). This view was reasonably typical of informed historical opinion at that time and later.

1.3 The Mandate of Heaven

The more religious often saw more truly mystical signs of imminent dynastic fall: omens and apparitions, strange animal births and plant productions, mysterious lights, fires, and comets. A white dog might appear with black spots spelling out characters for chaos, or an uncanny bird might land in an Imperial park. Dragons were seen in the skies (Brook 2010); an imaginative person can easily see dragons in the wildly rolling clouds of a South China typhoon. In Korea, a horse entered a Buddhist hall in the kingdom of Paekche around 655 and walked round and round as if performing the ritual clockwise circumambulation that Buddhist worshipers perform. The horse continued to walk until he fell dead of exhaustion. The kingdom fell soon after (Best 2007: 186, 476–477).

This omen-shopping has serious effects on records of climatic problems. Historic records tended to preserve accounts of disasters when those came just before a crash, and to preserve accounts of favorable omens when those came just before a glorious victory. For instance, to return to Paekche, In the 490s, just before a disaster, its annals reported a dismal series of climate disasters—floods, droughts, famines, snowfalls. "Remarkably...not one of these aberrant weather conditions seems to have afflicted the neighboring domains of Silla and Koguryŏ" (Best 2007: 109, citing data from the *Samguk Sagi*, the earliest Korean comprehensive history). Paekche was attending to small events, as omens of local disaster. The same events were too small to be noted by countries a few miles away that were not on the brink of collapse.

The same occurred in Japan: "From Emperor Heizei (r. 806–9) to Emperor Go-Sanjō (r. 1068–73), as many as 1686 natural occurrences of an ominous nature were officially recorded: 653 earthquakes, 134 fires, 89 instances of damage to the crops, 91 outbreaks of epidemics, 356 calamitous occurrences of a heavenly nature (volcanic eruptions, comets, eclipses, thunder in clear skies), and 367 ghostly events. In contradistinction, the same documents record a grand total of only 185 auspicious occurrences during the same period. The reign of Emperor Yōzei (876–84) seem to have been plagued by the largest number of calamities (...133...) and...the largest number of auspicious signs (...31).... This period corresponds exactly to the rise to ascendancy of the Fujiwara house...." (Grapard 1999: 551); "a formidable interest in astrology...an alarming increase of strange movements on the part of planets and constellations..." (Grapard 1999: 554).

Even quite early, many writers were aware that the fall of a state had less to do with Heaven than with administrative policy. A text attributed to Sun Wu of the sixth century BCE, but actually written perhaps around 300–400 BCE, says: "The farm systems of the Fan and the Zhongxing [are taxed] one-fifth. Their fields are narrow, but they support many active troops. With a tax of one-fifth, the government in rich. When the government is rich, many active troops are supported. Their lords are arrogant, their officials extravagant; they presume to have merit and constantly desire armed conflict. Thus I said they would [collapse] first" (Sun 2008: 261, tr. J.H. Huang, slightly amended). This is perhaps the first clear, concise statement of a view that lasted throughout the history of the Chinese empire—for the very good reason that it was correct. Often, the causal arrow went the other way: an ambitious ruler raised taxes to mount campaigns. But heavy taxation, large armies, and ambitious rulers made a deadly combination throughout Chinese history.

Unusual levels of flood and drought were seen as evidence that the mandate was being withdrawn—usually because Heaven was punishing the sovereign for ineptness or evil ways. Even plants growing poorly or withering unseasonably were bad omens (Métailié 2015: 279). Westerners encountered this in the nineteenth and twentieth centuries, and saw it as inscrutable Chinese mysticism. In fact, it shows the Chinese read environmental deterioration as proof that the government was not doing its job. This perception was correct, whatever spirit-beliefs may have accompanied it.

From early times, East Asians worked hard to control and manage their environment, especially its hydrology, and thus saw environmental failure as due to human failing as well as to fate. They knew what westerners are only coming to realize: "natural" disasters are largely man-made (Muir-Wood 2016). Floods were due to deforestation, poor levee management, and poor water control in general, as Chinese sources were clearly stating by Han times (Elvin 2004; Marks 1998, 2012). Droughts were less manageable, but alleviating human suffering through food aid and other means was always possible, and was well managed local from early times. By Qing, there was a whole elaborate famine-relief system that worked amazingly well (Zhang et al. 2016; Will 1990; Will and Wong 1991).

The imperial government had to respond to such challenges. If it responded effectively, it had recaptured or saved the mandate, but if it was ineffectual, intelligent people predicted the withdrawal of the mandate and the fall of the dynasty. Chinese historians often saw this in terms of morality: the morality of the emperor, the morality of the high courtiers and elites, and the morality of the general populace. One form was summarized by Hans Bielenstein: "Some authors...claimed that imperial lines inevitably degenerate. The founder possesses great ability and energy.... Later rulers, raised in a luxurious and intrigue-ridden court, and indulging in alcohol and sex, are likely to be weaklings.... Dynasty founders who received the Mandate of Heaven are depicted by the ancient historian[s] as men of extraordinary ability, head and shoulders above their contemporaries. Those unworthy of the mandate are described as libertines" (Bielenstein 1986: 259). The "bad last emperor" became a stereotypic figure, sometimes (as in the case of the last emperor of Sui) made the antihero of salacious stories. One recurrent moral story held that a failed government had come under too much control by women and eunuchs. These were *yin* forces, shady (*yin* originally meant the shady side of a hill) and potentially ill-omened. This belief spread to Korea. The unfortunate king whose loss of Paekche was presaged by the circumambulating horse was subjected to a whole flock of made-up stories about his salacious and dissolute ways (Best 2007: 185–186).

These theories of morality are dismissed today by many historians, especially western ones like Bielenstein. (Frederick Mote 1999 is a major exception.) This dismissal is perhaps too quick. We have seen in the twentieth and twenty-first centuries what morality can do for or to a nation. On the other hand, as this book will show, "last emperors" were less often libertines than they were helpless children, unable to control the forces of destruction.

The moral crises noted by Chinese historians in imperial times were failings of the sovereign, palace politics, over taxation, financial mismanagement, and military overextension. Modern scholars add structural weakness, autocracy and over centralization. These add to less directly moral problems: overpopulation, poor farming leading to soil erosion, and, of course, climate change. Currently popular is a version of the classic Chinese theory: poor management of environmental stresses brought down dynasties (e.g. Kidder et al. 2016).

1.4 East Asia's Climate Record

In general, East Asian climate history is straightforward. Most of the region is dominated by the East Asian monsoon, which brings wet, hot summers and very dry winters that are unbearably cold in the north and can be cold even in the south. (Hong Kong is the only place in the tropics that has recorded subfreezing temperatures on the sea coast.) Only the farthest west and northwest parts of Xinjiang are outside the monsoon cycle (Li et al. 2017c; Zhang et al. 2016).

The monsoon varies greatly in intensity. In East Asia, a weak monsoon means massive drought and famine. A strong one means plenty of rain and warm temperatures, and thus good farming prospects, but also flooding. Weakness and strength are governed by total solar input. It has been thought that insolation at high latitudes was critical, but recent analysis of north China's rainfall by Warren Beck et al. (2018) leads them to conclude that insolation at low latitudes is more important. If the sun is hot over the Asian tropics, warming the Pacific surface, it weakens the trade winds and the monsoon. It also strengthens the Tibetan high caused by cold temperatures and intense solar radiation on the Tibetan plateau, and that high further weakens the monsoon. Meanwhile, the southern Indian Ocean warms, driving a strong India monsoon, so China's loss is apt to be India's gain.

Weaker insolation of the northern hemisphere makes the northern Pacific surface cooler, the southern correspondingly relatively warm, and the monsoon stronger. India suffers. Global ice volume—itself determined partly by the monsoon, though largely by more cosmic forces—is correlated. Over tens of millennia, rainfall in north China has varied fourfold or more. Within history, variation at that level is not unknown in dry parts of China's northwest.

Warm climatic periods make the region wetter. They not only make the monsoon more powerful, but they shift northward the intertropical convergence zone, bringing the origin of the monsoon into the seas off Vietnam. The same move takes it farther from tropical Asia, a change that is associated with droughts in southern southeast Asia, including Cambodia and the Mekong Delta region. China derives almost all its rain from the south and southeast winds of the summer monsoon, though the cold, dry winter monsoon's north winds can pick up enough moisture over China to bring drizzling, chilling rain to the south. Warmer weather dries up central Asia and brings more drying conditions to Tibet, but, in East Asia, only western Xinjiang and the highlands of Tibet are much affected by this. The Walker circulation—the east-west airflow that brings El Niño and La Niña to the eastern Pacific (Timmermann et al. 2018)—has some slight influence on south China.

There is one important exception to the equation of weak monsoons with dry years: eastern central China can become wetter (Zhang et al. 2018a). This is because the same years are generally cold in the north Atlantic, forcing westerly winds south. A narrow, southward-displaced westerly belt picks up moisture (presumably from interior China or Siberia) and dumps it on central eastern China in late spring and early summer. This is the period of *meiyu*, "flowering-apricot rains," so named because they come when the fruits of the flowering-apricot (*Prunus mume,* often mistranslated "plum") ripen. Visitors to Japan, where the pronunciation is *baiyu*, will well remember the gentle, soft drizzle of that time of year. More important, this explains why cool, otherwise dry periods are sometimes associated with terrible flooding in central China and in Japan. Conversely, years of strong monsoon can be drier than average in that area, as the *meiyu* fail.

Temperatures vary from extreme cold in high Tibet and in winter in the north and west to the fully tropical heat of Vietnam. Rainfall varies from essentially zero in the Takla Makan Desert to over 200″ in the southwest mountains. In general, the west is cold and dry; the north is cool and moist; the central east (including central and east-central China, southern Korea, and most of Japan) is temperate to warm-temperate. Far southern Japan and most of southern China are subtropical. A small part of China—southern Yunnan and Guangdong—is tropical. Vietnam is tropical and rainy. Its low mountains reach to cooler heights but are not cold or icy like those of southwest China.

Good rain and pleasant temperatures in the central and southeast regions of all these countries make them ideal for cultivation. The same is true of the highlands of west China, wherever extensive level or terraced land allows.

In most of the region, almost all precipitation is in summer, but the extreme west of Xinjiang has thin winter precipitation, much of it as snow in the Altai ranges. The northeast part of Japan is exposed to winds that blow from Siberia over the seas, picking up moisture and dumping it on Honshu, often in the form of snow. The "snow country" there can be cut off from the outside world even today by enormous snowfalls; such isolation now lasts a day or two, but in historic times it lasted all winter. Hokkaido also has snowy winters and year-round rainfall. So do Korea and neighboring parts of northeast China. Vietnam is far enough into the wet tropics to have some year-round rain, but it too is dominated by the summer monsoon (Fig. 1.3).

The natural cover of the eastern half of East Asia was forest, grading from tropical to subarctic according to latitude and altitude. The western half is largely desert, with lush steppe only in the eastern parts of Mongolia and Inner Mongolia and in parts of northeast China. Long, linear oases follow rivers based on snow and ice melt from the mountains. Most of Tibet is a high-altitude desert, virtually uninhabitable until today. Very recently, mining and commerce have led to the emergence of small settlements. Lower parts of Tibet, however, have fertile soil, and humans adapted slowly to the altitude, eventually maintaining large populations of farmers and stock raisers as high as 14,000′ or more.

1.4 East Asia's Climate Record

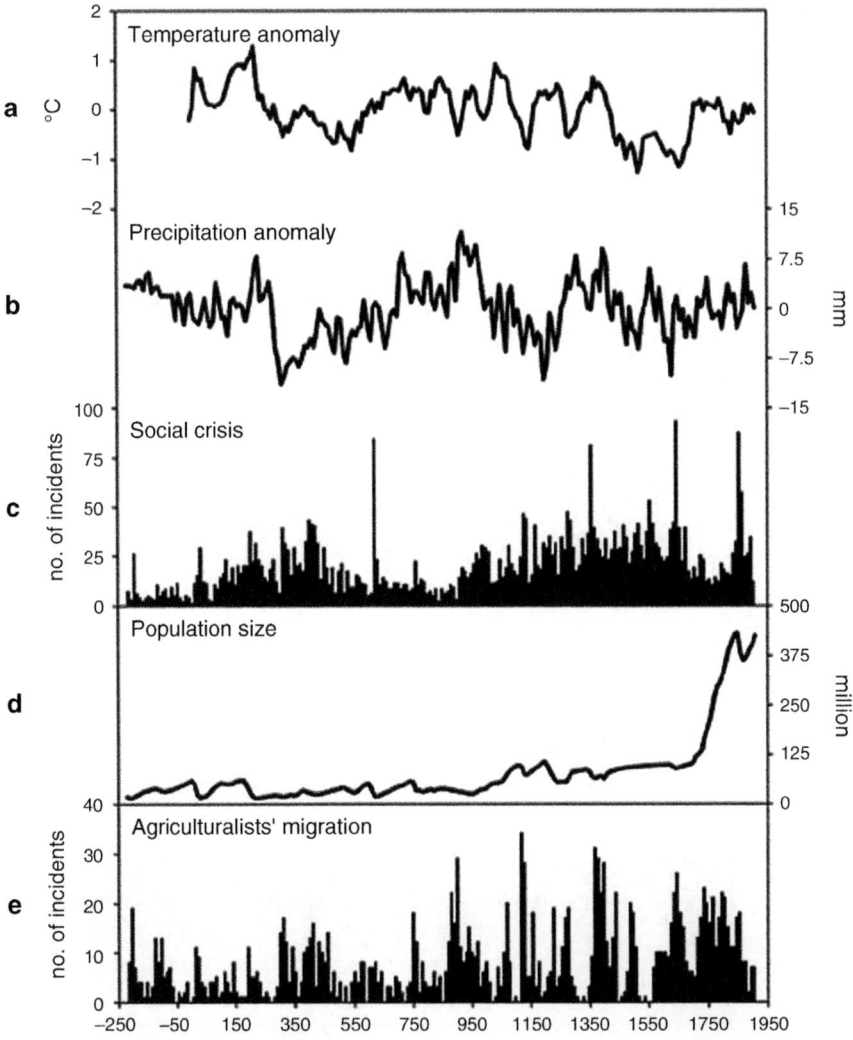

Fig. 1.3 Temperature and precipitation anomalies and agricultural migration. From Pei et al., *The Holocene*, 2018, by kind permission of Sage Publishers

China has the most diverse flora of any temperate land, with over 31,000 vascular plant species known; since over 100 are being discovered every year, the real total is probably over 36,000 (Lu and He 2017). Much of the diversity is in the southwest, where the Himalayas sweep down in vast ridges and canyons, with huge altitudinal gradients and many isolated sub-ranges. Vietnam and Japan are also stunningly biodiverse.

The last Pleistocene glaciation was one of the most intense in geologic time. It produced extremely cold temperatures throughout the mountain regions and far

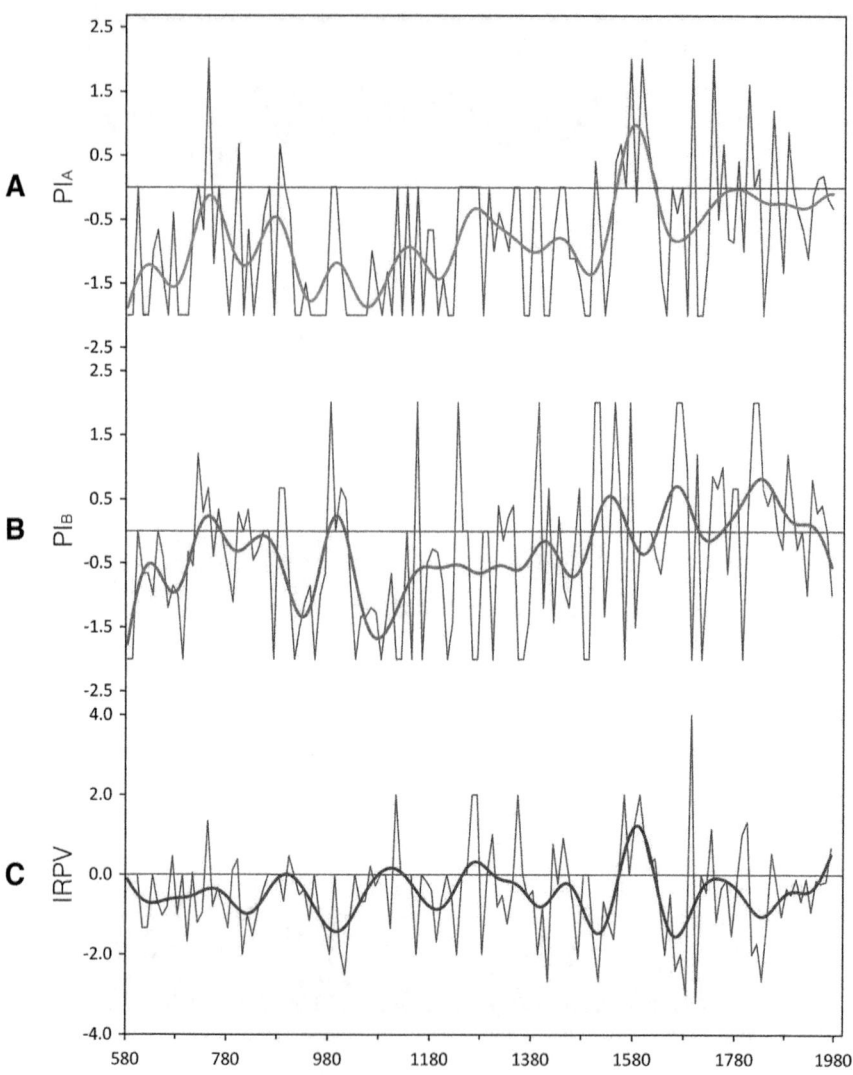

Fig. 1.4 Precipitation variability in northwest China, 580–1980 CE. PI, precipitation calculated (two ways) from drought and flood records; IRPV, precipitation variability calculated from that. Lee et al. 2015, *PLoS One,* open access per permission letter

north. Glaciation was extensive in the mountains, especially in Tibet, with maximum ice between 31,000 and 16,000 years ago. Japan's mountains were also glaciated. At the time, ice coverage was ten times that of today, and glaciers were much thicker. The ice disappeared rapidly after that. Temperature rather than lack of precipitation was the major cause (Hu et al. 2017). Temperatures did not rise as fast as today, but heating was rapid by geological standards (Fig. 1.4).

1.4 East Asia's Climate Record

The Ice Ages were caused by variations in the earth's orbit and tilt (when it tilts more on its axis, sunlight is more indirect, providing less heating), but some unexpected factors turn out to make a difference. The recent rise of Indonesia's spectacularly volcanic islands, for instance, had a role, because they poured forth an enormous amount of low-carbonate rock that could and did absorb CO2 in vast amounts, reducing global warming (Voosen 2018). Also, less surprisingly, the decline of vegetation during glaciations and its rebound during deglaciations provide a feedback effect, since vegetation also blots up large amounts of CO2.

One reason was that the potent greenhouse gas methane was released from wetlands and from melting permafrost and other sources. Tropical wetlands tend to dominate methane levels worldwide, and the drowning of much of the continental shelf by the South China Sea must have had an effect on this. Another major corollary of the end of the Pleistocene was a sharp rise in worldwide fires, caused by hotter, drier weather, but also at least in part by extinction of megafauna and consequent buildup of vegetation that the mammoths, mastodons, and other large animals previously ate (Bock et al. 2017). Wetland and fire dynamics obviously influenced many areas. North and central China had enormous wetlands, thanks to deposition of vast amounts of silt by winds and rivers in flat or hilly country with meandering rivers. The methane release must have been impressive; current emissions from rice fields can hardly be much bigger, and may be smaller, than the emissions from the swamps and floodplains that once covered much of north and central China.

The Ice Age ended around 13,000 years ago. A sudden, sharp reversal—the Younger Dryas interval—returned cold weather from about 12,900 to 11,700 years ago. It was ended by very rapid warming. The "Greenlandian" period of the Holocene is dated from this event to 8300 years ago. This was followed by a warm period sometimes known as the Altithermal, lasting from about 9000 to 5000 years ago; its peak at 8300 years ago ends the Greenlandian and initiates the Northgrippian period of the Holocene. This period was dry in much of inner Asia, but wet in north China, Korea, and Japan. The warm weather drove the monsoon farther north and made it more intense. Lake Dali in northern Inner Mongolia, now almost dry, was 60 meters deeper than today, showing that its basin received twice the precipitation it does now (Goldsmith et al. 2017). Slightly cooler weather 8200 years ago, and drier weather 6900 to 5900 years back, were registered in dry interior north China, but otherwise this period was optimal for that area (Li and Liu 2018). The cooling at 8200 was partly caused by the breaking of the Laurentian ice dam in eastern North America, releasing a vast glacial lake ancestral to the Great Lakes. So much cold, fresh water flowed into the Atlantic that the world was cooled for 150 years or more.

In what would become the Chinese heartland, around Luoyang near the middle Yellow River, forest gave way to steppe-meadow around 8800 years ago and progressively dried to steppe conditions by 7500 years ago; millet agriculture appeared around 8500 (Zhang et al. 2018c). The western end of China had a vegetation optimum, pursuant to substantial precipitation (Li et al. 2017c).

In the central Yangzi region, a wet period coincided with the very early Neolithic or pre-Neolithic, 8–9000 years ago. The climate dried, reaching a very dry period from 6500 to 3000 years ago. The pattern is quite different from northern China, reflecting more subtropical conditions and some influence from the Indian monsoon as well as from the South China Sea. The Mei-Yu Front, separating cool continental air from strongly monsoon-driven subtropical air, usually separates these regions (Zhu et al. 2017).

Arid Central Asia was kept dry in the Ice Age by the far-distant ice sheets of North America and stayed cold and dry even after those melted. The area did not warm and moisten until about 8000 years ago. A complex situation, caused by melting ice, northward or southward movements of warm water in the Atlantic, and wind circulation over western Eurasia led to dry conditions during very cold periods and warmer, wetter ones in the mid-Holocene, when moist air from as far as the Arabian Sea entered the region (Zhang et al. 2016, 2018d). It remained relatively dry even then, however, simply because of distance from oceans.

A sharp cooling followed, between 4500 and 4000 years ago. This was one of the most dramatic climate changes since the Ice Age. It was so important in climate history that it has recently been made the endpoint of the Northgrippian and the starting mark of a new geological period, the Meghalayan, 4200 years ago to present. However, the extent of the sharp drying and cooling event is debatable, and therefore the existence of a separate Meghalayan period is debated (Voosen 2018). What is certain is that the cooling and drying event was very widespread in the Northern Hemisphere, drastically affecting East Asia and much of North America. It was pronounced both in monsoon-dominated areas and in the Atlantic-dominated area of far northwest Xinjiang, where a dramatic peak in conifers and crash in *Artemisia* (sagebrush, wormwood) lasted from 4500 to 3500 years ago, signaling cool but dry times (Zhang et al. 2018d).

Archaeology reveals obvious effects on cultures of the time. This cooling hit the monsoon hard. It dried up north China, devastated early Neolithic cultures in the dry north and hit them hard throughout the East Asian region, and probably contributed to the rise of more complex and more southerly-centered Neolithic cultures such as Longshan in the southern China plain and the Yangzi valley (Goldsmith et al. 2017; Gao et al. 2018; see also Barnes 2015). Developing a more complex, hierarchic society with clear lines of control is a frequently-noted response to worsening conditions, though causal links have never been well worked out. A locally very wet period, with extensive flooding, happened around 4000–3600 years ago (Liu et al. 2017). It seems to have been particularly a problem in the middle Yellow River area.

The upper Yellow River basin participated in the cooling and drying after 4000 years ago, with a particularly sharp decline from 3800 to 3400 BP, impacting the neolithic Qijia Culture, which was replaced at 3600 by the less agricultural and more pastoral Kayue culture (Zhao et al. 2018). Central Asia was quite dry 2500–3500 years ago, with devastating effects on cultures there—largely west of what is now China, however.

1.4 East Asia's Climate Record

In at least one lake in south China, the early Holocene was very wet, due to strong summer monsoon, and cooling did not set in there till 3.4 thousand years ago (Li et al. 2018). This may indicate advantageous conditions in south China. Paradoxically, that may have discouraged the development of more complex society. In the upper Yangzi area, the warm wet optimum fell at 8 to 6 thousand years ago. Conditions then became dry until 3000 years ago, following which they became wet again, with fluctuating temperatures and rainfall over the last thousand years. This is quite different from other records in China and may owe something to the novel method of assessment: measuring fatty acids fixed by bacteria in a stalagmite.

Far south China, including central and southern Yunnan (Tan et al. 2018) and Hainan Island, as well as Vietnam, displays a different pattern. Here, the northward shift of rain in warm periods leads to drier times, as the Intertropical Convergence Zone moves north, pushing the monsoon with it. Cooler periods were correspondingly wetter. In Yunnan, the cool period from 310 to 570 was wet, the Medieval Warm Period (especially 1160–1245) and modern times were and are dry (Tan et al. 2018).

Tibet, affected by the Indian monsoon as much or more than by the East Asian, warmed with the rest of the world after 10,000 years ago, filling the large lakes of the Tibetan Plateau to their highest levels around 8000–9000 years ago, as the summer monsoon strengthened and brought more rain. Temperature, not rainfall, limited and still limits Tibet's vegetation (Li et al. 2017c). Some lakes shrank, others continued full, until the major cooling and drying trend around 4200–4000 years ago (Shi et al. 2017). By 3800, grain agriculture was sharply trimmed back, with many higher sites abandoned (D'Alpoim Guedes and Bocinsky 2018). Millets will not grow in the full harsh cold of the Tibetan Plateau, and their rather tentative extension by Majiayao cultivators into the east Tibetan valleys was cut short by this cooling period. Hunting and gathering persisted. Tibet had to wait for the coming of wheat and barley, around 3750 years ago, to get a successful farming livelihood. These crops were adopted in the north, but did not reach the south until later, limiting its success (D'Alpoim Guedes 2018; D'Alpoim Guedes and Bocinsky 2018).

Warm, wet conditions continued in Japan, which had a lush forest grading from subtropical in the south to cool-temperate in the north. The entire period from the Pleistocene to the coming of rice agriculture is known as the Jomon. Jomon economy was based on nuts and other natural crops, though some horticulture took place in at least the later millennia. The land provided so many acorns, chestnuts, hazelnuts, and other high-energy, high-nutrient wild crops, and the sea produced so many fish and shellfish, that agriculture did not seem to pay well. Vast shellmounds accumulated in favorable settings.

A significant trend over the last 4000 years has been a steady weakening of the summer monsoon and strengthening of the winter monsoon, overlain by a rather loose cycling with roughly 700–800-year periodicity. This tracks solar energy over the Northern Hemisphere. The cycles correlate with the Atlantic meridional overturning circulation (Kang et al. 2018) and wider worldwide changes. The gradual change in the monsoons has been too weak to show marked effects on history, but the cycles, described below, are related to the relatively warm and moist times in the glory days of Han, late Tang, and Song and the difficult cold periods in the Sui, Ming, and Qing dynasties, with more moderate climate in Zhou, most of Tang, and late Qing.

Geologists prefer to talk of "years ago" or "BP" (Before Present), but we now reach historic time, and it is convenient to shift to "BCE" reckoning. North China and Inner Mongolia were dry and cool from 2500 BCE onward until perhaps Han times, with some amelioration around 900–500 BCE (Tian et al. 2017a). A pleasant, warm, moist period, known in the west as the Roman optimum, helped both the Roman Republic and Roman Empire and the Chinese and Korean states between 200 BCE and 200 CE. (On this and later details, I follow Brooke 2014; Kang et al. 2018; Kidder et al. 2016; Lieberman 2003, 2009; Lin et al. 2016; Pei et al. 2016a, b, 2018; Pei and Zhang 2014; Wei et al. 2015, 2016; Yin et al. 2016; Zhang et al. 2007, 2008, 2011, 2014, 2016; Zhao et al. 2018.) Particularly useful is the work of David Zhang, Qing Pei, and Harry Lee at Hong Kong University. (I cite only a few of their many papers.) The rather sudden end of this optimum just after 200 CE is associated with the fall of Han in China and the onset of decline in the Roman Empire; nor is this the only case where cooling impacted both China and the Mediterranean (Harper 2017; William Thompson, email of March 3, 2018).

This gave way to cooler conditions, especially after 260 (D'Alpoim Guedes and Bocinsky 2018). As in Tibet earlier, North China had to turn more toward wheat and barley, leading to further spread of flour milling (the rotary mill had been introduced in Han) and a change of diet. Pastoralism increased, and central Asia turned more and more away from millet and toward the more cold-tolerant crops. The peak was a sharp cold and dry period from 536 to 650, the low point of the hypothesized 700–800-year cycle beginning in Zhou times. This sharp cooling followed on volcanic eruptions. The period from 536 to the 550 s may have been the coldest since the Younger Dryas, outdoing even the Little Ice Age (Harper 2017: 218).

After 650 the climate modified to conditions much like today's, or a bit more warm and wet, giving the Tang and Silla dynasties and the Heian period of Japan a golden opportunity to flourish. A weak monsoon seems to have occurred in 910–930 (Zhang et al. 2008). Otherwise, warming after 800 was visible. One record of warming in the 600 s, and of climate since, is the Japanese court record of when the court went to enjoy viewing the cherry blossoms. The peak of bloom tracks the first warmish days in spring and reflect the shifts in climate. Alas, duller fare—tree rings, cave speleothem records, and sea sediments—have replaced the court record, being more accurate. Still the cherry blossom record deserves note as the oldest accurate record to show climatic change. (See the great classic *Times of Feast, Times of Famine* by Emmanuel Le Roy Ladurie 1971.)

A major drought struck central Asia in 783–850, contributing to the fall of the Uighur empire in 840 and to the reduction of Silk Road trade. It probably weakened the Tang dynasty. In 950–1300 came the Medieval Warm Period, also known as the Medieval Climatic Anomaly, at the peak of another 700–800-year cycle. It brought warmer and wetter conditions to all the region except the far west, but with sharp fluctuations, especially in the eleventh and twelfth centuries, when sudden returns to more average (i.e. colder) conditions hit China hard. The early twelfth century seems to have been especially cool, the late twelfth particularly warm. Sharply cooler periods (as recorded in China Sea sediments) occurred in 640–670, 1030–1080, and 1260–1280, when major volcanic eruptions blocked some of the sunlight (Kong et al. 2017). South China endured very wet conditions (compared to earlier times) in the Medieval Warm Period and again in the last 600 years. Large floods occurred there, driven by El Niño conditions and solar fluctuations (Zhu et al. 2017).

Vietnam prospered weather conditions were cool or warm, with its water-based, rice-based economy, but drought could be devastating (Kiernan 2017: 158). Presumably south China did the same. The rest of China did better in the warmer spells, with good harvests widely reported. However, these periods were also wetter, and more rain also meant more flooding along major rivers.

Western Central Asia gets its rain mainly from the Atlantic. The Medieval Warm Period was often wet in some areas, tracking Atlantic storm and evaporation action; the Little Ice Age was cold and dry, as elsewhere. The mountains, especially east of center, reflect a more monsoon-influenced climate, but northwestern Central Asian mountains were dry in the Medieval Warm Period (Fohlmeister et al. 2017). Cool, dry conditions are shown by expansion of conifers in the northwestern Xinjiang Altai (Zhang et al. 2018d), but *Artemisia* did not decline at this time as it had earlier, showing that moisture was not so reduced. Since the MWP, the far northwestern Xinjiang Altai has seen a striking and fairly steady increase in grass and deciduous trees and a decline in conifers. Zhang et al. (2018d) interpret this as evidence for warm dry conditions but I suspect it to be heavily influenced by human manipulation of the landscape. Cutting conifers for fuel and houses, and burning for pasture, would explain this pattern better than a climatic influence rendered dubious by the fact that the Little Ice Age was universal worldwide for most of the time.

The Tianshan Mountains and the more easterly Qiling Mountains are on a climatic border, where the very last of the Atlantic moisture is lost. They show some erratic cycles compared to other parts of what is now China. The Taklamakan Desert is more or less the tripoint where the Atlantic moisture, East Asian monsoon, and Indian monsoon all fail. It is normally rainless. The little rain it gets may be from any of the three sources. In the early Holocene it was moister, probably from Indian and East Asian rainfall.

The boundary between Atlantic and monsoonal weather systems remains somewhat unclear. It is normally taken to be in central to eastern Mongolia and northwest Xinjiang, but Zhang et al. (2018d) think it was farther east, at least in earlier millennia.

A record from Lake Tian'E of the Qiling, just south of Dunhuang, shows wet periods from 1270 BCE to 400 CE, 1200–1350 CE, and after 1600, with dry periods between (Zhang et al. 2018b). Rouran declined after 220 and was abandoned by 645. Dunhuang was deserted twice, in 1372 and 1524. Other cities were abandoned later. The loose correlation with other dry times is evident.

After 1300, the Little Ice Age slowly came on, at the trough of a 700–800-year cycle, producing extremely cold, dry conditions, especially at certain periods in the 1400s, 1600s, and 1700s (Lee et al. 2016). Even the tropics suffered, not from cold but from locally sharp reduction of rain (e.g. Kiernan 2017: 177 for Vietnam). A dramatic cold peak occurred in 1420–1450, once again due to volcanic eruption (Kong et al. 2017). This ameliorated in the later fifteenth century (Kiernan 2017: 198), with another cold peak in the early seventeenth century. However, in Hainan and elsewhere in the south, the Little Ice Age brought *more* rain, because of southward compression of the Intertropical Convergence Zone where tropical air meets temperate air, and related activity of the Walker circulation, which brings warmer, moister air close to land in the west Pacific (Zhang et al. 2017). The Medieval Warm Period was correspondingly dry, bringing disastrous droughts to southern Southeast Asia.

In general, though, the 1500s were a period of high variability; the period from 1560 to 1660 was again brutally cold and dry. Drought ravaged China and Vietnam (Kiernan 2017: 223). Worldwide, climate was terribly cold, often stormy, and prone to be even drier in dry regions while often flood cold wet ones. This was followed by some amelioration in 1660–1800, but times were still not pleasant. The Zhang and Lee group has pointed out that the cold phase corresponds to a particularly horrific period of war, food shortage, high prices, and disease in Europe, followed by improvement tracking the warming after 1660 (Zhang et al. 2011). They remain climate determinists, but do not explain why the Manchus could pick the coldest period of the coldest phase to invade and conquer China from the frozen north, and then run it firmly and successfully for the rest of the LIA. Geoffrey Parker's book *Global Crisis* (2013) tells a long and harrowing story of the catastrophes that followed in the seventeenth century, from Europe and North America to China and Japan. He blames climate for a large share of the horrors of that bloody century. Indeed, one must grant climate a share in the ruin of the Ming Dynasty and the doldrums of Central Asia in that era. On the other hand, Japan flourished, and so did icy Korea, to say nothing of the Manchus. Climate is not destiny (Di Cosmo 2018; Di Cosmo et al. 2017).

The Little Ice Age was followed by a slow warming after 1800 or 1850. This gave way after 1900 to more steady warming as human-released greenhouse gases added themselves to natural warming and eventually took over the major warming role. (Alleged human-caused warming by rice agriculture in dynastic times is not demonstrated. Among other things, the allegers forget that the rice, at first, largely grew in marshes and wetlands that had always released methane.)

Southeast Tibet was rather different, with a sharply cool and dry period from 400 BCE to 700 CE, and little signature of the Medieval Warm Period or Little Ice Age (Bird et al. 2017). It was oddly cooler after 1700 also. Global warming appeared only after 1970 (Li et al. 2017b). Since the Little Ice Age, glaciers in Tibet have shrunk by approximately 1/3 (Qiao and Yi 2017).

A recent summary of the climate of Yelang Cave, Guizhou, can really stand for much of East Asia: "very cold and dry climates with poor vegetation cover during the Last Glacial Maximum [15,000–18,000 years ago]; cold and dry climate with poor vegetation during Younger Dryas; warming and wet climates with the best vegetation (mainly C3 plants) coverage during 9–11.5 ka BP corresponding to a maximum insolation; drying climate with decreasing vegetation intensity from 9 ka

1.4 East Asia's Climate Record

to 8 ka; relatively good vegetation cover during 5–6 ka; the summer monsoon strength decreased from 5 ka to 1.5 ka BP, but increased during the Medieval Warm Period to produce wet climates and abundant vegetation; and strongly decadal variations in moisture budget and strong deforestation due to human impact during the past 600 years" (Zhao et al. 2017: 102). Plants with the C3 carbon metabolic pathway—the vast majority of plants—are adapted to moderate conditions; some plants, mostly grasses, are C4, more tolerant of and faster growing under hot conditions. This is highly relevant to Asian history because the millets that were staple foods in early North China, Japan, and Korea are C4 plants, as is the maize that has now replaced them. These grow very well in north China's hot, rather dry summers. Most other Chinese staples, including rice, are C3.

Working largely with Ming and Qing data, Tian et al. (2017b) have shown that cool temperatures exacerbate problems of drought, locusts, and related problems, leading to famine and epidemic diseases becoming more frequent during such times. The north and west are especially prone to cool and dry extremes. Floods correlate with epidemics, so wet periods have their own problems, especially in the south, where too much water is more often a problem than too little.

Western Eurasia, and indeed the whole world, had rather similar fluctuations. The British Isles, for instance, hit a rough period from 3500 to 3000 BCE, another from 1000 to 400, a warm period during the height of the glory of Rome, and then the disasters of the 540–650 period and later the Little Ice Age, both of times which exacerbated the bubonic plague sieges that swept Europe (Bevan et al. 2017; Brooke 2014; Campbell 2016; Harper 2017; Parker 2013). However, the period around 2000 BCE, when China was hit with drier conditions, was good for Britain; warmer, drier times caused wheat to flourish, and produced a population peak not regained for many centuries.

Otherwise, minor to substantial fluctuations in the record appear, but are largely in the category of "weather" rather than "climate." They have lower amplitude and last much less long than the above events. Even brief stressful periods—especially cold, dry ones—may be associated with wars.

The major confounder in studying the effect of climate change on dynastic cycling is the point made above: China exacerbated or even created its own problems. Walter Mallory's classic study *China: Land of Famine* (1926) stressed the role of deforestation, erosion, badly managed river dykes, wetlands drainage, and other environmental ills on China's horrific history of droughts and floods. Recent studies have gone on to confirm this (Cressey 1955; Elvin 2004; Marks 1998, 2012). Leading economic historian Shiba Yoshinobu, for instance, has frequently made the point for the Song Dynasty (McDermott and Yoshinobu 2015). Song engaged in massive deforestation for iron smelting, ceramics making, and printing (pines were burned for soot to make ink). Enormous, uncontrollable floods devastated the country and threatened the dynasty (Zhang 2016).

The same could be said, *mutatis mutandis*, for the other parts of the region. However, Japan, Korea, and Vietnam, as well as southeast China, invested from early times in extremely intensive irrigated rice agriculture, based on massive modification of landscapes by dykes, terraces, canals, leveling of paddies, extending land by polders (called "sand fields"—*shatian*—in China), reservoirs, and catchment basins. This "landesque capital," as geographers call it, buffered the environment against change rather than making it more vulnerable. Deforestation in Korea and most of China canceled some of the benefits, but Japan and Vietnam managed their forests better. In Japan, the Tokugawa shoguns in the seventeenth-century realized what deforestation was doing to water cycles, and put a sudden stop to it (Totman 1989). Japan is two-thirds forested today, a level surpassed only by Finland (Von Verschuer 2016). China, especially the old core provinces of the north, remained the most prone to disasters related to human activity (Fig. 1.5).

The post-Pleistocene climate changes summarized above were not enormous. They were nothing like what is expected for the near future with human-caused global warming. The changes in temperature were one or 2 °C. Exact numbers are hard to find, but studies of the better-known Mediterranean reveal that historic changes were about 1.5 °C at most (Guiot and Cramer 2016). This means that the 2–4 °C expected in the next hundred years, and the far greater warming farther on, will have devastating effects worldwide.

Overall, the areas most exposed to disasters by climate and weather changes were north China and the Central Asian deserts and steppes. A center of recurrent disaster was the inner North China Plain and the westward mountains and steppes of Inner Mongolia, Shanxi, and Shaanxi. Here the monsoon reached its northwest limit. The smallest expansion caused flooding, though also lush crops. The smallest contraction caused utterly devastating droughts. Population expanded during moist cycles, contracted by the harshest of Malthusian mechanisms in dry ones. It is no surprise that this was always an area of instability, conflict, regime change, and sudden changes in mortality.

China's climate and plant cover are more stable in the south and in the far west (Chinese Central Asia). The vegetation is a rather young one, averaging Miocene in evolutionary origins, but with a large number of very ancient taxa, especially in the southeast (Lu et al. 2018). Some downright relics of the age of dinosaurs, such as the ginkgo and *Metasequoia*, survive only in China. These are balanced by rapid evolution of herbaceous plants in the east and the mountainous west and far northwest. The result is that China has by far the most diverse flora of any temperate country.

Japan is buffered from the changes affecting East Asia. All Japan is quite wet, with reliable year-round rainfall. The country ranges from subtropical in the southern islands to subarctic in the high mountains of northern Honshu and in Hokkaido, but most is warm temperate. Warmer times allow expansion north, cooler times are met by contraction to the south, and Japan's centers of power have always been in the temperate, well-watered middle.

Korea is less well protected, but it too has fairly reliable rainfall and a range of temperatures; more northerly than Japan, it is largely a country of cold winters. Warm periods were notably beneficial to the northern part (now North Korea). Droughts that afflicted north China were apt to afflict Korea as well.

Vietnam is always warm and rainy, but warmer periods could be devastatingly dry, as they were in Cambodia, and, farther afield, in southern Mexico in the

1.4 East Asia's Climate Record

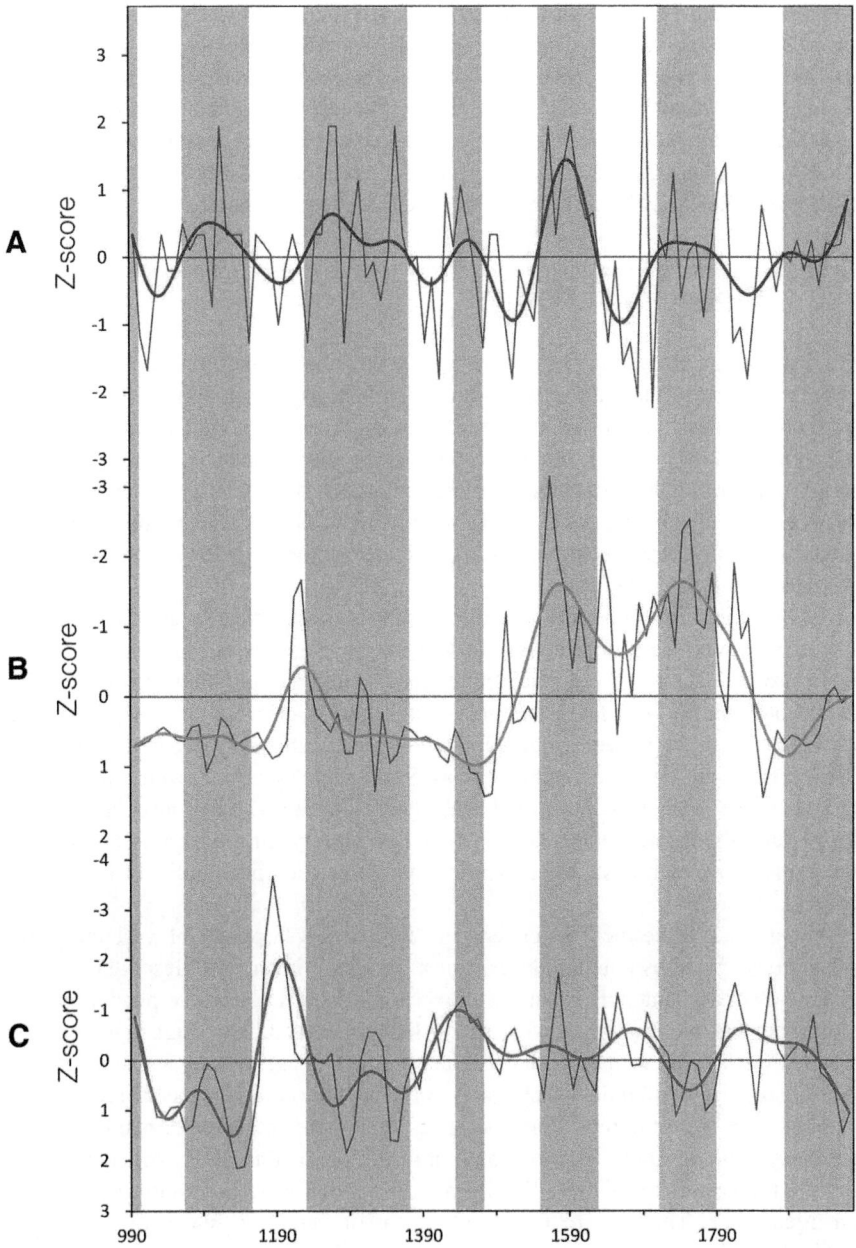

Fig. 1.5 Comparison of intraregional precipitation variability as above with other northwestern China indicators: sediments from Sugan Lake, Qaidam Basin, and cave records from Jiuxian Cave, Shaanxi. Lee et al. 2015, *PLoS One,* open access per permission letter

Medieval Warm Period. The Intertropical Convergence Zone shifts north, moving the rain belt north. China becomes wetter, but Vietnam, especially the southern part, can be drier. Conversely, cool periods are drier everywhere, and can move the Convergence Zone far south, again leaving Vietnam to potentially dry weather. Vietnam is thus exposed to variation to a degree rather surprising for a wet tropical country. Southeast Asia's climate vicissitudes correlate broadly with other regions (Lieberman 2003, 2009), but sometimes the direction of change is very different.

1.5 Geography as Destiny

Geography is not destiny, but some geographic facts condition human behavior closely enough to have quite specific effects. The Near East found itself afflicted with drought, and developed a system based on drought-tolerant wheat and barley supplemented by livestock that can hunt their food over distances. East Asia, with more rain and water, developed more intensive, sustainable, water-dependent systems, especially high-yield rice agriculture. The Near East's mobile sheep and goats are replaced by pigs and chickens. The luck of the draw led to economic lock-ins that conditioned the future.

Ecological unsustainability tracks cropping systems clearly: it is closely tied historically with wheat-barley-livestock systems everywhere, and with maize in many areas. Sustainability is much more often seen in rice-growing agroecosystems, but also maize in very favorable areas for it, and root crops. Unsustainability is almost universal in pioneer fringes, among nomads in chronic war zones (such as the Middle East), and in some mountain areas. Sustainable systems are quite widespread among hunter-gatherers and small-scale horticulturalists. They also characterize long-established states with relatively little opportunity for pioneering. Such states, including the major East Asian polities, are forced to find relatively sustainable sources of food.

Wheat-barley-livestock dependence facilitated development of vast monocrop plantations owned by elite landlords (priests or kings) in ancient Mesopotamia, then to elite-owned plantations in ancient Greece and Rome. Extensive plantation agriculture spread worldwide as western civilization spread. The huge estates of the western world went with less intensive and productive agriculture, which meant that cities had to be small unless they could draw by ship on large areas of farmland.

East Asia was different from the start, with more intensive agriculture and a tendency to have small owner-operated farms. Large landholdings were common at first, but gave way over time. Small, intensive farms were necessary. Large cities emerged early. They led to the development of yet more intensive agricultural methods, such that food could be provided nearby.

The idea of hating and fighting against nature and all things natural came from these monocrop plantation societies. This pernicious idea is reflected in ancient Near Eastern literature, including the Epic of Gilgamesh and the verse in the Bible that tells humanity to "subdue" the earth and "have donimion" over it (Genesis 1: 26). The view propagated especially in pioneer fringes, and thus became universal in

Americas and Australia, changing only very recently. The opposite extreme is found in rice areas of southeast Asia and Japan, and to a lesser extent in south China and southern India, with governments backing smallholders and conservation. Eastern Asians could not afford to hate and destroy nature. They depended on a healthy environment, and they knew it from early times (Anderson 2014a, b) (Fig. 1.6).

Resource-rich centralized countries that are easily ruled from a center tend to be centralized, autocratic, and imperial; China is the most striking case, but Spain and France occur to mind. Small regions well placed on sea trade routes can become thalassocracies. Greece and, later, the Netherlands are the best examples; in eastern Asia, only the Ryukyu Islands, long independent as the Kingdom of Liuqiu, managed to do this. Some states can support themselves by being central in overland trade, like Switzerland in Europe and Samarkand, Ferghana, and Kashgar in old central Asia.

A vast homogeneous tract of land, without obvious centers, will become a realm of shifting nomad empires if it is a steppe where high mobility is possible. Any disaffected group can take off over the steppe. Any nomad who could conquer a Central Asian river valley, or, better, North China, had a center to operate from. However, no Central Asian oasis valley had a clear advantage over any other, leading to frequent wars between them. No stable center lasted forever, though Balkh, Samarkand, Ferghana, and Bukhara had long runs as centers of empires. However, they never became powerful enough to challenge China. Its foreign conquerors were from the north and northwest, not the dry west.

A fertile region without good nomadizing or centralizing options naturally produces evenly-spaced statelets. These are spheres of influence, radiating out from cities. Because hexagons are the most economical way to pack divisions into a space (they minimize boundary lines), such statelets come to resemble bathroom tiling, though shapes vary with landscape. The concept was developed by the geographers Von Thunen and W. Christaller, but applied to anthropology and economics, and to China, by G. William Skinner and his students (Skinner 2001; Smith 1976). China's economy was always based on this pattern, with nested polygons all the way up from market villages to the great "macroregions" of Skinner (1977, 2001). As Skinner showed, when China broke up into separate kingdoms during periods of disunion, it broke along macroregion boundaries. When China was divided, the major divisions were always along the great ecological divide between wheat-farming north and rice-farming south. This occurred in the interregnum from 220 to 581 and again from 907 to 1279, with the exception of a brief period after 960, but even then the Liao Khitans held the far north and steadily expanded till they controlled all the north. When north and south broke up, they broke up along macroregion lines.

Japan, Korea, and Vietnam have similar polygons, though Japan's are extremely distorted by sea lanes; sea transport was faster and easier than land. Especially around Japan's Inland Sea, lords could rule long coastlines instead of neat hexagons of land. It was also easy to rule an entire small island, whatever its shape. Vietnam's linear, mountain-bounded shape was less distorting, but the center had long, narrow local regions. Central Asia, with its nomadic herders and settled linear riverine oases,

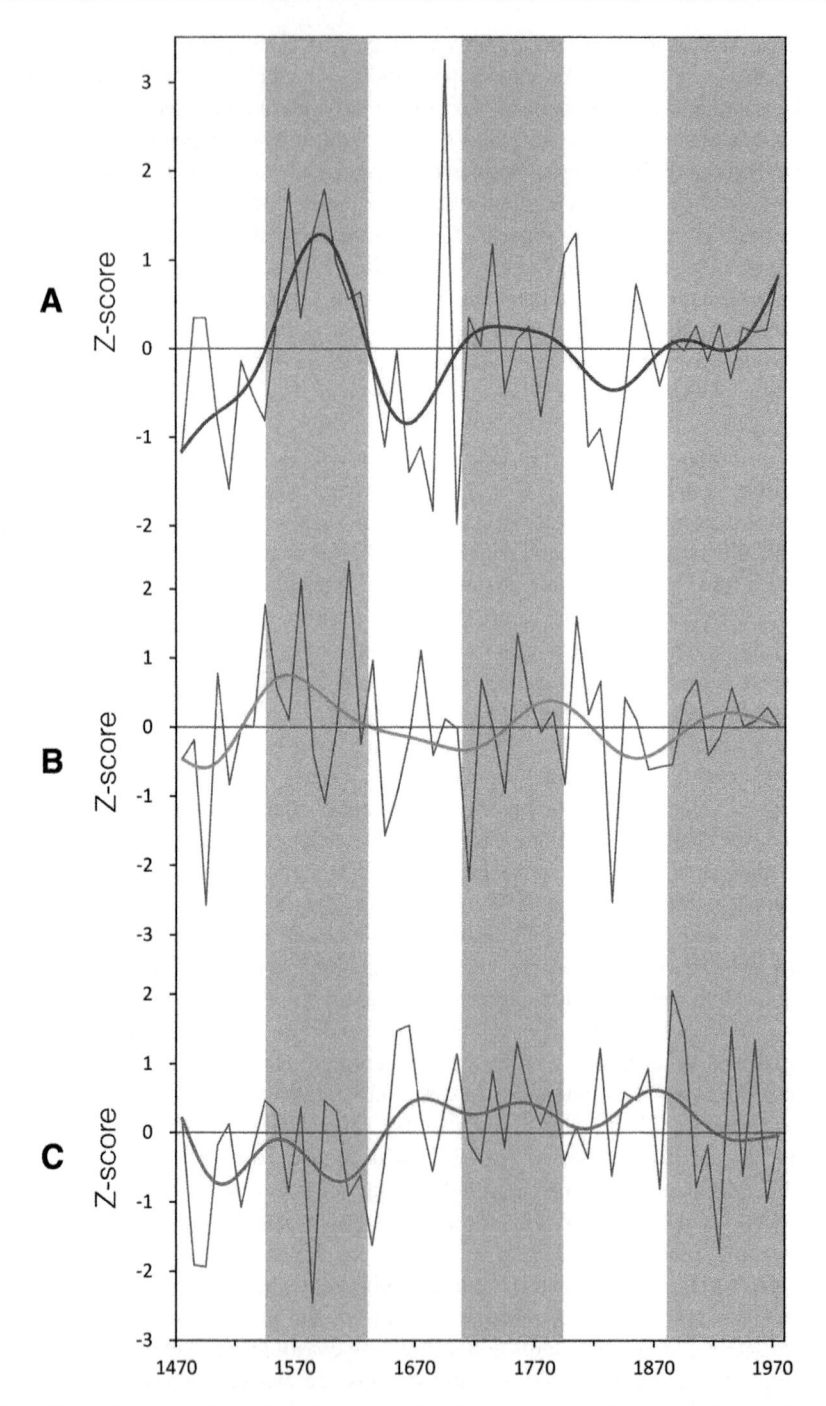

Fig. 1.6 Further comparison of precipitation variability in northwest China: with Palmer drought index and with precipitation in north-central China. Lee et al. 2015, *PLoS One,* open access per permission letter

1.5 Geography as Destiny

displays the same basic pattern, but the shapes of regions are distorted by river courses and stock routes. A given river valley would normally be the center to a widely-extended, but largely desert, region.

Certain defined areas keep emerging over and over as power centers. Robert Carneiro's conscription hypothesis (Carneiro 2003; Johnson and Earle 1987) works for these: they are lush, fertile valleys surrounded by deserts or semideserts, such that escape from them is difficult for significant numbers of people. A state can take over and assert control with relative ease. The classic valleys where civilizations began are central to vast regions. They are the places that are both fertile valleys and nodes through which trade and commerce naturally flow. In East Asia, the core valleys are China's Yellow, Wei, and Yangzi valleys, Japan's Kanto Plain, and Vietnam's Red and Mekong Deltas. Worldwide, we have the Valley of Mexico, the Cuzco and Titicaca Valleys in Peru and Bolivia, central Mesopotamia, the Nile Valley, the Indus and Ganges Valleys. By contrast, Rome and Athens were not very stable centers. Indeed, nowhere in Europe is an obvious central place. Paris is probably closest. Amsterdam was central to western Europe when waterborne trade was dominant. Europe's lack of an obvious center from which the rest of the continent could be conquered and held has long been regarded as critical to its maintenance of many small, varied states and societies (see e.g. Morris 2010).

Cultural copying goes along trade routes; the developing trade partners who get into value-added bootstrapping are the ones who do it most, the trade partners who just supply raw materials do it least. This partially explains the westward march of civilization from Mesopotamia to Syria to Greece to Rome and then Italy to France to Low Countries to England and Germany, and the rise of Japan to much greater prosperity than China in the nineteenth century. . East Asia has always been largely agricultural, but during its great ages in the past, it was often a leader in manufacturing and value-added work of all types; this characterized the golden ages of Tang and Song in China, and the Tokugawa shogunate in Japan.

The different fates of Rome, Byzantium/Istanbul, and China relate to the geography of conquest and trade. Rome ruled the world when the Mediterranean, which the Romans called "our sea," was central, and shipping trade on it was vital. But the rise of land powers ("barbarians" though they might be) and land trade were fatal to the Roman empire, and to the Greek thalassocracies and conquest states.

China shows a different pattern: a polity that survives as an intact empire, with roughly similar borders, through many changes of regime (for China's full economic history, see Von Glahn 2015). Disunited till 221 BCE, it conquered what was to be its core territory soon after that date. It slowly and erratically expanded and contracted thereafter, but the core stayed roughly the same: the land that became the so-called "18 provinces." Long periods of disunion never became permanent; far-flung conquests were rarely permanent either; and China now looks like the Qin Empire of 221 BC, with the major additions of Xinjiang, Tibet (with neighboring southwest China), and Manchuria. The reason is that China is naturally centripetal. The wealth and fertility are greatest in the center. The peripheries are rugged, poor, and thinly populated. China is indeed the Middle Country. (The word *Zhongguo* probably meant, originally, "Middle Countries," with reference to several core states in the Yellow River drainage, surrounded by peripheral states.)

The center of the realm, and the cockpit of power in early China, was the Wei River valley around Xian. The Wei Valley is like an arrow aimed at the center of China. Historically, whoever held it had tremendous power over the whole empire. China was conquered from or via the Wei Valley several times. The other conquests were all from the north or northeast, and usually involved securing the Wei very early in the game. Only the Ming Dynasty won as a genuine internal uprising from China's economic core area (see e.g. Mote 1999).

The Wei Valley is a rather small, dry region. It is not an economic powerhouse, and indeed is difficult to supply. Thus, the temptation was always to move the capital downriver to the Yellow River cities, or even farther afield to Beijing or Nanjing. All these were hard to defend and easy to conquer. Thus, when an empire succumbed to the temptation to move the capital closer to the economic center, a conqueror could move into the Wei Valley, secure it, and move on to the downriver towns. This was a frequent military aspect of China's dynastic cycles.

China suffered repeated conquests that followed the same military-geographic pattern. A fantastically rich center, surrounded by dismally poor borderlands, was continually conquered from the steppes and mountains, but never held from there. Conquerors moved to the center.

Rich, extensive agricultural lands like north China tend to produce totalitarian, reactionary states. The logic is clear. The easiest way to achieve wealth in such situations is by getting people to farm. This leads to a regime of landlords and servile workers—serfs, slaves, illegal immigrants. The chief concerns of the rich, who are typically absentee landlords, is making sure that the servile workforce does not revolt or decamp. Progress is scary to the elites; it gives the workers more opportunities. Better to fight progress and make authority ever more repressive. Rent-seeking is more profitable in the short run than development.

One wonders what would have happened if China had not had the geographic dominance of the North China Plain. Many have speculated that the trading cities of the lower Yangzi, Fujian, and Guangdong would have been free to develop their own versions of freedom and capitalism. Something like this happened in Japan, though agrarian government remained strong there. Japan is the exact opposite of China. Instead of a country neatly centered on a vast, level, farmable core, it is a chain of mountains with narrow coastal plains. The Kanto Plain is the only large and productive flat land. Japan thus naturally tended to break into a chain of quasi-independent lordships, integrated only during times when the shoguns were exceptionally powerful.

Korea and Vietnam had still another type of geography. They were mountainous, but had extensive, rich, productive plains for rice agriculture. Korea had the plains around Seoul and some other sizable rice lands. Vietnam centered on the Red River delta, the Hue-Danang plains, and later (when fully conquered) the Mekong delta. Both could have been independent progressive city-states, but the Chinese imperial model—introduced by invasion and early conquest—was too pervasive.

Overall, in the last analysis, the dominance of wet-rice agriculture allowed a centralized imperial agrarian regime to dominate even if it has a rather small land base, such as the Kanto Plain in Japan and the Red River Valley in Vietnam.

The deserts and mountains of central Asia were a very different matter. They could only be exploited through a complementary relationship of nomadic stock raisers and oasis farmers. It is necessary to explain here that the overused word "nomad" does *not* mean people wandering at will over a vast steppe. It means, usually, people who have a definite, fixed, long-known cycle of moving from winter pastures to summer ones. Nomads cannot afford to wander at will. They must know every spring, every stream, every pasture in their habitats, and when those will be reliably usable. Neither were the central Asian herders "barbarians" in the pejorative sense. They had, in the periods we consider here, highly sophisticated cultures, eventually maturing into organized states. They depended on agricultural societies for everything from metal and quality cloth to supplementary grain for their diets. Many societies were mixed, depending on both agriculture and extensive herding. This pattern goes far back in time (Chang 2018). More common still was complementary interactions, with herders trading with settled people (Barfield 1989, 1993). A tendency arose very early for nomads to conquer settled regions and draw agricultural supplies from them. This often brought nomad polities into conflict with agrarian states, especially on the China frontiers.

Ideology is indeed grounded in means and relations of production, as Marx said, but ideologies can last for thousands of years. The idea that ideology changes automatically and quickly in lockstep with superficial economic changes is not supported by either Marx or historical evidence.

"Natural" disasters tracked the rise of agriculture. China logged 1015 major droughts from Tang through Ming (920–1619; Mallory 1926: 41). These increased with time—not because the weather got drier, but because a combination of steady deforestation and other degradation, combined with steadily rising population, made droughts more catastrophic. In the north, where drought is more serious and population did not rise so much during the period in question, there was an economic decline in and after the 1100s and 1200s—not directly because of climate (though weather was stressful), but because of Song, Jin and Yuan Dynasty mismanagement of human-land relationships. Korea, Japan, and Vietnam had similar problems with drought and flood. Central Asia's worst problems were cold and snow in the north, cold and drought southward.

Floods track overpopulation and poor river-course management more than weather. Earthquakes and locust plagues also cause disasters in proportion to population density and vulnerability, providing some control on climate, since drought and flood really are subject to climate effects while earthquakes are not. Earthquakes are most serious in the north, where they are common, and where houses are frequently in loess caves or built of unreinforced adobe that can collapse and bury the inmates. In imperial times there were bad weather events, floods, droughts, and famines almost every year somewhere in China. Explaining dynastic cycling by disasters is perilously close to explaining a variable by a constant.

1.6   Cycles

"Men make their own history, but they do not make it just as they please; they do not make it under circumstances chosen by themselves, but under circumstances directly encountered, given and transmitted from the past. The tradition of all the dead generations weighs like a nightmare on the brain of the living" (Marx 1972: 15)

Since China is by far the largest polity in East Asia, and the first to develop urban and literate society, it must be the featured society in the present work. Other polities have their own distinctive histories, however.

Imperial China was united under nine successive dynasties, holding something close to the classic "Eighteen Provinces," for most of the time from 221 BCE till 1911. It was also divided for many centuries into smaller states, which had their own dynastic families. It has since been united briefly under the Republic and stably under the Communist regime. The present book deals only with the imperial dynasties.

The major dynasties included three very short ones, and six that ranged from less than 200 to over 400 years long. All these longer ones were punctuated by coups or partial conquests leading to territorial loss, imposing a shorter cyclicity of about 60–70 years on the long cycles. Periods of disunion, ranging from a few years to several centuries, separated the dynasties. Before 221 BCE, what was to be China was divided up among many small states or pre-states. Hegemonic central regimes holding the key central area from about 2000 to 771 BCE. They are considered early dynasties.

Most dynasties included periods of cultural brilliance, such as the middle period of Former Han, the middle years of Tang, and the late eleventh century in Song. The Chinese still recall these as glorious times. They were "golden ages" in A. L. Kroeber's sense (Kroeber 1944).

No other country has exemplified so perfectly the cyclic nature of history in agrarian empires. China is thus the best testing ground for cycle-based theories of change. Scholars have sharply questioned the cyclic nature of history in the world in general, but no one can ignore the recurrent patterns of China's dynasties.

Korea and Vietnam had much less clear cycles; reigns were very long in the former, and ranged from brief to extremely long in the latter. Patterns are hard to find. Japan, on the other hand, had cycles rather similar to China's, but they were not imperial; Japan has had only one imperial dynasty in its history. The cycles involved the shoguns, who were usually the de facto rulers.

The present book compares environmental issues with Ibn Khaldun's classic fourteenth-century model of dynastic cycles (Ibn Khaldun 1958; cf. Anderson and Chase-Dunn 2005; Chase-Dunn and Anderson 2005). Peter Turchin's recent reworking of Ibn Khaldun, using much more sophisticated biological and mathematical models than the medieval Tunisian had at his disposal, adds considerably to explanatory power (Turchin 2006, 2016).

Meanwhile, from another realm of learning, a very similar model, the resilience cycle, has been developed by C.F. Holling (Gunderson and Holling 2002; Gunderson et al. 2009). Leading Sinologist Stevan Harrell is currently working on applying this to Chinese history (personal communication). Holling's general theory can incorporate Ibn Khaldun's specific one, and I shall do so herein. This allows use of his convenient terms for phases of cycles.

The history of China is not simply one of cycles. Still less is it the endless repetition that Tennyson meant when he wrote "better fifty years of Europe than a cycle of Cathay" (line 184 of his poem "Locksley Hall"). China's real history is one of progress in feeding, clothing, housing, and caring for an increasing population. It is the story of the "silent gardeners" who patiently, lovingly, caringly tended the tree of humanity in eastern Asia, while the emperors and generals carved their names in the bark; as the Chinese observed, emperors and generals earned their reputations by killing.

There are always three things happening in any society: secular progressive change (such as China's agricultural progress), cycling, and conservation of tradition. Giovanni Arrighi (as cited by Moore 2015: 165) has put this in the form of three general questions to ask about a historic phase or civilization: "What is cumulative? What is cyclic? What is new?"

One can add "What is unchanging?" The "changeless" or "traditional" side of China has been vastly overemphasized in the past (as Richard von Glahn pointed out in 2003), but China does have some striking continuities: the same language (however changed by time) for over 3000 years, the same patrilineal kinship system with extended families and larger lineal groups, the same small-scale intensive approach to farming, the same beliefs in dragons and sky gods. Traditional medicine, philosophy, and literary forms continue. China has maintained for over 3000 years an autocratic imperial-bureaucratic system of government. Even today's Communist China might well be considered closer to its imperial past self than to the utopian visions of Karl Marx and Frederick Engels. Profound changes have occurred in all aspects of life, but China is still recognizably the same civilization that it was 2500 years ago. England, Germany, and Spain are not, let alone the United States. To some extent, the interaction of tradition and progress drove China's cycles.

I have written two books (Anderson 1988, 2014b) and many articles to celebrate China's heroic story of progress in science, technology, agriculture, and the arts of life. Others, both east Asian and western, have done far better than I at recording this great enterprise. I find recording progress and advance more rewarding than recording the fall of dynasties. Still, cycles have their own interest, and recently China's cycles have been attributed to climatic change. As a human ecologist who has done some work in China, I found the claims of climatic determinism hard to ignore.

1.7 Cause and Explanation

Claiming climate as a "cause" of anything requires defining the word. "Cause" is a vexed word, and often a weaseled one. Aristotle noted that the cause of something may be the purpose for which it was created, the act of creating it, the plan for creating it, or even the material of which it was made (Aristotle 1953, Greek original ca 350 BCE). Aristotle points out that parents cause a child in several ways; they plan for the child, engender him or her, provide the physical substances ("blood"—i.e. nutrition in the womb—and food), and so on. He further notes that a house can only be understood as something built for people to occupy, by builders trained in the art, according to an architect's plan, and using the appropriate stone, lime, wood, and tile.

Today, we are concerned about rising global temperatures, and so we look for the effects of the Medieval Warm Period—a very warm period worldwide, often called the Medieval Climate Anomaly—in the past. But what does heat really do? It makes us sweat: exposure to fierce sun directly causes sweating. It somewhat less surely makes us drink more and eat more salt; we can abstain (and possibly die). But it does *not* cause us to turn on the air conditioner; that happens only if we want to be chilly (Aristotle's ultimate cause), have planned for an air-conditioned house (Aristotle's planning and intention cause), a strong desire for comfort motivates the act (Aristotle's immediate cause), and an air conditioner in operational state is at hand (Aristotle's material cause). A strongly motivated environmentalist who refuses to use the energy, a person too poor to buy an air conditioner, and a person without electric power will have to suffer, however bad the heat. Still less does heat or cold "cause" the rise and fall of empires.

We today are often ambiguous about whether we mean ultimate purpose, ultimate predisposing factor, immediate plan, immediate agency, or some mix of the four. Andrew Vayda has sharply critiqued the naïve use of "cause" in anthropology, noting the tendency for anthropologists to use vast, vague things like climate change and capitalism to explain very specific "events" that have their own specific histories. He counsels us to look at the whole causal chain, from widest context to individual decisions to actual behaviors (Vayda 2009).

Vayda was particularly critical of using analytical abstractions as "causes." "Cultures," "ideologies," "social structures," "capitalism," and the like are social scientists' abstractions from data; they cannot cause anything. On the other hand, the data from which they are abstracted do have real-world relevance. Culture, for instance, is a general term for learned, socially shared ideas and behaviors. "Culture" cannot act, but the ideas and behaviors included in "culture" are considered and performed by real people. They therefore have real-world causal importance. Similarly, climate change cannot directly cause people to do any *specific* thing. It creates conditions in which people are motivated (or even forced) to do *something*.

Marx sought to link long, slow economic cycles to agents of revolution and change, but he was careful to point out that humans are impassioned actors—not passive pawns or coolly rational deciders. He wrote of people as emotionally reacting with anger, resignation, or deflection to economic crises (see notably Marx 1973), in some contrast to many of his modern followers who see "capitalism" as a direct causal agent (even though, unlike Marx, they cannot always define it). Max Weber (e.g. 1978) reminded us that only human individuals *act*. "Society" does not, "culture" does not, "economics" does not. The fate of societies, in the end, comes home to individual decisions and individual agency. These, summated, become social forces. C. Wright Mills, in *The Sociological Imagination* (1959), critiqued both abstract-idealist and narrowly-individualist views, arguing for a

1.7 Cause and Explanation

sociology in which real human agents deal with large-scale forces, and do so within changing systems. I follow Weber and Mills, and also Anthony Giddens in his concept of structuration (Giddens 1984): structure created and constantly changed by agency.

An interesting class of "actors" is Benedict Anderson's "imagined communities" (Anderson 1991). Once people have "imagined" nations, empires, and dynasties into existence, and accept them as real, those entities take on a life of their own. The Chinese themselves, except for confirmed skeptics like Wang Chong, believed that spirits of seas, mountains, rivers, and trees could also act willfully and sometimes importantly. (On paradoxical "actors," see Latour 2005). The gods may not be real, but people act according to their beliefs in those gods, and imagine their gods as being very much like human societies. This adds Durkheim, especially his views on religion (Durkheim 1995 [1912]), to Marx, Weber, and Ibn Khaldun, as the foundational theorists of the present book.

A similar virtual actor is "money." In China as in the west, money was once real: cattle, gold, shell. It soon became purely notional: an abstraction, symbolized rather than realized in coins and bills. Yet money "makes" people act, through their shared belief in it. On the other hand, concepts that are ill-defined by scholars and unknown to the rest of the world do not have agentive effect. Analytical abstractions from "categorical imperatives" to "neoliberalism" do not cause anything. On the other hand, belief in the specifics that those terms more or less capture may cause individuals to act.

Especially pernicious is the tendency to assume that humans will always react the same way to the same stimulus or situation. Common experience (and Akira Kurosawa's classic film *Rashomon*) teaches otherwise. The same stress breaks one person, strengthens another, drives a third to escape. We must link general causes to specific human acts. There are cases—notably, recent elaborations on Ibn Khaldun's theory of cycles—in which the links are well spelled out in current literature. There are others—notably, certain recent theories of climate change, ecological overshoot, and disease—in which the links are left so totally obscure that we must be highly dubious of the result.

Vayda's methods of analysis lead to constructing causal chains, or causal trees. In the case of climate, we may speak of a *back story*, including the climate and the long-standing desires and intentions of leaders; a *middle story* of immediate sociopolitical, economic, and military conditions; and a *front story*, consisting of the actions taken by those leaders in pursuit of their goals. The results may be stasis, collapse, innovative change, outright war, or other coping mechanisms. These may succeed or may not, depending on how well the leaders have calculated their strength.

To maintain that climate change caused culture change, one should show that the relationship of the former to the latter is not due to chance. This is not possible, but it is possible to show that the relationship is at least more than mere chance would predict. Ideally, one dreams of a probability of .01. Keith Kintigh and Scott Ingram (2018) decided to test alleged correlations of seven major climatic and cultural events in the late pre-Columbian American Southwest. With the aid of modern computing, they could generate 3,000,000 imaginary climatic regimes with equal proportions of drought to wet years and cold to hot years, and correlate all of them with the seven cultural changes. (Imagine doing that by hand). They paid special attention to long droughts (5 years or longer). It turned out that the average correlation of the 3,000,000 imaginary events was just as good as the real correlations observed, demolishing all proof that climate caused culture change. Of course, as they point out, it does not *disprove* such connection—merely makes it unprovable one way or the other. More and better research is needed to establish real causes.

1.8 Human Nature: Basics Needed for Understanding History

At this point it is necessary to specify the general human truths that lie behind cycle theories. Human behavior leads to processes that alternately strengthen and weaken social bonds. People need air, water, food, shelter from heat and cold, sleep and waking, protection from danger, medical care, and some degree of space in which to live. They also need to feel in control of their lives; this is a well-documented life-and-death need for higher animals. Above all, they need society; humans are extremely social animals. (Abraham Maslow's classic list of human needs [Maslow 1970] has been importantly updated by Douglas Kenrick et al. 2010. See also Bandura 1982, 1986; Langer 1983; Schulz 1976).

If the society is to last more than a generation, they also need to reproduce. The bottleneck in traditional societies like imperial China was infancy. Regularly throughout history, before the late twentieth century, from a fifth to a half of all children died before age five. Particularly striking in East Asian history is the number of emperors who lost all their children, despite having many wives and getting the best care the country could afford. In the 1200s, two consecutive Song Dynasty emperors died childless, creating chaos in the succession and hastening the fall of the dynasty to the Mongols (Davis 2009a, b).

The best explanation for the sociability and the constant group rivalries and group hatreds that characterize human life is that humans evolved in large kinship groups. Possibly this group orientation evolved through competition—including war—with other groups, the larger group tending to be the winner (Bowles 2006; Choi and Bowles 2007). On the other hand, it may have happened peacefully, through the advantage to wide-flung cooperation (González-Forero and Gardner 2018). Human groups are normally around 50 to 150, with larger, loose groupings of about 500; this is what one sees among hunter-gatherers and in modern face-to-face communities and friendship networks (Dunbar 2010).

The classic models of war of each against all, rational self-interest, Freudian id, and so on cannot predict anything about human sociability, let alone such complications as empathy, innate moral tendencies, and innate attention to social cues (all well demonstrated in infants). They cannot predict why genuine individual maximizers (psychopaths and sociopaths) are considered abnormal. Conversely, models of humans as innately good, moral, nonviolent, and the like cannot survive the slightest acquaintance with history or with crime statistics. Humans are social; they do all for the group, and above all they fight and often die for it. They maximize individual benefits, *but most of those benefits are social*: love, approval, status, acceptance, warm affection, support, care, sympathy, and—alas—the pleasure of working the most horrible and sadistic tortures on anyone who threatens one's group or one's own status within it. People are good or evil according to what they believe their society teaches and demands.

In cases of failure to satisfy survival needs, concern for them must take priority, and people must deal with those problems before going on to other matters (such as enjoyment or empire-building or governing). Knowing what to do most directly follows from experience, but often people must draw on cultural knowledge to survive (on society, culture, and survival, see Henrich 2016). Accumulated cultural wisdom has given us everything from successful bear-avoidance methods to effective medicines--increasingly efficient and satisfactory fixes to problems. The search may be regarded as "science," in a very broad sense of the term. It is, in the words of Laurence Flanagan describing the archaeological record, "the story of Man's attempts to keep the wolf from the door by means of better doors and better wolf-traps" (1998: 3). He adds: "[T]he fact that humankind itself is unpredictable is the quintessential stumbling-block for archaeologists. We have to assume that the people whose dwelling-places, artefacts [sic], lives even, we are dealing with were rational, integrated, sane, and sensible human beings. Then we look around at our own contemporaries and wonder how this belief can possibly be sustained" (1998: 5).

A common and often unhelpful way of dealing with threat is to escape into quietism. Buddhism and Daoism taught that approach in East Asia. It is not clear that meditation solves many problems. One suspects that the nameless, unremembered individuals—Heine's "silent gardeners"—who domesticated rice did a great deal more for East Asia than its famous "immortals" and meditators.

Other common methods include denial, rationalization, satisficing, and simply bearing it all. Proactive rational coping, the only successful way of dealing with most problems, seems to be often the last to be invoked. Asia, like the rest of the world, had a constant tension among ways of dealing with problems. Its superb scientific tradition, from domestication of rice and millet to modern satellite launching, never achieved clear priority over war and hate, or over quietism and escape. Until the late twentieth century, science there never became the vast, self-conscious, highly funded enterprise that it is in the world today.

1.9 Rationality

"Rational choice" theorists often get away with claiming people are rational because they assume certain human behaviors are innate and "rational." In fact, human choices inevitably involve more or less emotion, and often work against genuinely rational goals. These emotional motives are often the unpleasant ones so loved by Machiavelli: revenge and vengeance, aggression and rapine, jealousy and envy and the other Seven Deadly Sins. But one must also include love and sex. Most important of all is the need for society and social status, approbation, respect, recognition, acceptance, and the other intangible but all-important social goods. These notoriously make people happily court death, everywhere, every day. East Asian philosophers and historians were aware of this, and more concerned with moral choices than with rationality.

Not even extreme rationalists like Gary Becker (1996) could ignore emotional goals. We now well know that human minds are biased in certain "irrational" but understandable and usually useful directions (Gigerenzer 2007; Kahneman 2011). For one important example, humans display an irrationally extreme devotion to short-term interests, versus long-term and wide-flung concerns. Even truly terrifying long-term threats are downplayed or disregarded, relative to trivial immediate issues. In Asia, this led to a chronic pattern of neglecting such things as river dyke repair, irrigation work maintenance, and tax roll updating. Attention to them was displaced by immediate concerns such as local rebellions and nomad incursions. One story repeated over and over in Chinese history, and by Chinese historians, was the slow but steady encroachment on roadways by farmers desperate for just one more furrow. The roads shrank until some hard-nosed authority widened them again.

David Hume said: "Reason is, and ought only to be the slave of the passions, and can never pretend to any other office than to serve and obey them" (Hume 1969 [1740]: 462). Politics is notoriously dominated by emotion (Caplan 2007; Lakoff 2006; Westen 2007). Conflict is inevitable, and scary; it cannot be eliminated and cannot fail to arouse some stress. Alan Beals has argued that conflict resolution is the most basic and important thing that culture does (Beals 1967).

Individuals do try to maximize satisfaction of wants and needs, irrational though they often are in pursuit thereof. Often this comes at the expense of the common good, giving us the classic "tragedy of the commons" (Hardin 1968, based on Thomas Hobbes). Mancur Olson (1965) argued that people need "side benefits": individual rewards for sacrificing self-interest for collective good. Fortunately for humanity, a very important "side benefit" is the sense of pride and accomplishment that comes from being recognized as helpful, reliable, and a "nice person." Much of Chinese literature turns on that point. Other "side benefits" include the various aesthetic delights of festivals and ceremonies associated with the agricultural cycle or with maintaining social order. By 2000 years ago, statesmen were aware that these rewards were important, even necessary, to motivate people to work for the community.

Most people do not really try to maximize wealth. They do not maximize leisure, either; people love to be active and even to work. As the German proverb says, *arbeiten und lieben,* work and love, are the sources of life's real meaning. In a Nazi death camp in WWII, Viktor Frankl found that those who survived were those who had something extremely important to live for: usually family, sometimes career (Frankl 1959, 1978). In psychotherapeutic practice and psychological research after the war, he found more and more depending on these ultimate, deepest meanings.

Emotions include the familiar love, hate, fear, anger, disgust, and liking. Less often considered emotions, but important and related, are mood-states like interest, excitement, enthusiasm, boredom, annoyance, frustration, and irritation. Contrary to endlessly repeated claims, there is no believable evidence that emotions are profoundly different between cultures (Anderson 2011). There is, however, even within any one culture, quite enough complexity to attract philosophical attention (Elster 2007). In human agency, everything is mediated through self-efficacy (Bandura 1982, 1986). People strive to control the uncontrollable through prayer, ritual, and magic (Malinowski 1944, 1948). Such rituals in East Asia served the real function of bringing people together for common effort, but they would not have been able to do that if people had not believed they were effective at bringing rain, curing ills, or maintaining the state.

1.10 Human Good and Evil

Common from China to Europe is the Hobbesian idea—anticipated in China by Hanfeizi—that people are individuals locked in perpetual strife, or at least in selfish competition. They are calmed—if at all—only by a "social contract" (Hobbes 1950 [1651]), putting a king in charge, or by the more vague and general hand of culture and religion.

More accurate is the segmentary unity described in the old proverb: "I against my brother; my brother and I against our cousin; my cousin, brother and I against our village; and our village against the world." Kinship is not the only source of "brotherhood." Religion, ethnicity, personality, occupation, personal history, and sheer accident can all create bonds, and all can outweigh kinship on occasion. Another extension is personality. Asia has the same personality dimensions as the rest of the world, though they may be differently expressed (Bond 1986).

The proverb reminds us that one basic question we face daily is whether competition will be individual vs. individual, or—far more usual—subgroup vs. subgroup within the society, or society vs. society. Next most basic is whether the competition will be positive-sum, zero-sum or negative-sum. In declining times, group competition within a society tends to win out, because dominant groups become desperate to maintain their position, and think their best hope is to do it by taking down other groups. The worst problem in society is generally not simple greed but social hate. Most people will harm themselves to almost any degree, including sacrificing their lives in various painful ways, to harm enemies—especially genuine enemies of the group, but often personal enemies or even imagined ones.

Times of rapid change can have the same effect, but if the trajectory seems upward, hope will make people work together. Most people realize that they can do better working with others to improve the situation. There will always be zero-sum and negative-sum gamers, and they tend to get the power because they fight hard and without moral scruples. This must be outweighed by the hopes of the others, so the more clear and rapid the improvement, the more sociability wins out—it motivates the others to out-shout the hypercompetitive ones. Loving all is an ultimate ideal, but simply being solidary with others—mutual aid, mutual respect, or at least mutual consideration—sometimes occurs. This "human nature" is basic to cycle theories, and is amply confirmed by modern psychology and sociology.

Solidarity flourishes in folk communities, because everyone is dependent on everyone else. *The problem is that it is always easiest to unite people by hate, and against enemies.* It is always hardest to unite by asking for self-sacrifice for more pleasant ends. *Thus, all solidarity-building institutions, even if they start out advocating pure love, tend to create hate.*

Adam Smith's attempt to escape this by appealing to rational self-interest fails because, as he well knew (Smith 1910, 2000), a society must have a moral and institutional shell around the market or the market quickly collapses upward into a cartel or downward into robbery and chaos.

When people self-sacrifice, it is usually under the direction of leaders, and thus often for the rich and powerful. Within societies, there is a range of personalities from followers to leaders, with many more followers. There is also a range from charismatic leaders to not-so-charismatic ones. The leaders define the agendas, motivate people, and escape the dominance of abject fear, so even though people are mostly fearful, things get done.

1.11 Population Growth

East Asia's coping with climate took place against a backdrop of slowly increasing population and eventual high density. Population density interacted with climate change in such important ways that demographics must be considered at some length. Ester Boserup (1965) held that population growth causes intensification of agriculture. As she saw it, population growth puts increasing subsistence pressure on people, who must then work harder and harder to extract more and more product from a fixed supply of land. This is Thomas Malthus at his most dismal: population growth must always occur, it must always outrun food production, and people can only work ever harder.

Boserup's actual data came from rather unusual cases. Her data were from then-British parts of Africa, where British colonialism pressured people to work harder and prevented people from doing much else about population pressure.

As Chayanov pointed out long ago (Chayanov 1966; Durrenberger 1984), traditional farming families need a great deal of labor power, since most work is done by hand. They therefore have many children. This, not mindless breeding, is certainly the reason for large family sizes in traditional rural societies, as has been shown by countless anthropological studies using Chayanov's theories. However, families must balance the need for workers with the ability to raise them. Children require years of care and parental sacrifice before they can work. Parents thus limit births according to the likelihood of being able to raise children who will be able to have satisfying or at least productive lives (Feng et al. 2010).

History reveals that people do indeed often deal with population pressure by intensifying, as John Brooke (2014) documents at length. Only rarely, however, do they innovate when under pressure from real want. In the short run, the usual response to population pressure on the food base is first to migrate if possible, to "approach the food" as the Chinese say.

1.11 Population Growth

An important article by Oliver Sheehan et al. (2018) compared 155 Austronesian-speaking societies, ranging from the Cham to Easter Island, to see the effects of landscape modification and social stratification. Some of these societies are very simple, with no social stratification or major landscape projects. Others are highly complex and hierarchic socially, and have sophisticated land modification by terracing, creation of gardens involving heavy earth-moving effort, or other "landesque capital." Still others have one or the other. The comparison over space and time showed that social complexity and land modification increase in feedback, with neither one leading the other. However, political complexity does precede land modification, or drive it, when there is a driver. Most societies show a good correlation between complexity and landesque capital. The exceptions are, for the most part, either small islands that require terracing and extensive garden construction but have too few people to allow much complexity, or large, well-populated, but flat areas where terracing and other extensive earth-moving is unnecessary and impractical. It appears that population pressure by itself is not a driver. In fact, it seems to be more effect than cause; people increase when they have a reliable, intensive food production system.

People also trade manufactured goods and raw materials such as metals for food, and have been doing so since long before civilization. People can also rely on aid and charity for food; these were well developed in China by the 1600s.

Many models assume that population simply grows. This is not the case. Modern demographic research has amply proved that population growth rates are dependent variables (Bengtsson et al. 2004). Every culture encodes enough knowledge about birth control, abortion, and infanticide to allow its human bearers to plan their families. China and Japan used various techniques, from delayed marriage to infanticide, to prevent rapid population growth (Bengtsson et al. 2004; Sommer 2015). In fact, China's population growth rate, when documented, proved to be too low to be the result of natural uninhibited increase (Feng et al. 2010). Only infanticide (and, theoretically, abortion, but Sommer proved it was rare) can explain the low rates of fertility among married, co-residing couples. Migration of men to work far from home also kept childbirth rates low in many areas. Japan practiced infanticide even more extensively. Thomas Smith argued in his book *Nakahara* (1977) that at least some parts of Japan practiced infanticide reaching up to 50% of births. Tibet, medieval China, and probably other areas used monasticism; excess people were shipped off to monasteries, which could incorporate a large percentage of the population. War was also a release.

Population increasing steadily but not at "natural fertility" rates allowed China, Vietnam, and Japan to be leaders in agricultural intensification throughout their histories. In fact, they probably owed their ability to innovate to the *lack* of extreme pressure, among other factors.

Infanticide in China (but not in Japan) was largely (though not exclusively) of girls, which depresses population growth far more than infanticide of boys. The tendency of rich men to have several wives further removed women from availability to poorer men, such that about ten percent of men in old China were left to be "bare sticks"—hopeless bachelors (Moise 1977). Family structure, variant marriage patterns, long-continued breastfeeding, and simple abstinence all had their effect. The result was that East Asian populations did not simply grow.

Malthus (1960/1798) was aware that people did limit their birthrates, but believed that this did not usually help much (in that pre-"pill" age). The French were already limiting births by his time, and he knew it (Bengtsson et al. 2004), but he also knew that France had considerable hunger. John Brooke (2014) argues for a lack of Malthusian checks, but he misses the constant downward sift of people into poverty, disease, and finally death, a situation memorably described for China by Edwin Moise (1977).

Sadly, the commonest fate of those confronting Malthusian pressures in East Asia was to die of malnutrition, or at best to die unreproduced. It is possible that *most* deaths in China and Korea over the 2000-year run of empire were due to malnutrition. This was true even when the cause of death was labeled "illness"; weakened bodies could not survive even minor diseases. Vietnam and Japan, blessed by better food production capacity and rather thin population through much of history, were less troubled, but had their famines and Malthusian checks (see Kiernan 2017 for Vietnam). Central Asian polities seem usually to have dealt with population pressures by sending their children to war.

James Z. Lee has spent many years studying surviving Chinese records, and finds quite long life spans, robust maintenance of living standards, and population regulation through various mechanisms (Bengtsson et al. 2004). His Manchurian sample and some other sampled lineages elsewhere in China had longer life expectancies than the Imperial family, a fact which he explains by pointing out that the Imperial family lived in polluted and disease-ridden Beijing (Bengtsson et al. 2004: 295). With life expectancies ranging upward from 40, Manchuria was certainly exceptional by premodern agrarian standards. The Qing Dynasty imperial lineage had life expectancies at birth ranging from 25 for women to 33–35 for men, with life expectancies at 10 reaching 35 and 38 respectively (Bengtsson et al. 2004: 296). Rich Zhejiang was more like Manchuria; China as a whole, in the early twentieth century, more like the imperial lineage. Comparable figures for Japan indicate very respectable longevity, at least in the richer areas.

These worlds were separate, demographically, through most of history, but many Central Asians did settle in China over time. China became a safety valve (sometimes through conquest, often through migration) for the steppes and deserts. Conversely, China sent vast numbers of people, mostly from Guangdong and Fujian, to southeast Asia, where they populated whole regions of Thailand and Laos, and later a great deal of Malaysia and Indonesia. Back and forth migration across the shifting China-Vietnam boundary was also a regular feature of interaction through time.

A volume of archaeological and historical studies (Marcus and Stanish 2006) shows that people intensify for all sorts of reasons, and even manage to make their lives better off. Frequently, they manage to raise production per unit area while reducing, not increasing, their work load. One way is irrigation.

Perhaps most amazing are the long water tunnels known as *karez* in Xinjiang and elsewhere in eastern Central Asia; they are the same as the *qanats* of Iran and the

Middle East. They were probably invented in Iran or Mesopotamia, and have been known for millennia. A karez is a tunnel extending back into an alluvial fan, to tap the underground flow. Desert rivers typically rise in the mountains, and sink into the sands of their alluvial fans, the water disappearing underground. Water tunnels tap this flow. They were common in what is now Xinjiang. They are elsewhere unknown in East Asia, but remain a visible part of the scene in Iran and Afghanistan.

Famines happened almost every year in East Asia. There were huge ones in north China in late Ming and in north and central China in late Qing. China dealt with the problem by employing innovation, intensification, migration to newly opening lands, or successful famine relief. Local famines were common (every other year on average), but widespread famines were usually due to war or similar unrest. These slowed population growth long before population pressed on natural limits. Probably the greatest famine in all history was Mao's Great Leap Forward, in which 30 to 50 million people died (Dikötter 2010); it was caused by one man's behavior. As Amartya Sen showed long ago, and as subsequent studies confirm, there have been *no* famines since the 1930s that were not caused by war or politics as opposed to absolute lack of food (Sen 1973, 1982, 1984, 1992, 1997, 2001).

A stagnant population upgrading from grain to meat, vegetables, and fruit produces far more food market growth, and thus more agricultural intensification, than a rising population getting hungrier and hungrier. China's history provides a superb example (Anderson 1988; Smil 2004).

Actual agricultural *innovation*, as opposed to mere intensification of existing methods, is genuinely rare. Moreover, the great innovations in history—the domestication of the key plants and animals, the development of irrigation technology, the development of effective means of food storage and pest control, and the like—are anonymous. We have no idea who developed bread, wine, beer, pickling, salting, smoking, or anything else we depend on for our lives today.

In East Asia, agricultural innovation was concentrated in Former Han and other "golden age" periods (Anderson 2014b). Markets, government policies, and opportunities for self-betterment motivated farmers to work harder and smarter.

Food demand often goes *up* when birth rates decline and population growth stops short. Instead of having more children, people eat "higher on the hog" (this classic American idiom refers to eating pork chops instead of sow belly). China today is a dramatic case: people are eating more, and especially more animal protein, which involves turning grain into meat rather than into human food. Grain consumption has skyrocketed (Smil 2004). But this tracks economic *progress*, not want. Bruce Lerro (2000: 46–48) sees the usual western belief in "progress," fueled by insatiable curiosity and a strong desire to improve one's lot, as rare indeed in human history.

East Asian societies rarely came up against absolute Malthusian limits, but very often lost vast numbers of people to malnutrition. The explanation is that the government was often too weak to guarantee law and order, infrastructure maintenance, security of life and property, and smooth operating of governance systems ranging from enforcement of contracts to protection of roadways and dykes.

1.12 Growth and Non-growth

"In reflecting upon the poverty of Turan and Arabia, I was at first at a loss to assign a reason why those countries had never been able to retain wealth, whilst, on the contrary, it is daily encreasing [sic] in Hindustan. Timur carried into Turan the riches of Turkey, Persia, and Hindustan, but they are all dissipated; and during the reigns of the four first Caliphs, Turkey, Persia, part of Arabia, Ethiopia, Egypt, and Spain were their tributaries; but still they were not rich.... Hindustan has been frequently plundered by foreign invaders, and not one of its Kings ever gained for it any acquisition of wealth; neither has the country many mines of gold and silver, and yet Hindustan abounds in money and every other kind of wealth...." 'Abd al-Karim Kashmiri (d. 1784), historian in Delhi (quoted Levi and Sela 2010: 263–264).

'Abd al-Karim Kashmiri has a good question. The short answer is that India could constantly renew its wealth by farming and manufacturing, while Central Asia ("Turan") and Arabia maintained their wealth all too often by plundering; plundered wealth is soon taken by new plunderers, with much burning and destruction in the process.

The long answer involves a deeper understanding of why some societies develop in one way, some in another. In the above material on agriculture, and by extension other economic matters, a very basic model that works and is highly predictive is the concept of *induced development,* elaborated by Hayami and Ruttan 1985 building on earlier work by Theodore Schultz (1968). They noted that innovation efforts concentrate in areas where clear "bottlenecks" exist and can be removed. The "bottlenecks" are specific areas of relative backwardness that limit economic growth.

They showed that agriculture historically developed to eliminate bottlenecks in factors of production. Denmark, long on capital but short on land, invested in and developed capital-intensive agriculture with extremely high productivity per acre. Taiwan, long on labor and short on land, lavished incredible amounts of labor on tiny plots. The early United States, blessed with abundant land and capital but scarce labor, developed a very extensive agriculture, using a great deal of land and a lot of machinery (capital). As land has gotten scarcer and capital more abundant, the US has poured more and more capital into agriculture.

East Asian agriculture, based on cheap labor but scarce and limiting supplies of land, developed fertilizing, planting methods, crop selection for high yields, recycling, and other organic-based innovations to a uniquely high degree (Anderson 1988; Bray 1984; Elvin 1973; Hayami and Ruttan 1985). When food supply and food security became limiting to the global economy in the 1960s, all traditions were used, but the East Asian system proved most effective on a worldwide scale. Breeding high-yield varieties, using more fertilizers, and improving cultivation techniques launched the Green Revolution, climaxing in China in the amazing work of Yuan Longping (e.g. 2002), who used traditional breeding techniques to increase yields of rice. Rice in Han days produced 1500 pounds per acre per crop; by Qing it was yielding 2500; breeding in the 1950s and 1960s brought it up to 5000–7000; it now reaches 15,000. Modern GMO technology has joined traditional methods in developing new (and, be it noted, quite safe) rice varieties. Korea reached comparable levels, and Japan under the intensification-oriented Tokugawa was even better off. There is little doubt that Japan in the eighteenth and nineteenth centuries had the most productive agriculture of any nation in the world, in terms of production per acre of farmland (see e.g. Von Verschuer 2016). China's Pearl River Delta and Vietnam's Red River Delta were comparable (Gourou 1955).

Bottlenecks in transportation led to East Asian innovations, made by various ethnic groups around the China Seas, that included the compass, watertight compartments, fore-and-aft rigging that would allow sailing close to the wind, the sternpost rudder, and indeed most of the key inventions that lifted sailing from hugging the shore to voyaging over oceans.

These developments did more than anything else to tie the East Asian region into a world-system from early times (cf. Chase-Dunn and Hall 2011). The early and impressive development of shipping made trade, commerce, communication, seaborne invasion, pilgrimage, migration, and all forms of travel easier and more effective. All this explains why East Asia progressed and developed steadily over thousands of years, but suggests one reason why it eventually lost the "progress" sweepstakes to the west—"for now," as Morris (2010) rather ominously reminds us.

1.13 Power

Dynastic cycling is ultimately about getting and holding power, and thus a brief note is needed on what I mean by "power." Power is the common currency of history and theories of history, but is sometimes inadequately discussed in works such as the present one, so some details are needed. [What follows is based largely on Max Weber (1946, 1951, 1967, 1968, 1978), including interpretations by Randall Collins (esp. 1986).]

Power, in society and in sociological usage, is the ability to make people do things they do not want to do. Put another way, it is the ability to decide for others, and to prevent them from deciding for themselves. It can be exercised by direct force, by indirect force, or by persuasion. Paying people to do something is not really power if they could go elsewhere for their money, but being able to cut off people's incomes and thus starve them to death is real power. Telling people to act morally is not power (they could choose otherwise), but being able to ostracize them and deprive them of social life if they do not follow a moral rule is real power (cf. Henrich 2016).

Elites, in this book, are defined as those who can make large numbers of people follow the elites' decisions; common people are those who cannot decide much beyond their own households. Contra Michel Foucault and James Scott (1985, 1998), power is not intrinsically bad; it is neutral. The problem is that "bad" people want it and abuse it. This is especially true of coercive and force-based power, though power to persuade is routinely abused by demagogues and con artists. Competition for power leads to the anarchic, cut-throat jockeying often seen in politics. Critical here is the hard-to-define, indeed almost intangible, concept of *respect*. Some politicians have a combination of courage, social skills, and knowledge that gives them prestige (Henrich 2016) and makes them *respected*.

The classic sociological definition of a state is that it is a society in which established elites can legitimately call out the use of force, but ordinary individuals cannot (except in self-defense). (Elites may delegate force, e.g. to sheriffs, but they retain legal authority over it.) A state must trade on legitimacy. If most people do not respect the elite, or a specific law, there is not much the elite can do about it.

China learned the lessons of legitimacy early. In a premodern world, Chinese elites could not police every village and house. By Zhou Dynasty times, they had learned to cultivate legitimacy through ritual, song, history, monumental architecture, and the other displays that establish a government (Anderson 1991). Through these techniques, the Chinese imperial government usually commanded an astonishing amount of loyalty and support due to wide belief in its legitimacy. When it lost that, it collapsed. It had lost the intangible but critical legitimacy that kept even bandits and rebels in awe, and, more to the point, kept the officials loyal and honest enough to run the country.

Threats of social sanction—public shame, dishonor, or ostracism—are worst of all; many, if not most, soldiers and suicide bombers are motivated by supporting their comrades (Atran 2010) and by avoiding shame. Revolutionaries often focus on the failure of the elites to deliver tangible rewards: stability, economic growth, food and water. This implies that "legitimacy" is a mix of fairness and rewards, with a component of blind loyalty.

In imperial times, when people were forced by harsh conditions to be tough, it was the simple fact that war was the general way of life, and one fought or was overrun. The Chinese dynastic founders were reasonable, talented men who were not psychopaths (with the possible exception of Zhu Yuanzhang of Ming), but they fought hard, because they would have been killed otherwise. *Only strict laws, equally enforced, can prevent destructive negative-sum games from starting,* since psychopaths and similar amoral persons will always succeed otherwise, changing the game inevitably to a mutual-destruction game. This is the truth behind Hobbes' "warre."

East Asia knew three forms of autocracy: Literal one-man rule, rule by elites (aristocracy), and bureaucratic rule. The East Asian states had the first of these at the very beginning, but soon converged on the third. None ever had democracy or republics, unlike ancient Greece and Rome.

One form of authority derives from religion. Religion is the collective representation of the community (Durkheim 1995 [1912]), and especially the vehicle for indoctrinating the citizenry with the basic, fundamental ideas and ideals of the community. All the respect and authority due to the community should ideally adhere to the religion, though in fact it rarely commands that much respect. Religious authority depends on the belief of the faithful. True believers in the religion very often doubt the credibility of its practitioners and spokespersons. The result is that, in practice, an ecclesiastical potentate must act much like a secular one, a point made clear by Machiavelli (2005 [ca. 1515]: 73–75).

From time immemorial, secular rulers have tried to take advantage of religion. The crudest way is by simply claiming divine right. This, however, runs a risk: since the ruler cannot deliver rain or health any better than anyone else, a long drought or a plague is apt to get him ousted. This was one way climate change unquestionably affected East Asian politics: turns for the worse clearly delegitimized several monarchs, including the last Mongol emperors of China.

Even in the past, and certainly today, religion was not the only collective representation of community. Politicians, in China and central Asia perhaps more than elsewhere, early learned ways to make loyalty to the state, its laws, and its culture transcend religion. China successfully commanded loyalty despite or because of extreme religious pluralism.

1.14 States

States emerged in the best spots for farming and trade. There is a good correlation between the points of origin of highly developed agriculture and the points of origin of states. The fall from perfect correlation is easily explained by geography. Agriculture in the Near East was earliest in the Jordan River-Orontes River trench, but the state arose in Mesopotamia, where vastly more agricultural land exists, and where trade is easier. Similarly, agriculture in Mexico seems to have emerged in the Balsas River drainage, but states arose not far off in the Grijalva-Usumacinta Delta and the Valley of Mexico, where—again—there is better farmland and more opportunity for developing trade routes. New Guinea had early agriculture, but lacks a central area where a state could reasonably form. It is still fractioned by its rough geography.

The link to civilization should now be clear. The same ecological advantages hold, assuming only that people moved down out of the mountains into ecologically favorable alluvial plains—vast areas of fertile soil, easily conquered and held. Trade and raid do the rest. The correlation between the nodes of great trade routes and the urban centers of early civilizations is perfect. Archaeological finds show the trade came first; high-quality stone and other commodities were traded widely, and the early cities developed at the nodes where routes converged and fertile soil and water made agriculture productive. The more central an area is to the great trade routes, the earlier and more dramatically fast-rising the civilization, with China (as well as Mesopotamia and Egypt) in prime place. Civilization and states developed at what has always been the geographic center of cultural and political China, and spread outward. Conversely, civilization conspicuously failed to emerge in early times in ideal ecological situations that were marginal to long-distance trade, such as California, Argentina, South Africa, and southeastern Australia.

Trade, contact, and communication are basic to states. The origin and rise of states depends on wide-flung trade and communication (Wengrow 2011). It is not mere coincidence that the earliest wheeled vehicles emerge rather suddenly over a large region from Mesopotamia to Germany (Mischka 2011) in the period 3300–3600 BCE, just before true states begin (3100–3000 BCE).

People farmed, traded, intensified their agriculture, smelted metals, and lived in circumscribed valleys for thousands of years before the state arose. The actual trigger for forming states was probably warfare (Engels 1942 [1892]; Turchin et al. 2013). Peter Turchin and collaborators noted that the rise of horse domestication, chariots and horsemanship in central Asia occurred just before, or just as, state formation was occurring in the core areas, and that horse and chariot warfare arose and spread in fairly clear lockstep with the rise of states. Turchin developed an agent-based model that predicted the rise of states in the nearest fertile regions to steppe warfare—the regions the steppe warriors were most fond of raiding. The model predicted 65% of the variance (Turchin et al. 2013). The other 35% is explained largely by trade and soil fertility. We have no idea whether this was true for the Chinese case—we lack the archaeology—but it seems likely.

Nomad raiding pressure on settled, rich, densely populated agricultural societies forced the latter to organize defensively, which in turn forced the nomads to organize better themselves (Engels 1942 [1892]). Eventually the settled societies invent states as a defensive system, and the nomads develop what are now called "secondary" or "reflex" states in response. One need not assume, however, that the pressure was from nomads. Settled but aggressive neighbors produced the same result. China, as well as Mexico, Peru, and other areas, formed states without horse-riding nomad raiders.

Early states and empires were patrimonial: they were run by dynasties, defined by kinship. Wars were between states each governed by a dynasty, or were between factions within a dynasty. Patrimonial states became more bureaucratic over time, and ultimately the bureaucrats could generate their own factions, leading to factional fights that often involved dynastic rulers. Throughout history, most states have been hereditary monarchies, ruled by descendants of some original conqueror. It was virtually universal in Eurasia throughout most of history. The Maya, Aztec, Inca, African, and other remote states independently invented this model; it was not confined to Eurasia. Kings ranged from benevolent and caring to tyrannical. The most tyrannical—the Tamerlanes and Neros—played the state as a negative-sum game: they made everyone worse off, and made people jockey for power to see who could make the other groups worse off faster, thus sparing themselves temporarily.

There were often a few states that were republics or even democracies, and this too is seen almost worldwide, but not in East Asia, where all well-documented states were dynastic. The democratic form of rule is largely identified with ancient Greek innovators, though local democracies and republics appeared in many parts of the world. Democracies and republics were rare, and largely a phenomenon of the Mediterranean and northwest Europe, until the Enlightenment. Since then, they have rapidly grown, especially through colonial and neocolonial pressures on former colonies. They are, however, hothouse flowers, rapidly wilting as resources run out and dictators consolidate control.

The modern dictatorial state is a third form: a state ruled by a democratically elected leader or one who rises through a coup, or occasionally through conquest. Such dictatorships are often negative-sum games. The ruler rules by setting groups against each other. At times, especially if dictatorships survive their first one or two leaders, they can evolve into more growth-oriented polities, as has happened in contemporary China.

A theory often mentioned in the same cultural-ecological breath as Boserup's, and examined even more thoroughly in Marcus and Stanish's volume (2006), is Karl

Wittfogel's theory of "Oriental despotism" (Wittfogel 1957). Wittfogel held that early states in dryland alluvial valleys find it necessary, or at least expedient, to develop large-scale irrigation works. This gives them control over the populace, who must labor on these vast projects if they want to eat. Dryland alluvial valleys are the most productive of all agricultural settings, but require irrigation to produce much—in many cases, to produce anything at all. Wittfogel assumed that irrigation above minimal level required top-down organization.

This has turned out to be wrong. Irrigation has to be managed at the local level, because it has to be fine-tuned according to constantly fluctuating local conditions, and because individual farmers constantly struggle over it. Far from being conducive to despotism, irrigation is conducive to local control. The world's great authoritarian and "despotic" empires usually started in rainfed areas, or in areas like the Nile Valley where natural flooding made irrigation a minor concern. These empires often had a hard time conquering irrigated lands, because of the fierce localism bred by local irrigation management.

On the other hand, once an empire consolidates control of an irrigated valley, dryland or otherwise, its leadership often commissions huge irrigation and water-management works, and uses this to leverage political control (Marcus and Stanish 2006; see also Zhang 2016). We see this in today's world, especially in the case of big dams (Scudder 2005), and there is every reason to assume it in the past.

Wittfogel was misled by his studies of China's Liao Dynasty. His work on Liao was brilliant and pathbreaking, but suffered from poor availability of data. Research has shown that Liao tried hard to be despotic but failed dismally. They depended on rainfed agriculture. Irrigation. Water works were important but not critical. The previous several dynasties had been quite open societies—astonishingly open, in fact, for a premodern regime—and had succeeded brilliantly for long periods, while Liao lasted only about a century.

All this could not have been known to Wittfogel, who had only late, official sources to draw on. Wittfogel's analysis and theory were thoughtful and incisive. Julian Steward, an early convert to Wittfogel (Steward 1955), lived to see the hydraulic hypothesis disproved; Steward (1977) argued that it had been a wonderful theory, because it stimulated so much attention and because it could indeed be disproved, leading to considerable progress in understanding state formation. In fact, the ways of organizing irrigation and water control in ancient societies were many and various (Scarborough 2003). Egalitarian societies did well. So did "heterarchic" ones, with various complementary power systems.

The rise of writing is dramatically correlated with the rise of states. Writing begins in the advanced-chiefdom phase, usually as a means of recording goods paid into the center, sometimes as a ritual or ceremonial tool (see Baines et al. 2008; Houston 2004).

Chinese characters are an important factor in melding East Asia into one world-system. Korea, Japan, and Vietnam all used Chinese characters, and Korea usually used the Chinese language for important writing. Japan and Korea early invented phonetic scripts to write their own languages, or at least the grammatical particles they had to add to Chinese-character texts in Japanese or Korean. But these phonetic scripts remained minor components of the language until twentieth-century Korea converted entirely to *hangul*. Vietnam converted slowly but surely to the Roman-letter script devised by Jesuit missionaries in the seventeenth and eighteenth centuries. Learning about older history and literature, however, still requires knowledge of Chinese and its script.

Central Asia never used Chinese characters. Among other reasons is the complex and intricate grammar of the languages there. They cannot use a script developed for a language with a word-order grammar in which prefixes and suffixes are few. The Turkic and Mongol people in what is now China used scripts based ultimately on Sanskrit writing. Arabic-based scripts dominated further west. Today, most of the languages have fallen under enough Soviet influence to use Russian letters, variously adapted to write sounds that do not occur in Russian. The Russians under Lenin and Stalin promoted this precisely to draw the Mongol, Tadzhik, and Turkic people into the Soviet cultural orbit.

1.15 World-Systems

This book is written from a world-systems perspective. World-system theory, originally elaborated by Immanuel Wallerstein (1976), deals with the relationships between polities (see Chase-Dunn and Anderson 2005; Chase-Dunn and Hall 2011; Inoue and Chase-Dunn 2018). A given group of polities will typically have a core of particularly powerful or central polities that dominate economy and information flows, and often dominate the region militarily; a semiperiphery of middle-level polities fringing the core or closely related to it; and a periphery of more remote, less powerful polities less involved in trade, commerce, and action, but often supplying raw materials to the core. These last are normally repressed or raided by the core and semiperiphery. They are thus kept in relations of dependency or relative weakness.

Terms of trade are one way of keeping this system going. A typical case is the trade reported for Han-dynasty China with the tribal realms that were to form Korea: "Bronze mirrors, silk brocade, jade, vermillion, and gold seals from China were exchanged for the hardwood timber, fish, salt, iron, and agricultural produce of the region.... Many of these goods [from China] were of symbolic nature: caps, robes, seals, and precious items...were status goods enhancing the prestige and authority of native elites" (Seth 2011: 19). Korea was giving up its true wealth in exchange for baubles, doubtless extremely overpriced ones. Moreover, they were for the elite, to mark and shore up their status, while the goods exported were taken—doubtless at low cost—from the ordinary people. Similarly, in Tang times, Japan exported to China and Korea "such natural products as gold and gold dust, mercury, pearls, sulfur, pine, cryptomeria, and hinoki cypress, and also various handicraft items.... They imported from China brocades, damasks and other rich silks, ceramics, writing implements, books, paintings, and copper coins; from Koryŏ came ginseng and saffron..." (McCullough 1999: 95–96). The same pattern is evident. Later, as Korea and Japan moved to semiperiphery status and Japan became core of its own little world-system, the manufactures shifted more and more toward quality items and the "hi-tech" of the era. Extremely sophisticated ceramics, for instance, were an important Korean export by Song times.

This is typical of core-periphery relations, and, far more generally, of primary producers everywhere when they deal with urban makers of expensive goods. Farmers and miners starve, while rulers buy overpriced ceremonial trappings from the core countries. The pattern endures today. Heads of peripheral states from Arabia and Africa to Latin America and southeast Asia are now corrupted by multinational firms that encourage them to buy yachts and jewels, even as the Korean chieftains bought mirrors and seals. (Modern economists have confronted this issue: Bunker and Ciccantell 2005; Humphreys et al. 2007; Ross 2012. Their more general comments on the nature of the problem apply about as well to Korea 2000 years ago as to Africa today.)

Seimperipheral countries are those that lie near the core, but are inhibited by political or military weakness from developing into core status. Korea and Vietnam were semiperipheral countries through most of China's imperial history; they simply could not challenge the massive Chinese monolith. Japan played a semiperipheral role vis-à-vis China, but was the core of its own small nested world-system, including the Ainu and, later, the Ryukyu Islands.

The problems of being a core polity are many. It must maintain its economic and military power. This often leads to its becoming overextended Economic overextension makes it vulnerable to shocks anywhere in the system. Military overextension leads to insupportable expenses, above all in manpower, though also in wealth. A core polity attracts pressure from neighbors. It thus often finds itself beset by enemies on all sides, forcing military overextension. This occurred in China's Han Dynasty. Conversely, attempts to stay peaceful, and thus avoid the costs of militarism, may backfire, by making the state more vulnerable to militant neighbors; this led to the fall of the Song Dynasty. All these and more feed into the cyclic decline and collapse of major states.

Often, the polity that finally takes down a core polity is a semiperipheral one. Core powers tend to maintain peace with each other, or at least distance. Very rarely does a peripheral power rise high enough to defeat a core polity, though we shall examine below the most spectacular case of peripheral rise in all history: the triumph of the Mongols under Chinggis Khan. Semiperipheral powers have the geographic position (near the core), the incentive, and often the ability to mobilize military resources quickly. About half the dynastic transitions in Chinese history involved semiperipheral powers conquering the Chinese Empire from its borders. The other half involved takeover of the Chinese state by internal coup or rebellion.

China was fortunate in that, through most of history after the formation of the Chinese Empire in 221 BCE, no other power could even remotely challenge its core status. Except for part of the period from 1127 to 1279, when the Mongols rivaled and eventually conquered it, China was the richest and most powerful state in eastern Asia. For most of that time, the gap between China and the next most powerful realm was enormous. Yet, semiperipheral states flourished, and often succeeded in taking on China and conquering it.

China regularly warred with neighboring semiperipheral states. We have no idea of the languages spoken by the states conquered early by Shang and Zhou, before 200 BCE. They were so thoroughly absorbed that no record remains. Once the Chinese empire was formed in 221 BCE, China conquered and incorporated semiperipheral marcher states, especially in early centuries. The Xiongnu empire in the northwest, the Nanzhao state in what is now Yunnan, and the kingdom of Yue in the southeast are footnotes in history books today.

Typical is a story in which a semiperipheral state is caught between two expanding cores, or, perhaps worse, between a core and a rising peripheral polity. One example is the progressive disappearance of the Serbi: the Xianbei, Northern Wei, and Khitan, who spoke languages related to Mongolian, now extinct without trace. They were ground down between peripheral Mongol and Tungus powers and the Chinese core. The "Yue" and Nanzhao states were similarly ground between expanding China and semiperipheral southeast Asian states. China often used this fact of statecraft to advantage, "using barbarians to control barbarians" (euphoniously put as *yi yi zhi yi*—the term "Yi" referring, originally, one grouping of non-Chinese on the northern frontiers). They set more remote groups against closer ones. This backfired badly when the Mongols or the Tungus were the more remote groups in question. Those hardened warriors simply incorporated the conquered armies and used them to roll over China itself.

The shifting story of languages on the western frontier of China is important to show world-system dynamics in borderlands (marchlands) close to an imperial core. In that situation, ethnic groups could arise, rapidly gain tremendous power when the empire is weak, lose that power when it strengthens again, and become extinct. The same fate befell the Xiongnu, Xixia (though their language-group survives), and other once-powerful groups that are now lost.

Farther away, the Sogdians who once ruled western central Asia were squeezed between peripheral Turks and core Persians, and their language survives in one tiny mountain village. European equivalents are the rise, enormous power, and total disappearance of the Scythians, Goths, Huns, and Vandals. (At least the Vandals' name survives today, with a new meaning that they would have perversely enjoyed.) The Byzantines played the same game as the Chinese, backing the more remote against the nearer, with similar success or lack of it. Of course, success meant that the formerly remote ally was now at Byzantium's walls, and thus a rival or outright enemy. Then the Byzantines had to deal with that—usually by playing the same game over again. This combined with Byzantium's convoluted and murderous imperial rivalries gave us the term "Byzantine politics." The same conversion from remote ally to next-door enemy happened in China's dealings with the Jurchen.

Societies farther from the center, more protected by mountains or seas, are more stable; the Koreans, Mongols, and Japanese are examples in East Asia. These proved impossible to conquer and hold over time. Vietnam and Korea held out, despite occasional reduction to colony status. Mongolia was ruled by the Qing Dynasty as part of the Manchu Empire, but what is now Mongolia (as opposed to Inner Mongolia) was never actually Chinese. Thailand, Burma, and other southern states were theoretically "tributary." Japan was never challenged from outside, except for a brief and unsuccessful Mongol foray. It was, in fact, the core to a small world-system of its own, before its rather late incorporation into the East Asian world-system.

Peripheral polities were often tribal societies: the Siberian Indigenous groups, the upland peoples of southeast Asia, the Taiwan Aboriginals, the Ainu. They provided minor trade goods, and in southern China they were slowly incorporated in an expanding Chinese empire.

China remained at the center of its own East Asian world-system until the nineteenth century, when the European (or Western) world-system expanded and incorporated it into what has now become a single network that is, literally, a "world" system.

China and the west interacted from earliest times. They were moderately involved with each other, indirectly, by 1 CE. Contact grew rapidly, reaching the point of armed clashes in the eighth century, at which time trade was extensive. The two world-systems finally merged in the Mongol Empire, which conquered all Eurasia from Korea to Hungary and from Siberia to India. The empire slowly disintegrated, leaving the two ends of Eurasia separate again, but Portuguese expansion and sea trade quickly ended that. The two systems remained fused at the hip, so to speak, by trade, commerce, and communication, but largely separate in military and political systems until the nineteenth century (Chase-Dunn and Hall 2011).

More important is the way world-systems theory can integrate all the above issues—climate, power, contingent history, geography, foodways, population dynamics, disease, and all—in one comprehensive view. World-systems thinking integrates across regions and time periods, rather than stopping short at inevitably arbitrary national boundaries. It is well adapted to studying the effects of climate on history.

East Asia, for many centuries, has been a single world-system, with China as core, Korea, Japan, and Vietnam (and to some extent other southeast Asian countries) as semiperiphery, and central Asia as periphery, sometimes locally rising to semiperipheral status. Korea and Vietnam were incorporated early, by brutal military invasion and continued political and military pressure. Southeast Asia was intensely involved in a wider world-system, linked by the trade network stretching from China to Rome (Chew 2018).

World-systems are linked together by four subsystems: information networks, prestige-goods networks, political-military networks, and bulk goods networks (Chase-Dunn and Hall 2011; cf. Abu-Lughod 1989). Usually, these develop in the order given. Information travels fastest, either with or closely followed by prestige goods. In Asia, we see the coming of prestige goods with the very first long-range contacts of peoples, and then the rise of very extensive prestige-goods trade, especially along the Silk Route, from well before the Common Era. The most significant thing, however, was the spread of Chinese culture, which became the defining high culture in all areas except farthest Central Asia. Japan, Korea, and Vietnam formed their own Sinicized high cultural traditions, variously localized; all retained tremendous local cultural dynamism and independence, shown in their distinctive languages.

Political and military networks are commonly taken as established by the time a major military confrontation occurs; by this standard, what is now the core provinces of China were largely a single world-system by 700 BCE, entirely so by 200 BCE. Central Asia, Vietnam and Korea were brought in by the Han Dynasty. Tibet attacked China in the 700s. Japan was somewhat integrated by large-scale missions in the 600 s, but not militarily attacked until the 1200s. Japan was in fact its own little world-system for most of history, only loosely integrated with the rest of East Asia (Batten 2003). It came within the Chinese purview in Han times, when envoys appeared from the Wa (i.e. some polity or polities in Japan). A Queen Pimiko or Himiko of Wa sent missions in the 200 s to the state of Wei. Japan and Korea were in rather more touch. Japan imported Korean iron very early, and otherwise drew heavily on the peninsula (Batten 2003: 188).

Bulk goods were routinely flowing back and forth at about the same time periods, but central Asia was not really integrated into a bulk goods network until the Mongol period or later; parts of it never were until the twentieth century. Bulk shipping of everyday commodities, notably including coins, metal, and ceramics, reached Korea and Japan by the 600 s and were extensive by the 1200s. However, remote parts of Chinese Siberia and northern (Ainu) Japan were not involved in bulk goods trade till the seventeenth century or later. By this time, sea trade was extremely extensive everywhere, but Japan and China both pulled back sharply on it, and Korea did so later. Of course, food was extensively shipped within regions, and to some extent between the major countries. World-system dynamics had an up-and-down career in the east.

1.16 Conclusion

This chapter has set the stage for more specific theories. The climate and geography of East Asia has been summarized. The needs of humans and the resulting structure of society, including dynamics of population and innovation, have been considered. The next task is to look at direct theories of dynastic cycling.

References

Abu-Lughod, J. (1989). *Before European hegemony: The world-system A.D. 1250–1350*. New York: Oxford University Press.
Amrith, S. (2018). *Unruly waters: How rains, rivers, doasts, and seas have shaped Asian history*. New York: Basic Books.
Anderson, E. N. (1988). *The food of China*. New Haven, CT: Yale University Press.
Anderson, B. (1991). *Imagined communities* (2nd ed.). London: Verso.
Anderson, E. N. (2011). Emotions, motivation, and behavior in cognitive anthropology. In D. Kronenfeld, G. Bennardo, V. C. de Munck, & M. D. Fischer (Eds.), *A companion to cognitive anthropology* (pp. 331–354). New York: Wiley-Blackwell.
Anderson, E. N. (2014a). *Caring for place*. Walnut Creek, CA: Left Coast Press.
Anderson, E. N. (2014b). *Food and environment in early and medieval China*. Philadelphia, PA: University of Pennsylvania Press.

References

Anderson, E. N., & Chase-Dunn, C. (2005). The rise and fall of great powers. In C. Chase-Dunn & E. N. Anderson (Eds.), *The historical evolution of world-systems* (pp. 1–19). New York: Palgrave Macmillan.

Aristotle. (1953). *Metaphysics* (R. Hope, Trans.). New York: Columbia University Press.

Atran, S. (2010). *Talking to the enemy: Faith, brotherhood, and the (un)making of terrorists*. New York: HarperCollins.

Baines, J., Bennet, J., & Houston, S. (Eds.). (2008). *The disappearance of writing systems: Perspectives on literacy and communication*. London: Equinox.

Bandura, A. (1982). Self-efficacy mechanism in human agency. *American Psychologist, 37*, 122–147.

Bandura, A. (1986). *Social foundations of thought and action: A social cognitive theory*. Englewood Cliffs, NJ: Prentice-Hall.

Barfield, T. J. (1989). *The perilous frontier: Nomadic empires and China, 221 BC to AD 1757*. Cambridge, MA: Blackwell.

Barfield, T. J. (1993). *The nomadic alternative*. Englewood Cliffs, NJ: Prentice-Hall.

Barnes, G. L. (2015). *Archaeology of East Asia: The rise of civilization in China, Korea and Japan* (3rd ed.). Oxford: Oxbow Books.

Batten, B. (2003). *The end of Japan: Premodern frontiers, boundaries, and interactions*. Honolulu: University of Hawai'i Press.

Beals, A. (1967). *Culture in process*. New York: Holt, Rinehart and Winston.

Beck, J. W., Zhou, W., Li, C., Wu, Z., White, L., Xian, F., Kong, X., & An, Z. (2018). A 550,000-year record of east Asian monsoon rainfall from ^{10}Be in loess. *Science, 360*, 877–881.

Becker, G. (1996). *Accounting for tastes*. Cambridge, MA: Harvard University Press.

Bello, D. (2016). *Across forest, steppe, and mountain: Environment, identity and empire in Qing China's borderlands*. Cambridge: Cambridge University Press.

Bengtsson, T., Campbell, C., & Lee, J. A. (Eds.). (2004). *Life under pressure: Mortality and living standards in Europe and Asia, 1700–1900*. Cambridge, MA: MIT Press.

Best, J. W. (2007). *A history of the early Korean kingdom of Paekche, together with an annotated translation of the Paekche annals of the Samguk Sagi*. Cambridge, MA: Harvard University Press.

Bevan, A., Colledge, S., Fuller, D., Fyfe, R., Shennan, S., & Stevens, C. (2017). Holocene fluctuations in human population demonstrate repeated links to food production and climate. *Proceedings of the National Academy of Sciences, 114*, E10524–E10531. http://www.pnas.org/content/114/49/E10524.full.

Bielenstein, H. (1986). Wang Mang, the restoration of the Han dynasty, and later Han. In D. Twitchett & M. Loewe (Eds.), *The Cambridge history of China. Vol. 1. The Ch'in and Han empires, 221 B.C.-A.D. 220* (pp. 223–290). Cambridge: Cambridge University Press.

Bird, B., Lei, Y., Perello, M., Polissar, P. J., Yao, T., Finney, B., Bain, D., Pompeani, D., & Thompson, L. G. (2017). Late Holocene Indian summer monsoon variability revealed from a 3300-year-long Lake sediment record from Nir'pa co, southeastern Tibet. *The Holocene, 27*, 541–552.

Bock, M., Schmitt, J., Beck, J., Seth, B., Chappellaz, J., & Fischer, H. (2017). Glacial/interglacial wetland, biomass burning, and geologic methane emissions constrained by dual stable isotopic CH4 ice Core records. *Proceedings of the National Academy of Sciences, 114*, E5778–E5786.

Bol, P. K. (2003). Neo-Confucianism and local society, twelfth to sixteenth century: A case study. In P. J. Smith & R. von Glahn (Eds.), *The song-Yuan-Ming transition in Chinese history* (pp. 1241–1283). Cambridge, MA: Harvard University Asia Center.

Bond, M. (Ed.). (1986). *The psychology of the Chinese people*. Oxford: Oxford University Press.

Boserup, E. (1965). *The conditions of agricultural growth*. Chicago: Aldine.

Bowles, S. (2006). Group competition, reproductive leveling, and the evolution of human altruism. *Science, 314*, 1569–1572.

Bray, F. (1984). *Science and civilisation in China. Vol. VI: Biology and biological technology. Part 2: Agriculture*. Cambridge: Cambridge University Press.

Brook, T. (2010). *The troubled empire: China in the Yuan and Ming*. Cambridge, MA: Harvard University Press.
Brooke, J. L. (2014). *Climate change and the course of global history*. Cambridge: Cambridge University Press.
Bunker, S., & Ciccantell, P. (2005). *Globalization and the race for resources*. Baltimore, MD: Johns Hopkins University Press.
Campbell, B. M. S. (2016). *The great transition: Climate, disease and Society in the Late-Medieval World*. Cambridge: Cambridge University Press.
Caplan, B. (2007). *The myth of the rational voter: Why democracies choose bad policies*. Princeton, NJ: Princeton University Press.
Carneiro, R. (2003). *Evolutionism in cultural anthropology: A critical history*. Boulder, CO: Westview.
Chang, C. (2018). *Rethinking prehistoric Central Asia: Shepherds, farmers, and nomads*. Oxford and New York: Routledge.
Chase-Dunn, C., & Anderson, E. N. (Eds.). (2005). *The historical evolution of world-systems*. New York: Palgrave Macmillan.
Chase-Dunn, C., & Hall, T. D. (2011). East and west in world-systems evolution. In P. Manning & B. Gills (Eds.), *Andre Gunder Frank and global development* (pp. 97–119). London: Routledge.
Chayanov, A. V. (1966). *The theory of peasant economy*. Homewood, Ill: Richard D. Irwin.
Chen, Q. (2015). Climate shocks, dynastic cycles and nomadic conquests: Evidence from historical China. *Oxford Economic Papers, 67*, 185–2024.
Chew, S. (2018). *The Southeast Asia connection: Trade and polities in the Eurasian world economy, 500 BC-AD 500*. New York: Berghahn.
Choi, J.-K., & Bowles, S. (2007). The coevolution of parochial altruism and war. *Science, 318*, 636–640.
Collins, R. (1986). *Weberian sociological theory*. Cambridge: Cambridge University Press.
Cressey, G. B. (1955). *Land of the 500 million: A geography of China*. New York: McGraw-Hill.
D'Alpoim Guedes, J. (2018). Did foragers adopt farming? A perspective from the margins of the Tibetan plateau. *Quaternary International, 489*, 91–100.
D'Alpoim Guedes, J., & Bocinsky, R. K. (2018). Climate change stimulated agricultural innovation and exchange across Asia. *Science Advances, 4*, 10, eaar 4491, October 31, http://advances.sciencemag.org/content/4/10/eaar4491/?fbclid=IwAR1JRsDxkdZ6w13-kXl4auW9Hyr1dUCpMfyPlimgmGszK6ES6q6_9gEtCr0
Davis, R. L. (2009a). The reigns of Kuang-Tsung and Ning-Tsung. In D. Twitchett & P. J. Smith (Eds.), *The Cambridge history of China. Vol. 5. The sung dynasty and it precursors, 907–1279* (pp. 756–838). Cambridge: Cambridge University Press.
Davis, R. L. (2009b). The reign of Li-Tsung. In D. Twitchett & P. J. Smith (Eds.), *The Cambridge history of China. Vol. 5. The sung dynasty and it precursors, 907–1279* (pp. 839–912). Cambridge: Cambridge University Press.
Di Cosmo, N. (2018). *The scientist as antiquarian: History, climate, and the new past*. Institute for Advanced Studies, Posting. https://www.ias.edu/ideas/2018/di-cosmo-history-climate-and-new-past
Di Cosmo, N., Oppenheimer, C., & Büntgen, U. (2017). Interplay of environmental and socio-political factors in the downfall of the eastern Türk empire in 630 C.E. *Climatic Change, 145*, 383–395.
Dikötter, F. (2010). *Mao's great famine: The history of China's Most devastating catastrophe, 1958–1962*. New York: Walker and Co.
Dunbar, R. (2010). *How many friends does one person need? Dunbar's number and other evolutionary quirks*. Cambridge, MA: Harvard University Press.
Durkheim, E. (1995[1912]). *The elementary forms of religious life* (K. E. Fields, Trans.). New York: Free Press.
Durrenberger, E. P. (Ed.). (1984). *Chayanov, peasants, and economic anthropology*. New York: Academic.

References

Elster, J. (2007). *Explaining social behavior: More nuts and bolts for the social sciences.* Cambridge: Cambridge University Press.
Elvin, M. (1973). *The pattern of the Chinese past.* Stanford, CA: Stanford University Press.
Elvin, M. (2004). *The retreat of the elephants: An environmental history of China.* New Haven, CT: Yale University Press.
Engels, F. (1942[1892]). *The origin of the family, private property and the state, in the light of the researches of Lewis H. Morgan.* New York: International Publishers.
Feng, W., Campbell, C., & Lee, J. Z. (2010). Agency, hierarchies, and reproduction in northeastern China, 1789 to 1840. In N. O. Tsuya, W. Feng, G. Alter, & J. Z. Lee (Eds.), *Prudence and pressure: Reproduction and human Agency in Europe and Asia, 1700–1900* (pp. 287–316). Cambridge, MA: MIT Press.
Flanagan, L. (1998). *Ancient Ireland.* New York: St. Martin's Press.
Fohlmeister, J., Plessen, B., Dudashvili, A. S., Tjallingii, R., Wolff, C., Gafurov, A., & Cheng, H. (2017). Winter precipitation changes during the medieval climate anomaly and little ice age in arid Central Asia. *Quaternary Science Reviews, 178,* 24–36.
Frankl, V. (1959). *Man's search for meaning: An introduction to Logotherapy.* Boston: Beacon.
Frankl, V. (1978). *The unheard cry for meaning: Psychotherapy and humanism.* New York: Simon & Schuster.
Gao, L., Xiong, S., Ding, Z., Jin, G., Jiabing, W., & Ye, W. (2018). Role of the mid-Holocene environmental transition in the decline of late Neolithic cultures in the deserts of NE China. *Quaternary Science Reviews, 190,* 98–113.
Giddens, A. (1984). *The constitution of society.* Berkeley, CA: University of California Press.
Gigerenzer, G. (2007). *Gut feelings: The intelligence of the unconscious.* New York: Viking.
Goldsmith, Y., Broecker, W. S., Hai, X., Polissar, P. J., DeMenocal, P. B., Porat, N., Lan, J., Cheng, P., Shou, W., & An, Z. (2017). Northward extension of east Asian monsoon Covaries with intensity on orbital and millennial timescales. *Proceedings of the National Academy of Sciences, 114,* 1817–1821.
González-Forero, M., & Gardner, A. (2018). Inference of ecological and social drivers of human brain-size evolution. *Nature, 557,* 554–557.
Gourou, P. (1955). *The peasants of the Tonkin Delta: A study in human geography.* New Haven, CT: HRAF Press.
Grapard, A. G. (1999). Religious practices. In D. Shively & W. H. McCullough (Eds.), *The Cambridge history of Japan, Vol. 2, Heian Japan* (pp. 576–643). Cambridge: Cambridge University Press.
Guiot, J., & Cramer, W. (2016). Climate change: The 2015 Paris agreement thresholds and Mediterannean Basin ecosystems. *Science, 354,* 465–468.
Gunderson, L., & Holling, C. S. (2002). *Panarchy synopsis: Understanding transformations in human and natural systems.* Washington, DC: Island Press.
Gunderson, L., Allen, C. R., & Holling, C. S. (2009). *Foundations of ecological resilience.* Washington, DC: Island Press.
Hardin, G. (1968). The tragedy of the commons. *Science, 162,* 1243–1248.
Harper, K. (2017). *The fate of Rome: Climate, disease, and the end of an empire.* Princeton, NJ: Princeton University Press.
Hayami, Y., & Ruttan, V. (1985). *Agricultural development* (2nd ed.). Baltimore, MD: Johns Hopkins University Press.
Henrich, J. (2016). *The secret of our success: How culture is driving human evolution, domesticating our species, and making us smarter.* Princeton, NJ: Princeton University Press.
Hinsch, B. (1988). Climatic change and history in China. *Journal of Asian History, 22,* 131–159.
Hobbes, T. (1950 [1651]). *Leviathan.* New York: Dutton.
Hofman, J., Sharma, A., & Watts, D. J. (2017). Prediction and explanation in social systems. *Science, 355,* 486–488.
Houston, S. (Ed.). (2004). *The first writing: Script invention as history and process.* Cambridge: Cambridge University Press.

Hu, G., Yi, C., Zhang, J., Dong, G., Liu, J., Xu, X., & Jiang, T. (2017). Extensive glacial advances during the last glacial maximum near the eastern Himalayan Syntaxis. *Quaternary International, 443B*, 1–12.
Hume, D. (1969[1739–1740]). *A treatise of human nature*. New York: Penguin.
Humphreys, M., Sachs, J., & Stiglitz, J. (Eds.). (2007). *Escaping the resource curse*. New York: Columbia University Press.
Ibn Khaldun. (1958). *The Muqaddimah* (F. Rosenthal, Trans.). New York: Paragon.
Inoue, H., & Chase-Dunn, C. (2018). *A multilevel spiral model of sociocultural evolution: Polities and interpolity systems*. In Paper, international studies association, annual conference, San Francisco.
Johnson, A., & Earle, T. (1987). *The evolution of human societies: From foraging group to agrarian state*. Stanford, CA: Stanford University Press.
Kahneman, D. (2011). *Thinking, fast and slow*. New York: Farrar, Straus and Giroux.
Kang, S., Wang, X., Roberts, H. M., Duller, G. A. T., Cheng, P., Lu, Y., & An, Z. (2018). Late Holocene anti-phase change in the east Asian summer and winter monsoons. *Quaternary Science Reviews, 188*, 28–36.
Kenrick, D., Griskevicius, V., Neuberg, S. L., & Schaller, M. (2010). Renovating the pyramid of needs: Contemporary extensions built upon ancient foundations. *Perspectives on Psychological Science, 5*, 292–314.
Kidder, T. R., Haiwang, L., Storozum, M. J., & Zhen, Q. (2016). New perspectives on the collapse and regeneration of the Han Dynasty. In R. K. Faulseit (Ed.), *Beyond collapse: archaeological perspectives on resilience, revitalization, and transformation in complex societies* (pp. 70–98). Carbondale, IL: Southern Illinois University Press.
Kiernan, B. (2017). *Viet Nam: A history from earliest times to the present*. New York: Oxford University Press.
Kintigh, K. W., & Ingram, S. (2018). Was the drought really responsible? Assessing statistical relationships between climate extremes and cultural traditions. *Journal of Archaeological Science, 89*, 25–31.
Kleiner, K. (2009). I know that nose. *Scientific American Mind*. April-May, 7.
Kong, D., Wei, G., Chen, M.-T., Peng, S., & Liu, Z. (2017). Northern South China Sea SST changes over the last two millennia and possible linkage with solar irradiance. *Quaternary International, 459*, 29–34.
Kroeber, A. L. (1944). *Configurations of culture growth*. Berkeley, CA: University of California Press.
Lakoff, G. (2006). *Whose freedom? The Battle over America's Most important idea*. New York: Farrer, Strauss and Giroux.
Langer, E. (1983). *The psychology of control*. Beverly Hills, CA: Sage.
Latour, B. (2005). *Reassembling the social: An introduction to actor-network-theory*. Oxford: Oxford University Press.
Le Roy Ladurie, E. (1971). *Times of feast, times of famine* (B. Bray, Trans.). Garden City, NY: Doubleday.
Lee, H. F. (2018). Internal wars in history: Triggered by natural disasters or by socio-ecological catastrophes? *The Holocene, 28*, 1071–1081.
Lee, H. F., Pei, Q., Zhang, D. D., & Choi, K. P. K. (2015). Quantifying the intra-regional precipitation variability in northwestern China over the past 1,400 years. *PLoS One*. https://doi.org/10.1371/journal.pone.0131693.
Lee, H. F., Zhang, D. D., Pei, Q., Jia, X., & Yue, R. (2016). Demographic impact of climate change on northwestern China in the late Imperial era. *Quaternary International*. http://www.sciencedirect.com/science/article/pii/S1040618216303998.
Lerro, B. (2000). *From earth spirits to sky gods: The socioecological origins of monotheism, individualism, and hyperabstract reasoning from the stone age to the axial iron age*. Lanham, MD: Lexington Books.

References

Levi, S. C., & Sela, R. (Eds.). (2010). *Islamic Central Asia: An anthology of historical sources*. Bloomington, IN: Indiana University Press.

Li, J., & Liu, X. (2018). Orbital- and suborbital-scale changes in the east Asian summer monsoon since the last Deglaciation. *The Holocene, online*, (May 7). https://doi.org/10.1177/0959683618771479.

Li, J., Dodson, J., Yan, H., Zhang, D. D., Zhang, X., Xu, Q., Lee, H. F., Pei, Q., Cheng, B., Li, C., Ni, J., Sun, A., Lu, F., & Zong, Y. (2017a). Quantifying climatic variability in monsoonal northern China over the past 2200 years and its role in driving Chinese dynastic changes. *Quaternary Science Reviews, 159*, 35–46.

Li, J., Kong, L., Yang, H., Wang, Q., Yang, X., Ji, S., & Dao, C. (2017b). Temperature variations since 1750 CE inferred from an alpine Lake in the southeastern margin of the Tibetan plateau. *Quaternary International, 436*, 37–44.

Li, Y., Lingmei, X., Zhang, C., Liu, Y., Zhu, G., & Zhou, X. (2017c). Temporal and spatial evolution of vegetation and Lake hydrological status, China. *The Holocene*. https://doi.org/10.1177/0959683617744260.

Li, Y., Chen, X., Xiao, X., Zhang, H., Xue, B., Shen, J., & Zhang, E. (2018). Diatom-based inference of Asian monsoon precipitation from a volcanic Lake in Southwest China for the last 18.5 ka. *Quaternary Science Reviews, 118*, 109–120.

Lieberman, V. (2003). *Strange parallels: Southeast Asia in global context, c. 800–1830. Vol. 1: Integration on the mainland*. Cambridge: Cambridge University Press.

Lieberman, V. (2009). *Strange parallels: Southeast Asia in global context, c. 800–1830. Vol 2: Mainland mirrors: Europe, Japan, China, South Asia, and the islands*. Cambridge: Cambridge University Press.

Lin, Z., Ma, C., Tang, L., Liu, K.-b., Mao, L., Zhang, Y., Lu, H., Wu, S., & Tu, Q. (2016). Investigation of peat sediments from Daiyun Mountain in Southeast China: Late Holocene vegetation, climate and human impact. *Archaeobotany and Vegetation History, 25*, 359–373.

Liu, B., Wang, N., Chen, M., Wu, X., Mo, D., Liu, J., Xu, S., & Zhuang, Y. (2017). Earliest hydraulic Enterprise in China, 5100 years ago. *Proceedings of the National Academy of Sciences, 114*, 13637–13642.

Loewe, M. (1986). The religious and intellectual background. In D. Twitchett & M. Loewe (Eds.), *The Cambridge history of China. Vol. 1. The Ch'in and Han empires, 221 B.C.-A.D. 220* (pp. 649–725). Cambridge: Cambridge University Press.

Lu, M., & He, F. (2017). Estimating regional species richness: The case of China's vascular plant species. *Global Ecology and Biogeography*. https://doi.org/10.1111/geb.12589.

Lu, L.-M., Mao, L.-F., Yang, T., Ye, B.-F., Liu, B., Li, H.-L., Sun, M., Miller, J. T., Mathews, S., Hu, H.-H., Niu, Y.-T., Peng, D.-X., Chen, Y.-H., Smith, S. A., Chen, M., Xiang, K.-L., Le, C.-T., Dang, V.-C., Lu, A.-M., Soltis, P. S., Soltis, D. E., Li, J.-H., & Chen, Z.-D. (2018). Evolutionary history of the angiosperm Flora of China. *Nature, 554*, 234–238.

Machiavelli, N. (2005). *The prince* (W. J. Connell, Ed. and Trans.). (Italian Original ca. 1515). Boston, MA: Bedford/St. Martin's.

Malinowski, B. (1944). *A scientific theory of culture*. Oxford: Oxford University Press.

Malinowski, B. (1948). *Magic, science and religion*. Glencoe, IL: Free Press.

Mallory, W. H. (1926). *China: Land of famine*. New York: American Geographic Society.

Malthus, T. (1960[Orig. 1798]). *On population*. New York: Random House.

Marcus, J., & Stanish, C. (Eds.). (2006). *Agricultural strategies*. Los Angeles: Cotsen Institute of Archaeology, UCLA.

Marks, R. B. (1998). *Tigers, Rice, silk, and silt: Environment and economy in late Imperial South China*. New York: Cambridge University Press.

Marks, R. B. (2012). *China: Its environment and history*. Lanham, MD: Rowman and Littlefield.

Marx, K. (1972). *The eighteenth Brumaire of Louis Bonaparte* (C. P. Dutt and International Publishers, Trans.). (German Orig. 1869.). New York: International Publishers.

Marx, K. (1973). *Grundrisse*. Baltimore, MD: Penguin.

Masahide, B. (1991). Thought and religion, 1550–1700 (K. Wildman Nakai, Trans.). In J. W. Hall, & J. L. McClain (Eds.), *The Cambridge history of Japan. Vol. 4: Early Modern Japan* (pp. 373–424). Cambridge: Cambridge University Press.
Maslow, A. (1970). *Motivation and personality* (2nd ed.). New York: Harper and Row.
McCullough, W. H. (1999). The Heian court, 794-1070. In H. Japan, D. Shively, & W. H. McCullough (Eds.), *The Cambridge History of Japan, vol. 2* (pp. 20–96). Cambridge: Cambridge University Press.
McDermott, J. P., & Yoshinobu, S. (2015). Economic change in China, 960-1279. In J. W. Chaffee & D. Twitchett (Eds.), *The Cambridge history of China. Vol. 5, part two: Sung China, 960–1279* (pp. 321–436). Cambridge: Cambridge University Press.
Métailié, G. (2015). *Science and civilisation in China. Vol. 6, Biology and biological technology. Part IV: Traditional botany: An ethnobotanical approach* (J. Lloyd, Trans.) (Xli, 748 pp). Cambridge: Cambridge University Press.
Mills, C. W. (1959). *The sociological imagination*. New York: Grove Press.
Mischka, D. (2011). The Neolithic burial sequence at Flintbek LA 3, North Germany, and its cart tracks: A precise chronology. *Antiquity, 85*, 742–759.
Moise, E. (1977). Downward social mobility in pre-revolutionary China. *Modern China, 3*, 3–32.
Moore, J. (2015). *Capitalism in the web of life: Ecology and the accumulation of capital*. New York: Verso.
Morris, I. (2010). *Why the west rules...for now*. New York: Farrar, Straus and Giroux.
Mote, F. W. (1999). *Imperial China 900–1800*. Cambridge, MA: Harvard University Press.
Muir-Wood, R. (2016). *The cure for catastrophe: How we can stop manufacturing natural disasters*. London: Oneworld.
Olson, M. (1965). *The logic of collective action*. Cambridge, MA: Harvard University Press.
Parker, G. (2013). *Global crisis: War, climate change and catastrophe in the seventeenth century*. New Haven, CT: Yale University Press.
Pascal, B. (2005). *Pensées* (R. Ariew, Ed. and Trans.). Indianapolis: Hackett.
Pei, Q., & Zhang, D. D. (2014). Long-term relationship between climate change and nomadic migration in Central Asia. *Ecology and Society, 19*(2), 68–84.
Pei, Q., Li, G., Zhang, D. D., & Lee, H. F. (2016a). Temperature and precipitation effects on agrarian economy in late Imperial China. *Environmental Research Letters, 11*, 064008.
Pei, Q., Zhang, D. D., & Lee, H. F. (2016b). Contextualizing human migration in different agro-ecological zones in ancient China. *Quaternary International, 426*, 65–74.
Pei, Q., Lee, H. F., & Zhang, D. D. (2018). Long-term association between climate change and agriculturalists' migration in historical China. *The Holocene, 28*, 208–216.
Qiao, B., & Yi, C. (2017). Reconstruction of little ice age glacier area and equilibrium line altitudes in the central and Western Himalaya. *Quaternary International, 444A*, 65–75.
Ross, M. L. (2012). *The oil curse: How petroleum wealth shapes the development of nations*. Princeton, NJ: Princeton University Press.
Scarborough, V. (2003). *The flow of power: Ancient water systems and landscapes*. Santa Fe: SAR Press.
Schultz, T. (1968). *Agriculture and economic growth*. New York: McGraw-Hill.
Schulz, R. (1976). Some life and death consequences of perceived control. In J. S. Carroll & J. W. Payne (Eds.), *Cognition and social behavior* (pp. 135–153). New York: Academic.
Scott, J. C. (1985). *Weapons of the weak*. New Haven, CT: Yale University Press.
Scott, J. C. (1998). *Seeing like a state*. New Haven, CT: Yale University Press.
Scudder, T. (2005). *The future of large dams: Dealing with social, environmental, institutional and political costs*. London: Earthscan.
Sen, A. (1973). *On economic inequality*. Oxford: Oxford University Press.
Sen, A. (1982). *Poverty and famines: An essay on entitlement and deprivation*. Oxford: Oxford University Press.
Sen, A. (1984). *Resources, values, and development*. Cambridge, MA: Harvard University Press.

References

Sen, A. (1992). *Inequality reconsidered*. Cambridge: Harvard University Press and Russell Sage Foundation.
Sen, A. (1997). *Hunger in the contemporary world*. London: London School of Economics.
Sen, A. (2001). *Development as freedom*. New York: Oxford University Press.
Seth, M. J. (2011). *A history of Korea, from antiquity to the present*. Lanham, MD: Rowman and Littlefield.
Sheehan, O., Watts, J., Gray, R. D., & Atkinson, Q. D. (2018). Coevolution of Landesque capital intensive agriculture and sociopolitical hierarchy. *Proceedings of the National Academy of Sciences, 115*, 3628–3633.
Shi, X., Kirby, E., Furlong, K., Meng, K., Robinson, R., Lu, H., & Wang, E. (2017). Rapid and punctuated late Holocene recession of Siling co, Central Tibet. *Quaternary Science Reviews, 172*, 15–31.
Shun, K.-L. (1997). *Mencius and early Chinese thought*. Stanford, CA: Stanford University Press.
Sima, Q. (1993). *Records of the Grand Historian* (B. Watson, Trans.). (Revised edition). New York: Columbia University Press.
Skinner, G. W. (Ed.). (1977). *The City in late Imperial China*. Stanford: Stanford University Press.
Skinner, G. W. (2001). *Marketing and social structure in China*. Ann Arbor, MI: Association for Asian Studies.
Smil, V. (2004). *China's past, China's future: Energy, food, environment*. New York: RoutledgeCurzon.
Smith, A. (1910 [1776]). *The wealth of nations*. New York: E. P. Dutton.
Smith, C. A. (Ed.). (1976). *Regional analysis*. New York: Academic.
Smith, T. C. (1977). *Nakahara: Family farming and population in a Japanese Village, 1717–1830*. Stanford, CA: Stanford University Press.
Smith, A. (2000 [1759]). *The theory of moral sentiments*. Amherst, NY: Prometheus Books.
Sommer, M. (2015). *Polyandry and wife-selling in Qing dynasty China*. Berkeley, CA: University of California Press.
Steward, J. (1955). *Theory of culture change*. Urbana: University of Illinois Press.
Steward, J. (1977). *Evolution and Ecology: Essays on Social Transformation*. Ed. In *Jane Steward and Robert Murphy*. Urbana: University of Illinois Press.
Sun, W. (Sun Zi). (2008). *The art of war: Sun-Tzu* (J. H. Huang, Trans.). (2nd ed.) (Orig. 1993). New York: HarperPerennial (Orig. Pub as Sun-Tzu, by William Morrow).
Tan, L., Cai, Y., Cheng, H., Edwards, L. R., Lan, J., Zhang, H., Li, D., Ma, L., Zhao, P., & Gao, Y. (2018). High resolution monsoon precipitation changes on southeastern Tibetan plateau over the past 2300 years. *Quarternary Science Reviews, 195*, 122–132.
Tian, F., Wang, Y., Liu, J., Tang, W., & Jiang, N. (2017a). Late Holocene climate change inferred from a Lacustrine sedimentary sequence in southern Inner Mongolia, China. *Quaternary International, 452*, 22–32.
Tian, H., Yan, C., Lei, X., Büntingen, U., Stenseth, N., & Zhang, Z. (2017b). Scale-dependent climatic drivers of human epidemics in ancient China. *Proceedings of the National Academy of Sciences, 114*, 12970–12975.
Timmermann, A., An, S.-I., Kung, J.-S., Jin, F.-f., Cai, W., et al. (2018). El Niño-southern oscillation complexity. *Nature, 559*, 535–545.
Totman, C. (1989). *The green archipelago: Forestry in preindustrial Japan*. Berkeley, CA: University of California Press.
Turchin, P. (2003). *Historical dynamics: Why states rise and fall*. Princeton, NJ: Princeton University Press.
Turchin, P. (2006). *War and peace and war: The life cycles of Imperial nations*. New York: Pi Press.
Turchin, P. (2015). *Ultrasociety: How 10,000 years of war made humans the greatest cooperators on earth*. New York: Beresta Books.
Turchin, P. (2016). *Ages of discord*. Chaplin, CT: Beresta Books.
Turchin, P., & Nefedov, S. (2009). *Secular cycles*. Princeton, NJ: Princeton University Press.

Turchin, P., Currie, T., Turner, E., & Gavrilets, S. (2013). War, space, and the evolution of Old World complex societies. *Proceedings of the National Academy of Sciences, 110.* https://doi.org/10.1073/pnas.1308825110.

Vayda, A. P. (2009). *Explaining human actions and environmental changes.* Lanham, MD: AltaMira (division of Rowman & Littlefield).

Von Glahn, R. (2003). Imagining pre-modern China. In P. J. Smith & R. von Glahn (Eds.), *The song-Yuan-Ming transition in Chinese history* (pp. 35–70). Cambridge, MA: Harvard University Asia Center.

Von Glahn, R. (2015). *The economic history of China.* Cambridge: Cambridge University Press.

Von Verschuer, C. (2016). *Rice, agriculture, and the food supply in modern Japan.* New York: Routledge.

Voosen, P. (2018). New geological age comes under fire. *Science, 361,* 537–538.

Waley, A. (1996). *The book of songs* (New edn., ed. with J. Allen, Trans.). New York: Grove Press.

Wallerstein, I. (1976). *The modern world-system: Capitalist agriculture and the origins of the European world-economy in the sixteenth century.* New York: Academic.

Wang, C. (1907). *Lun-Heng* (A. Forke, Trans.). Leipzig/London/Shanghai: Otto Harrassowitz/Luzac & Co/Kelly and Walsh.

Weber, M. (1946). From max Weber: Essays in sociology (H. Gerth, & C. Wright Mills, Eds. and Trans.). New York: Oxford University Press.

Weber, M. (1951[1915]). *The religion of China: The sociology of confucianism and taoism* (H. Gerth, Trans.). New York: Free Press.

Weber, M. (1967). Max Weber on law in economy and society (M. Rheinstein, Ed., E. Shils, & M. Rheinstein, Trans.). New York: Simon and Schuster.

Weber, M. (1968). In S. N. Eisenstadt (Ed.), *Max Weber on Charisma and institution building.* Chicago: University of Chicago Press.

Weber, M. (1978). In G. Roth & C. Wittich (Eds.), *Economy and society: An outline of interpretive sociology* (Vol. 1–2). Berkeley, CA: University of California Press.

Wei, Z., Rosen, A. M., Fang, X., Su, Y., & Zhang, X. (2015). Macro-economic cycles related to climate change in dynastic China. *Quaternary Research, 83,* 13–23.

Wei, X., Wu, C., Ni, P., & Mo, W. (2016). Holocene Delta evolution and sediment flux of the Pearl River, southern China. *Journal of Quaternary Science, 31,* 484–494.

Wendell, P. (1967). *Oman: A history.* London: Longmans.

Wengrow, D. (2011). *What makes civilization? The ancient near east and the future of the west.* Oxford: Oxford University Press.

Westen, D. (2007). *The political brain: The role of emotion in deciding the fate of the nation.* New York: Public Affairs.

Will, P-E. (1990). *Bureaucracy and famine in eighteenth-century China* (E. Forster, Trans.). (Frech Orig. 1980). Stanford.

Will, Pierre-Etienne, and R. Bin Wong, with James Lee. 1991. Nourish the people: The state civilian granary system in China, 1650–1850. Ann Arbor, MI: University of Michigan Press.

Wittfogel, K. (1957). *Oriental despotism.* New Haven, CT: Yale University Press.

Yin, J., Su, Y., & Fang, X. (2016). Climate change and social vicissitudes in China over the past two millennia. *Quaternary Research.* http://www.sciencedirect.com/science/article/pii/S0033589416300369.

Yuan, L. (2002). *The second generation of hybrid rice in China.* In 20th International Rice Commission meeting, Bangkok. Retrieved 2013 from FAO website.

Zhang, L. (2016). *The river, the plain, and the state: An environmental Drama in northern song China, 1048–1128.* New York: Cambridge University Press.

Zhang, D. D., Zhang, J., Lee, H. F., & He, Y.-q. (2007). Climate change and war frequency in eastern China over the last millennium. *Human Ecology, 35,* 403–414.

Zhang, P., Cheng, H., Lawrence Edwards, R., Chen, F., Wang, Y., Yang, X., Liu, J., Tan, M., Wang, X., Liu, J., An, C., Dai, Z., Zhou, J., Zhang, D., Jia, J., Jin, L., & Johnson, K. R. (2008).

A test of climate, sun, and culture relationships from an 1810-year Chinese cave record. *Science, 322*, 940–942.

Zhang, D. D., Lee, H. F., Wang, C., Li, B., Pei, Q., Zhang, J., & An, Y. (2011). The causality analysis of climate change and large-scale human crisis. *Proceedings of the National Academy of Sciences, 108*, 17296–17301.

Zhang, D., Pei, Q., Lee, H. F., Zhang, J., Chang, C. Q., Li, B., Li, J., & Zhang, X. (2014). The pulse of Imperial China: A quantitative analysis of long-term geopolitical and climatic cycles. *Global Ecology and Biogeography*. https://doi.org/10.1111/geb.12247.

Zhang, Y., Meyers, P. A., Liu, X., Wang, G., Li, X., Yang, Y., & Wen, B. (2016). Holocene climate changes in the Central Asia Mountain region inferred from a peat sequence from the Altai Mountains, Xinjiang, northwestern China. *Quaternary Science Reviews, 152*, 19–30.

Zhang, W., Yan, H., Liu, C. C., Cheng, P., Li, J., Lu, F., Ma, X., Dodson, J., Heijnis, H., Zhou, W., & An, Z. (2017, September 20). Hydrological changes in Shuangchi Lake, Hainan Island, tropical China, during the little ice age. *Quaternary International, 487*, 54–60. https://doi.org/10.1016/j.quaint.2017.09.007.

Zhang, H., Griffiths, M. L., Chiang, J. C. H., Kong, W., Wu, S., Atwood, A., Huang, J., Cheng, H., Ning, Y., & Xie, S. (2018a). East Asian Hydroclimate modulated by the position of the Westerlies during termination I. *Science, 362*, 580–583.

Zhang, J., Huang, X., Wang, Z., Yen, T., & Zhang, E.'y. (2018b). A late Holocene pollen record from the Western Qilian Mountains and its implications for climate change and human activity along the silk road, northwestern China. *The Holocene, 28*, 1141–1150.

Zhang, J., Xia, Z., Zhang, X., & Storozum, M. J. (2018c). Early-middle Holocene ecological change and its influence on human subsistence strategies in the Luoyang Basin, north-Central China. *Quaternary Research, 89*, 446–458.

Zhang, Y., Yang, P., Tong, C., Liu, X., Zhang, Z., Wang, G., & Meyers, P. A. (2018d). Palynological record of Holocene vegetation and climate change in a high-resolution peat profile from the Xinjiang Altai Mountains, northwestern China. *Quaternary Science Reviews, 201*, 111–123.

Zhao, H., Huang, C. C., Wang, H., Liu, W., Qiang, X., Xu, X., Zheng, Z., Hu, Y., Zhou, Q., Zhang, Y., & Guo, Y. (2018). Mid-late Holocene temperature and precipitation variations in the Guanting Basin, upper reaches of the Yellow River. *Quaternary International, 490*, 74–81.

Zhao, M., Li, H.-c., Shen, C.-c., Kang, S.-c., & Chou, C.-y. (2017). delta-18 O, delta-13C, elemental content and depositional features of a stalagmite from Yelang Cave reflecting climate and vegetation changes since late Pleistocene in Guizhou, China. *Quaternary International, 452*, 102–115.

Zhu, Z., Feinberg, J. M., Xie, S., Bourne, M. D., Huang, C., Hu, C., & Cheng, H. (2017). Holocene ENSO-related cyclic storms recorded by magnetic materials in Speleothems in Central China. *Proceedings of the National Academy of Sciences, 114*, 852–857.

Cycles and Cycling

2.1 Ibn Khaldun's Theory of Dynastic Cycles

Every nation has built-in feedback mechanisms that keep it stable. Agrarian empires had to have a particularly effective feedback system, since they were huge, understaffed with government workers, easy to invade, and easy to destabilize. They thus tended to go on and on indefinitely with no basic changes. They almost all became hereditary monarchies. Most progressed, very slowly, by increasing their productive ability at about the same rate as their population.

Most agrarian states, and many modern states, also show cycles. The ancient Greeks saw a regular evolution driven by power consolidation: a regime begins with democracy, hardens into oligarchy, then monarchy, then tyranny, and finally a revolution or war causes it to return to democracy (see Aristotle 2013). Other theories of cycles have appeared since.

The most productive cycle theory in terms of prediction and research has been that of Ibn Khaldun (1332–1406), a Tunisian of Andalusian ancestry who eventually taught in Egypt (Dale 2015; Ibn Khaldun 1958). Ibn Khaldun's theories of cycles continue to flourish and inspire further research and refinement. His house, and the tiny neighborhood mosque where he spoke in his student days, survive today in Old Tunis.

He has been followed by an ever-increasing number of historians and social scientists. Thorough reviews of his and other cycle theories, by Joshua Goldstein (1988), George Modelsky (1987), Peter Turchin and Sergey Nefedov (2009), Chase-Dunn and Anderson (2005), and Hiroko Inoue and Christopher Chase-Dunn (2018), spare the trouble of reviewing an enormous and often repetitive literature. These authors have demonstrated beyond reasonable doubt that economic and political cycles do exist even in Europe—let alone in China.

Small, subtle shifts in culture and attitude can produce enormous results. *Historical cycles thus act as deviation-amplifiers. Common humanity and common culture predispose societies to cyclic fluctuations, not to stability over time.*

Often, the initial subtle shifts and deviations are too small for contemporaries to notice, or for future historians to pick out easily. Timing and intensity of cyclic change is apt to be determined, at fine scales, by contingent events. The assassination of Archduke Ferdinand in Sarajevo might have been just another political killing; in the event, it launched World War I. Chinese equivalent cases include the frequent points at which, in seriously troubled times, the only imperial heir was a very young boy. At such points, a leading general often took over and started a new dynasty. Thus, predicting *exactly* when a cycle will begin or end is quite difficult.

Ibn Khaldun was inspired by the ancient Greek ideas noted above. He certainly knew his Aristotle, and no doubt agreed with that philosopher that "…man is by nature a political animal…much more a political animal than any kind of bee or any herd animal…. nature does nothing in vain; and man alone among the animals has speech" (Aristotle 2013: 4—a better theory for the origin of language than many a modern one!).

In simplest form, Ibn Khaldun's theory runs as follows (Dale 2015; Ibn Khaldun 1958). A dynasty weakens and falls, providing an opportunity for an outside group to take over. The leader of this group must have enough charisma to attract followers. Then the new, powerful regime expands and grows, consolidating its hold, expanding through military success, and developing the economy. At first, almost everyone benefits, and the rulers can reign with a light hand. If this succeeds, a prosperous, tranquil time follows. However, population is increasing, economic expansion is limited, and the rulers and other elites are increasingly distant from the ordinary people. These rulers expand their take, through taxes and rent-seeking, and the ordinary people begin to get worse off rather than better. This leads to discontent among the people and increasing rivalry for power among the elites. The state is then rapidly weakened, and soon falls.

To some extent, all this is a sheer cost of bigness; as the government gets bigger, it gets more out of touch with the people, and powerful individuals in or out of government get more power that they can abuse. In game-theory terms, the state has evolved from a positive-sum game to a zero-sum game, and then to a negative-sum game in which many people harm their own interests to harm others more. The drivers are understandable human emotions: solidarity and mutual aid at first, then autocracy and rent-seeking by the elites, and finally increasing disaffection among an increasing number of people. Humans can fix the problem by growing the economy, as long as they are not prevented from doing so by rapacious, autocratic elites or by lack of necessary resources.

Ibn Khaldun's predictions are an example of agent-based modeling (*avant la lettre*). This art involves creating virtual people populating a virtual city or state, making them act as realistically as possible or necessary, and seeing how they generate processes, systems, and structures (Reardon 2018). It is the opposite of the top-down approach in which one takes the structures and systems and hypothesizes how those affect people on the ground. Today, agent-based modeling is done by computer, but Khaldun was perfectly capable of doing it in his head. He started with people as he knew them. He considered how their interactions within an ecological and political framework would generate particular processes.

2.1 Ibn Khaldun's Theory of Dynastic Cycles

Climatic determinists represent the opposite end of the spectrum. They assume or theorize that climate strongly encourages people to act in certain ways, and that once one has explained and recorded climate, one can predict what people will do—at least as far as one cares to predict at the time. Marx' model of determinism by modes of production represents a more mixed model: it is basically top-down, but Marx recognized that people are agents who make their own history. He thought they would respond predictably and uniformly to political-economic circumstances. By contrast, Ibn Khaldun saw people as less predictable, and prone to act in ways that produce unforeseen and unwanted consequences, such as the fall of a dynasty. In the present book I shall take an agnostic position, exploring top-down and bottom-up mechanisms, evaluating climate as cause and agents as actors on the ground. People are normally the agents, but climate can act directly, as by freezing people or killing through extreme heat.

Khaldun realized that cycles are the product of many individual decisions. He held that "God put good and evil into the nature of man" (Ibn Khaldun 1958: 261). He saw that group feeling brings out both. For Ibn Khaldun, society is based on kinship and extension of kinship, and "when common descent is no long clear and has become a matter of scientific [i.e. book-learned as opposed to experiential] knowledge, it can no longer move the imagination and…has become useless" (p. 263). People need each other and enjoy society, but they want control over their means of livelihood, too. Thus, a typical human "will stretch out his hand for whatever he needs and…take it, since injustice and aggressiveness are in the animal nature. The others, in turn, will try to prevent him…. This causes dissention…. People…cannot persist in a state of anarchy and without a ruler" forever, so in times of chaos a strong, charismatic leader arises to get people out of the mess (Ibn Khaldun 1958: 381). This differs from Hobbesian theory in that Ibn Khaldun sees people as naturally social and familial, in spite of competition. They are never in a state of "each against all." They do not need a social contract; they naturally group together and find a leader.

There is, by inspection, a distribution of humans from the very good to the very bad, the average being about halfway between. Other things being equal, the bad tend to take over, because they are more ruthless than the rest. Thus, whether one assumes people are selfishly rational or innately good, one should be prepared for the disproportionate success of the psychopaths and sociopaths. They not only succeed well themselves; they are often incredibly successful at launching and guiding mass movements. However, Ibn Khaldun noted that such people cannot usually start a cycle; they are too cruel and disruptive to put together a new dynasty. Dynasties and other governmental forms have to be started by people with an ability to build social solidarity.

In Ibn Khaldun's theory, a cycle begins when an empire has declined so much that a "barbarian" group can conquer it. Such borderland polities are usually semiperipheral marcher states, and modern research confirms Ibn Khaldun's contention they are often the conquerors of decadent empires (Turchin 2003, 2006). There has been a misconception that the "barbarians" had to be nomad herders, but Ibn Khaldun's prime examples were the Moroccan Almoravids and Almohads and the

Tunisian Fatimids. All these were Berber groups based in the town-village-field world of the Berbers, not on nomadic herding. Ibn Khaldun's theory requires "barbarians," but, very often, the conquerors are not external at all, but are generals or rebels who have developed the same deeply loyal band of warrior followers that "barbarians" traditionally have.

Barbarians, good generals, and successful statespersons build up '*asabiyah:* loyalty based on obligation and gratitude to the leader. The height of such 'asabiyah was the powerful lineage bonds of the Bedouin—the desert Arabs—but it can be developed by a charismatic leader, with or without kinship.

It is extremely difficult to start a snowball rolling—to get people to join more and more, and build up a sustained, growing force. Genghis Khan and the Chinese dynastic founders are among the very few in history who did it on a large scale. Ibn Khaldun anticipated Max Weber in seeing charisma as negotiated between leader and followers, not simply a property of leaders. He says such a leader must be brave, generous, willing to fight and suffer with his followers, and expert at social relationships. Success is also a matter of timing and luck. The "barbarians" must be in the right place at the right time.

Ibn Khaldun uses something like Latour's actor-network theory (Latour 2005): he postulates agents deciding on what to do, and producing cycles through such action. He saw that the interplay of individual choices vastly increases the amplitude of variation between highs and lows, so only a small shift in individual mentalities is needed to produce the enormous difference between a Golden Age and a violently nihilistic interregnum.

What matters on the ground is not barbarian status or marchland status, but having a solid, loyal, supportive following. Religious movements might do this, but in most of the world, seem rarely to succeed in taking and holding power. The Middle East is an obvious exception, but China is not; its religious movements never won. Zhu Yuanzhang, the founder of Ming, utilized a religious movement, but nobody seems to have believed he was deeply religious. He was out for his own power, not for the greater glory of God.

One must emphasize that 'asabiyah is not just social solidarity (contra the implications in Turchin 2003, 2006). It is not ordinary loyalty (mere followership), either. It involves people acting independently, on their own initiative, out of devotion to the common leader, group, or cause. 'Asabiyah is a special type of social solidarity that is ultimately based on personal loyalty reinforced by leaders' generosity to their loyal followers. Ideally, it comes not just from being in the same social group, but from fighting together, feasting together, sharing loot together, traveling and sharing hard times together—true bonding, reinforced by generosity and mutual self-sacrifice. At the least, it is a real personal dedication. It is systematically created, even manipulated—not just something spontaneous in groups.

Ibn Khaldun held that a major driver of the cycle is growth of population. Long before Malthus, he noted that as a country fills up, it becomes more difficult for the ordinary people to get ahead. Meanwhile, another key dynamic takes place: the elite get more and more efficient at taxing the people and otherwise appropriating wealth, thus generating more and more inequality. Everyone is caught in a squeeze between increasing competition for scarce resources and increasing offtake by the rich. The government officials are more and more easily corrupted; the people under them are more and more desperate. Eventually, selfishness of corrupt officials and desperation of masses lead to massive illegal behavior: deadly competition of elites, banditry among the masses.

The numbers of the elites are growing, too, a point emphasized by Turchin in his hypothesis of "elite overproduction." In the modern world, and to some extent even in old China, more and more of the economy is paper wealth—speculative bidding up of land and houses, risky loans, hoarded wealth, and the like. Today, that includes offshore numbered bank accounts, stocks and stock buybacks, and money manipulation in general. The elite come to serve themselves and each other, at the expense of the rest. Then the dynasty falls.

2.2 State and Family

Ibn Khaldun postulated a four-generation cycle in a family, which he also observed in the state. (Actually, the model works better for the state than for a family, so I suspect Ibn Khaldun started with the state in mind, and then applied his model to the family.) "Nobility...and prestige...reaches its end in a single family within four successive generations.... The builder of the glory...knows what it cost him to do the work, and he keeps the qualities that created his glory and made it last. The son who comes after him had personal contact with his father and thus learned those things from him....he is inferior...in as much as a person who learns things through study is inferior to a person who knows them from practical application. The third generation must be content with imitation and, in particular, with reliance upon tradition.... The fourth generation, then, is inferior to the preceding ones...." The leaders forget the real virtues and assume their people were always superior "by virtue of the mere fact of their (noble) descent...." (Ibn Khaldun 1958: 279–280) They thus fail to keep up the qualities of courage, loyalty, solidarity, generosity, cooperation, modesty, self-sacrifice, and leadership, and may even consider such virtues lowly. People then get dissatisfied and rebellious.

In a state, the same cycle happens. "The first generation retains the desert qualities, desert toughness, and desert savagery...privation...sharing... Therefore, the strength of group feeling continues to be preserved among them.... Under the influence of royal authority and a life of ease, the second generation changes...from privation to luxury and plenty... People become used to lowliness and obedience.... The third generation...has forgotten the period of desert life and toughness... They become dependent on the dynasty... Group feeling disappears completely" (Ibn Khaldun 1958: 344–345). In the fourth generation it all collapses.

Indeed, not only families but almost any community or neighborhood goes through cycles of integration and disintegration. They are usually much shorter than dynastic cycles, but not dissimilar. Communes in the hippie days regularly succumbed to buildup of factions. Businesses often fail as managers lose solidarity and come into conflict. Marriages go from honeymoon days to boredom to buildup of tensions, and often to divorce. The human condition is to oscillate from loyalty to factionalism, and to change reference groups accordingly.

There is probably a power law distribution of social group sizes and longevities, from birdwatching clubs to imperial dynasties (see Goldstein 1988; Turchin and Nefedov 2009). The common theme—the driver, in the final analysis—is that people must think they are getting more social and emotional satisfaction or more physical and economic security from the group than from leaving it.

2.3 Ibn Khaldun on Timing and Dynamics

Ibn Khaldun reckoned a generation as 40 Islamic years, i.e. 36 solar years (p. 346). A three-generation cycle would thus be just about 100 solar years, a four-generation cycle 144 years. In China, and elsewhere, the typical crisis-point is nearer 75 years than 100. This rough 75-year periodicity is found quite widely, including in the United States, with its meltdowns in the Civil War (78 years after full independence in 1783), the Great Depression (68 years after the start of the Civil War), and the Trump presidency (88 years after the start of the Depression). These three events involved major changes in governance, as opposed to the world wars, which did not.

Most Chinese dynasties lasted one 75-year cycle or less, but the great ones lasted three or four cycles. However, they were always broken by coups or attempted coups, and these tended to average close to 75 years apart. At the end, either the dynasty revived under very different leadership and policies, or it succumbed to the next wave of barbarians. Interregnal periods of varying length were characterized by chaos, and are not part of the 75–100 year cycle.

In spite of generations being fairly regular in timing, none of this gives a precise date. We can be fairly sure that a crisis of some sort will come about somewhere between 50 and 150 years into the regime—probably around 75–100 years.

A poorly-integrated empire like the USSR falls apart totally after one cycle (1917–1989: 72 years). A better-united one like China or the United States can survive a 75-year crisis, or even two, but rarely a third and—so far—never a fourth one. A small, well-integrated country like Korea or Switzerland can go on for centuries. On the other hand, huge crises that occur near the beginning of a cycle, such as the takeover by Empress Lǚ in Han and by Empress Wu in Tang, or the serious climatic changes in Japan shortly after the Tokugawa takeover, can be rather easily countered and the dynasty restored. (US parallels include the War of 1812 not long after independence, and World War II after the consolidation of the New Deal.)

A new leader, riding a wave of 'asabiyah, passes new laws, usually ones that reform the system and correct the problems that the new leadership believes to have caused the downfall of the previous dynasty. East Asian dynastic founders almost always did this. These laws, which normally move the system toward more equality and fairness (or at least are intended to do so), maintain themselves through the dynasty, but slowly become dead letters as enforcement wanes. A second cycle of "good" laws comes during the "golden ages" that typically occur in the middle of the dynasty. These laws tend to concern themselves more with quality of life:

environmental protection, reform of agrarian issues, institution of palace benefits and luxuries, institution of education, and the like. Mid-dynasty reanimation of dynasties, with consolidation and passage of new beneficial legislation, was common enough to inspire a Chinese term, *zhongxing,* "middle revival."

However, bad policies can be institutionalized then too. If these bad ideas become institutionalized, they can harm the dynasty further on (Alexis Alvarez, email of May 29, 2017). An example in Chinese history include the fateful if understandable choice by the Ming rulers after 1420 to give up seaward expansion and focus on Central Asia. Less a matter of policy, more a matter of irresponsibility, is the virtually universal tendency—well described by Ibn Khaldun—for the court to become addicted to luxury during such periods, and then find the habit hard to shake when times grow worse.

On the other hand, East Asian historians often noted that a new leader will take over a decimated government. Officials are dead, discredited, or in flight. The new leader can fill posts with his loyalists, but inevitably they are not numerous enough. If they are the traditional "barbarians" or are popular rebels, they do not have the training. The new leader must depend on those older officials who are willing to switch sides. Thus, a radically new regime has a huge handicap at the start. This is avoided when the change of dynasty results from a simple coup, but such a change was rare in old East Asia, and most dynasties had a hard time finding the trained manpower.

2.4 Ibn Khaldun on Decline

Simple sociality does not inspire action. It leads to passivity, from sheer conformity, often accompanied by fear of repression of any proactive accomplishment. As a state ages, the economy is concentrated in fewer and fewer, richer and richer, hands: the nobility in traditional societies, giant firms today. People become weaker as self-efficacy declines. Self-discipline, responsibility, and rationality suffer accordingly. The fight-flight-freeze response leads to mad violence, escapism, or passivity, conformity, and depression. *In fact, progressive loss of control and self-efficacy* (Bandura 1982, 1986), *and consequent loss of responsibility, as the cycle winds down, is perhaps the most important finding in Ibn Khaldun's work.* The boldness and courage of the original leader's troops has been replaced by weakness—either from elite degeneracy (one of Ibn Khaldun's favorite points) or mass impoverishment, dispiritedness, and alienation (a point echoed by many since, from Marx to American politicians).

However, the rapid meltdown of a unified society into one in which everyone is seeking his or her own survival, at any cost, is also disastrous. The unity that formerly kept people at least somewhat loyal to the state evaporates. Elite individuals and mass movements do what will benefit them at the polity's expense.The regime tends to deal with this by getting more repressive and reactionary. During declining years, vested interests—landlords, giant productive cartels, merchants, venal government agencies—suppress change and development.

Throughout East Asia, the innovations developed in each country would have comfortably fed all its citizens, if they had been propagated widely and fine-tuned for local use. After all, China and Vietnam fed over a billion people in the late twentieth century, and early Communist agricultural development was more a matter of rationalizing existing practices than of adding new ones. (Major modernization happened only after the mid-1970s.) China has lost a great deal of its farmland to urbanization, erosion, pollution, and desertification. China today has perhaps half of the food production potential of China 3000 years ago. It has lost a quarter of its farmland since the 1960s (see Shi 2010; Smil 1984, 2004: 125–128). Japan and Korea had modernized earlier. Tokugawa agricultural development was outstanding. Japan probably had the world's highest yields per acre from early Tokugawa times well into the twentieth century.

James Tong (1991; see also Parsons 1970) argues cogently that the choice to become bandits in the late Ming dynasty was quite rational, the most rational decision for many farmers. A bandit in good times may be just a deviant, but banditry in late Ming was a necessity for millions of ordinary people who simply wanted to save themselves and their families from starvation and violence caused by governmental breakdown. Highly successful banditry also characterized the China-Vietnam borderlands in late Qing (Davis 2017). Perhaps less rational, or perhaps rational in their way, were the huge religious movements that contributed to the fall of Han, Tang, Yuan, and Qing.

Ibn Khaldun saw that there is often a progressive change from fighting against strong enemies (in the building phase of the cycle) to oppressing and bullying weaker groups (elite or mass). These groups inevitably fight back, and eventually coalitions of them may become strong enough to take down the polity, or at least to ally themselves with the new conquerors.

This connects Ibn Khaldun cycles with the cycles of wars postulated by Arnold Toynbee and several others. These often assumed that a generation that had started life during a war would be unwilling to fight, but the next generation would think nostalgically of glory, giving us alternate generations. This proves not to be the case—World War II, coming on the heels of WWI, disproved it quite thoroughly. Still, some degree of alternation of violent and less violent wars does appear in the record (Goldstein 1988: 111–122, 264–271). Contrary to popular belief, wars are generally bad for the economy. Too much militarism brings down states and empires. Therefore, generations that have not known war are more apt to fight.

Thus arise "father-and-son cycles" (as Goldstein calls them), which turn on the tendency of sons to do the opposite of what their fathers did, especially in regard to war and peace. This simple mechanism gives us 25-year to 50-year cycles, often noted by anthropologists in studies of tribal war (notably Warner 1937, who compared Australian Aboriginal cycles of war with America and Europe). Sometimes the cycles lengthen to 75 years and approximate Ibn Khaldun cycles.

Turchin and Nefedov (2009) test this cycle theory in Ancient Rome, England, China, France, and Russia, finding it works well. Turchin and Nefedov give credit to Malthus and rather less to politics. Working largely with premodern civilizations they assume that a given period has a stagnant agricultural technology, allowing them to compute a "carrying capacity," as for animals.

Ibn Khaldun anticipated Durkheim (1995) in seeing religious unity as important—something to be constructed and highlighted. It does not arise without prior 'asabiyah (pp. 320–322). Ibn Khaldun saw "luxury and…a life of prosperity" (p. 286) as fatal. A ruling elite "claims all glory for itself and goes in for luxury and prefers tranquility and quiet" (p. 336). It can give way to a "stage…of contentment and peacefulness," but then to "one of waste and squandering" (p. 355). This makes people too soft to fight or even to take care of their interests. "Meekness and docility to outsiders" (p. 287) are the tragic end. (Compare the passivity and conformity that have grown in American society over the decades.)

Ibn Khaldun observed that the conquered come to imitate their conquerors, eventually losing their languages and customs (pp. 299–300). This is indeed the rule in the west, but China tended to be different. China notoriously absorbed her conquerors, except for the intrepid Mongols. Also, large states last longer, because decline tends to spread from the margins inward; the sheer size of the polity protects it, even though it has correspondingly more borders to guard.

In the declining phase of a dynasty, the country has overextended itself militarily. The ruler is apt to weaken the military by diverting resources to himself or the elite (Ibn Khaldun 1958: 355). This dynamic occurred frequently in China. A ruler may even hire mercenaries against his own rebellious people.

2.5 Generalizing the Features of the Theory

There was a strong class dimension to Ibn Khaldun's theory. Downfall comes when the elites are so cut off from the masses that government services are inadequate and the masses are sinking down into poverty and unrest. Very common, historically, has been the dodge in which a corrupt but united elite deliberately encouraged religious and ethnic ("racial") divisiveness in the body politic, in a "divide and rule" strategy. The Roman Empire perfected this; our term is simply a translation of their *divide et impera*.

A factor implied but not precisely stated by Ibn Khaldun is that during upward-bound parts of the cycle, people are hopeful enough to devise and implement hopeful agendas and forward-looking solutions, whereas during the down phase, even the most well-meaning people are prone to look to dealing with immediate problems by short-term alleviation or by going back to an imagined Golden Age in the past. The less well-meaning realize that things are deteriorating, and try to take others down to avoid being taken down themselves.

Thus, upward phases select for moral, helpful people who contribute to solidarity and innovation; the downward phase selects for bullies; and the violent, unsettled period at cycle's end selects for military leaders—at first bullies and thugs, later the charismatic individuals who will take over to start a new cycle.

From a completely different line of research, we have confirmation: a major study by Hemant Kakkar and Niro Sivanathan (2017) showed that leaders who are "overbearing...aggressive...and often exhibit questionable moral character" take over—with popular support—during "situational threat of economic uncertainty (as exemplified by the poverty rate, the housing vacancy rate, and the unemployment rate)" and also when "participants' psychological sense of a lack of personal control" (Kakkar and Sivanathan 2017: 6734).

For a society to survive, people have to have enough 'asabiyah to be actively working and self-sacrificing for progress, improvement, and seeing problems as opportunities. If people become happily conformist, living by deadening themselves with mass media, society will soon collapse. If the society is truly facing major invaders, *if it cannot mobilize most of its citizens to sacrifice their immediate comfort for the good of the system, that society will die.* This is the core of what happened in the ending periods of East Asian dynasties. It may be Ibn Khaldun's greatest realization.

Ibn Khaldun was also well aware of the age-old and universal tendency of religious movements to become corrupt and evil to the extent that they become successful. His brilliant sense of how reasonable individual acts summate into unexpected and unplanned systemic changes allowed him to realize how all these natural reactions would grow into a collapse. If people were rational, society would survive; people would work it out. However, even in the best of times, people's first response to challenge is always the fight-flight-freeze response.

The endless religious wars that ravaged Europe (see e.g. Te Brake 2017) were the product of monotheistic religions that monopolized—or tried to monopolize—the discourses on power, politics, morality, and community. In East Asia, these were always diffuse, and widely distributed over several religions. Another relevant point is that Europe was fractioned into countless small countries, principalities, free cities, and the like, and this led to constant local fights and squabbles. East Asia, even in periods of disunion, never had anything comparable. A general agreement to tolerate variety was region-wide. Ironically, this meant that the west wound up inventing liberty of conscience, a product of the religious wars (Te Brake 2017). East Asia was never forced to face this issue, and stays totalitarian even now.

Quasi-autonomous institutions and communities within societies go through their own Khaldunian cycles. Institutions such as politics and law, for instance, have their institutional entrepreneurs, who innovate and build movements and coalitions that create and change these institutions, using 'asabiyah just as "barbarians" do.

2.6 Barbarians

Ibn Khaldun's idea of "barbarians" was reborn by world-systems theorists in the twentieth century, in the concept of the "semiperipheral marcher states" as economic and military challengers of established power. The earliest known description of semiperipheral "barbarians" is as good as any. It is an anonymous account from the early second millennium B.C., describing an Amorite, member of a tribe from west of Mesopotamia who invaded and took over much of the Fertile Crescent at the time.

2.6 Barbarians

> ...he is dressed in sheep's skins;
> He lives in tents in wind and rain;
> He doesn't offer sacrifices.
> Armed...in the steppe,
> He digs up truffles and is restless.
> He eats raw meat,
> Lives his life without a home,
> And, when he dies, he is not buried according to proper rituals.
> (Van de Mieroop 2007: 83.)

Needless to say, the Amorites were not nearly as ragged and scruffy as here portrayed. The tropes, however, continue today. Truffles are still a poor desert nomad food in Arabia; apparently those desert truffles are not up to French ones. *Plus ça change, plus ça même chose:* the description of the Amorites is recognizably similar to American cartoon stereotypes of Muslims, and to the medieval Chinese stereotypes of steppe nomads.

Korea and North China were conquered by steppe-nomad or northern-forest invaders on several occasions. Only two of the resulting dynasties—the Mongols of Yuan and the Manchus of Qing—took over the whole country. By contrast, most of China's and Korea's dynasties, and all of Japan's shogunates, were started by generals or local warlords. No steppe nomads conquered China until they had previously set up Chinese-style states, and stabilized these over appreciable time. Only the Mongols conquered Korea, and no outsiders conquered Japan until 1945. Vietnam was safely removed from steppe nomads, though the Mongols—using a Chinese army—invaded.

Barbarians can invade and conquer a weakened state, but not a strong one. As Joseph Tainter said: "The overthrow of a dominant state by a weaker one is an event to be explained, not an explanation in itself" (Tainter 1988: 89). Ibn Khaldun was ahead of his time in recognizing that these "barbarians" are products of the state, not wild folk from nowhere. They are people, or peoples, that have always existed on the margins of the state, living by trade, raid, harassment, and—when possible—takeover. They depend on the state for grain, loot, and so forth; the state depends on them for things like horses, wild medicinal herbs, furs, gold, and, very often, slaves and women. Barbarians trade when weak, raid when strong, take over when *very* strong. They form their own "reflex states" whenever they can (Chase-Dunn and Anderson 2005; Turchin 2006; Turchin and Nefedov 2009).

Civilizations *produce* "barbarians," and interact intensely with them, creating a shifting "perilous frontier" (Barfield 1989, 1993) that guarantees Ibn Khaldunian cycles by guaranteeing the presence of highly sophisticated, knowledgeable would-be-conquerors always waiting in the wings. Modern scholars of barbarian-state relationships, like Thomas Barfield (1989, 1993) and James Scott (2009), restate an old truth.

The dangerous semiperipheral marchers in East Asia were the Central Asia states. The Mongols in 1100 were wandering forest tribes; by 1300 they ruled most of the civilized world; by 1400 they were merging into various local polities, having leaped from extreme periphery to world-system core to fragmented peripherality again. The complex and dynamic history of China's peripheral and semiperipheral states in Central Asia has been analyzed in detail in several important recent works (Allsen 2001; Barfield 1989; Di Cosmo et al. 2009, and others).

2.7 Collapse and War

Almost all regime changes in East Asia were most directly due to a succession of weak leaders giving way ultimately to a strong general (invader, warlord, qan, emperor). By contrast, in the west, religion was often the cause of wars and thus of regime changes: Christians, Muslims, Crusades, Wars of Religion, secular-humanist revolutions, nationalism (which learned its tricks from religion; Anderson 1991), and finally the full-blown hate ideologies such as fascism. Wars are usually over land, loot, and national interest, often including ethnicity—Hitler's "blood and soil," a phrase memorably used by Ben Kiernan in his magisterial history of genocide and mass killing (Kiernan 2007). Rarely, if ever, does a war have only one cause; it usually takes three or four of the above concerns to make a country risk war. The immediate trigger for war and most other violence, however, usually is jockeying for sociopolitical or personal power and control. This applies all the way from domestic violence to imperialist world wars. Individuals and nations are also sensitive to slights and "dishonor." Leaders that succeed do it by uniting followers against real human opponents. These could be higher, equal, or lower in power and control. In traditional times it usually meant more or less equals—the enemy polities. Obviously, a sensible war leader will try to find weaker enemies. However, rival states in ancient, medieval, and early modern centuries fought whenever they felt they had even a slight advantage. Since they were often deluding themselves, warring states in early times were often fairly equally matched.

The major East Asian dynasties survived child emperors (as in Han China), mentally deranged emperors (Ming), or emperors increasingly out of touch with the realities of change (Tang, Qing) before they fell. Frederick Mote in his great work *Imperial China 900–1800* (1999) stresses the classic Chinese historians' point that the emperors late in a dynasty were raised in the inner palace by eunuchs and palace women—the worst possible preparation for dealing with real-world problems. He agrees with traditional Confucians that the dynasties need not have fallen, had superb, brilliant, worldly-wise, authoritative emperors been available. Late-dynasty conditions made such a person almost impossible.

This provides us with the makings of a check list, oversimplifying Ibn Khaldun's theory to make it useful for assessing dynastic cycles:

1. *Rise* with leader who can command widespread loyalty; conquest and consolidation.
2. *Expansion*. Economic dynamism; growth of wealth and population, and bureaucracy. Military power leads to enlarging the polity.
3. *Widening gap* between elite and mass; widening gap in income and power leads to increasing gap in lifestyle and ideology, breaking down 'asabiyah

4. Larger and larger Imperial family and bureaucratic elite families: *"elite overproduction"* (see below)
 5. *Increasing corruption.*
 6. Economic *decline*: Malthusian squeeze combined with inequality.
 7. *Widespread unrest* as result.
 8. *Military overextension* as empire must defend borders against outsiders.
 9. *Collapse*, due to popular disaffection, economic decline, and military overstretch making empire vulnerable to attack from within and without.
10. *Fall,* typically to semiperipheral marcher state.

2.8 A Wider Theory of Cycles: Resilience

Ibn Khaldun anticipated modern "resilience theory" by several centuries. Resilience theory takes a cyclic view of the world, in which a period of expansion and exploitation is followed by a more stabilized one and then by decline and eventual renewal. It depends on ecological theories of populations. This theory derives from ecology.

C. S. Holling (Gunderson and Holling 2002) devised a cycle of ecosystem functioning that he has applied to human societies. He took the standard cycle of ecological succession, which begins with early regrowth, passes through slower maturation, and goes into a steady state and eventual decline from overmaturity and senescence, then a down phase while the population or system begins to recover but remains at a low level. He labeled these by following ecological practice to some extent. First comes the r phase (rapid reproduction). This leads to the stable, flourishing K phase: constant, conserving, at carrying capacity. (Cognate words start with K in German, and German ecologists invented the concept, hence the K.) Holling added the omega ("last!"), and alpha (first—the phase-before-the-restart). The senescent decline of the omega phase opens the way to a disruptive event—fire, flood, hurricane, insect pest outbreak, or the like—which renews the cycle.

A threshold may be passed that make catastrophic change to another ecological state inevitable. Application of this to human affairs is not facile Spenglerism. Holling points out that human economies are part of ecosystems, and subject to ecosystem dynamics.

Brian Walker and David Salt (2006) provide thoughtful human applications, especially to the overmaturity phase, which they characterized by:

"...increases in efficiency being achieved through the removal of apparent redundancies (one-size-fits-all solutions are increasingly the order of the day);
"Subsidies being introduced are almost always to help people *not* to change (rather than *to* change);
"More 'sunk costs' effects in which we put more of our effort into continuing with existing investments rather than exploring new ones....
"Increasing command and control (less and less flexibility);

"…more and more rules…
"Novelty being suppressed…." (Walker and Salt 2006: 85–87).

Clearly, this fits neatly with Ibn Khaldun's thinking. His four generations are *exactly* equivalent to the r, K, omega, and alpha phases of Holling—an amazing parallel, considering the difference of both the origin of the thinkers and the subjects in which they were trained. The one major difference is that Holling provides no mechanism, assuming the cycles are more or less properties of systems, whereas Ibn Khaldun nails the entire process very tightly to individual agency. This is because Holling is describing biological cycles in general, whereas Ibn Khaldun is sticking to human society.

Holling describes the ways that cycles are nested, as when the cycles of bacteria and fish in a pond are dependent on and nested within the cycles of the larger pond system, which in turn is nested in its whole drainage area. Human cycles are similarly nested—communities within polities within world-systems. This he calls *panarchy*. Others have described essentially the same phenomenon in other realms under such terms as "scaling up" (management) and "recursion" (linguistics and cognition).

Holling's model and terminology have been used to study the connection of climate and dynastic cycles by Wei et al. (2015), and are clearly useful. I will refer to the phases of East Asian dynasties by Holling's terms. Similar models exist in world-systems theory, recognizing ascending hegemony, hegemonic victory, hegemonic maturity, and declining hegemony (Goldstein 1988: 135, 144). These appear to correspond to Holling's r, K (two divisions), and omega. These cycles are about as long as Khaldunian cycles.

Joshua Goldstein argued for the reality of long waves in European economic activities and wages, clustering of innovations in rising and falling times, and, in general, war (Goldstein 1988: 204, 230–274). Goldstein's long wave begins with high prices and stagnant wages, causing a production and investment trough. Wages peak, prices drop, in a golden age. Then production and investment peak, outrun consumption, and lead to an innovation trough and often more war, which causes inflation and hurts the economy. This cycle takes about 50 years (Goldstein 1988: 259) or more. Much longer cycles of 130 to 300 years are postulated, defined by succession of hegemonic power, but they fail to stand up under the test of history. Indeed, cycles of roughly these time frames do exist in European history, but it remains to be seen what drives them and how stable they are.

A notable point about Goldstein's cycles, and indeed many of the cycles in European historical thinking, is that they have no people in them. "Production," "investment," "war," "prices," and "political power" interact without visible means of action. Not only are human motives left out, there are no humans mentioned at all. The contrast with Ibn Khaldun's meticulously actor-based model is truly striking. The cycles are thus hard to apply to China. In fact, in spite of some protestations to the contrary, the cyclic tradition from Hegel to Marx to Spengler to Toynbee to Goldstein essentially confines its attention to Europe. Marx treated "feudalism," a west-European institution only loosely paralleled in East Asia, as if it were a worldwide stage. Goldstein and those he follows disregard even the most obviously important neighboring areas such as Turkey and North Africa, let alone East Asia.

2.9 Coupled Cycles?

Many cyclic changes are correlated across Eurasia (Chase-Dunn, ongoing research; Chase-Dunn et al. 2007; Hall and Turchin 2007; Lieberman 2003, 2009; Parker 2013). At least some major cyclic correspondences correlate with climate (Chase-Dunn et al. 2007). The tight coupling of Old World economies after 1200, and especially after 1500, meant that chaos in one area refracted on others. The 1200s saw the Mongol impact throughout Eurasia. The fourteenth and fifteenth centuries brought the Little Ice Age, with major declines of population in both west and east; the west had huge plague epidemics.

East Asia had many epidemics, but they are described in a vague manner. This is not a result of poor medical writing; smallpox was well described, and handled by variolation, which the Chinese invented. Other diseases are sometimes recognizable in the record. The real problem is that East Asia's commonest serious epidemic disease was probably influenza emanating from the pig-duck-rice agriculture of the south—the source of most flu epidemics today. These influenza surges take a different form every year, and all tend to be long-lasting, with general signs of fever, respiratory problems, digestive upsets, weakness, and muscle aches. Worse, leptospirosis, another disease endemic in the south, has rather similar and equally vague and hard-to-define symptoms. Thus, the commonest diseases in the East have amorphous and inconsistent symptoms. This clearly put a damper on medical description.

The seventeenth century, especially the middle part, was quite violent all across Eurasia (Parker 2013). The enormously traumatic fall of Ming and rise of Qing in the 1640s, for instance, correlated with the Thirty Years' War (began 1618), the English Civil War (began 1638), and a number of other religious wars and similar troubles in Europe and India. All this was correlated with one of the coldest, harshest periods of the Little Ice Age. On the other hand, Korea and Japan were tranquil, and Central Asia had no more than its usual quota of wars, so once again climate is not destiny. The 1840s—another cold, bleak time—saw rebellions in Europe and China. In 1857 major rebellion struck India. By 1900, the world-system was unified enough to produce a "world" war in 1914–1918; this world conflict really started earlier, with imperial expansion, and with Russia's, Japan's, and China's wars in the early twentieth century. These coupled disasters do seem to imply climatic drivers as well as political ones, but the climate drives collapse only when the political stars are also aligned.

Many rises and falls after 1500 have been conspicuously *un*coupled. The fifteenth and sixteenth centuries was bloody in Europe and Japan, peaceful in China. The seventeenth was violent in China, peaceful in most of the rest of East Asia. The eighteenth century saw several brushfire wars in Europe, climaxing in the American and French Revolutions, while China was peaceful. In the nineteenth century, the

Napoleonic Wars did not affect East Asia. Even within East Asia, where China's cycles drove political events across the region, dynastic changes in Japan and Vietnam were often uncoupled from China's. Korea, far more dominated by China, saw dynastic changes determined by Chinese events.

2.10 Turchin, Nefedov, and Longer Cycles

China's longest-lived dynasties that are historically well attested were Han at 426 years, Tang 286, Song 319, Ming 276, Qing 267, for an average of about 315 years, but note that the very long but twice-interrupted Han raises the average. These dynasties all managed around four Khaldun cycles of 75 years each, or three or four cycles if we count 100 years to a cycle. Korea's were generally longer, climaxing in more than 500 years for the Yi dynasty.

A better fit is with Peter Turchin and Sergey Nefedov's (Turchin 2006; Turchin and Nefedov 2009) long cycles. They have identified 200-to-300-year cycles widely, and find them to be typical of agrarian societies. In East Asia, these typified countries with high levels of loyalty to the regime or government. They had highly bureaucratized governments, reaching down into local levels, such that rebellion was difficult to organize. They had fairly successful delivery of basic law and order and food security. Thus, the cost of loyalty was rather low. Reform seemed much more attractive than rebellion.

Modern research on civil wars shows that they track weakness in states (Collier and Sambanis 2005). Both economic and political weakness is a danger, in proportion to how serious it is and whether it is getting worse or better. There are always opportunists waiting for a chance to attack; Paul Collier maintains this is "greed," but it appears from the record that it is often psychopaths, sociopaths, and pathologically violent people that initiate it, not people who are simply greedy. In any case, as a state weakens, it supplies less and less in the way of public goods, and loses legitimacy. Then more and more people rise against it (see below). Sometimes it falls, and a new regime takes over, often consolidating its hold by mass political killings of its opponents. Sometimes the older regime finds new strength and wipes out the rebels.

Turchin and Nefedov base their thinking in part on Ibn Khaldun and in part on the same ecological cycles that independently inspired C. S. Holling. (Turchin, like Holling, was trained as a population biologist.) They are also somewhat influenced by Kondratieff waves, roughly 50-year cycles of business that were identified by Nikolay Kondratieff in 1922 (Goldstein 1988; Grinin et al. 2014; these waves have a shadowy existence, and have not been applied to China, so I shall not consider them here). They see the same general forces playing over the longer term. Some of their cycles, as in the United States case (Turchin 2016), are close to the length and configuration of Ibn Khaldun cycles. However, they also recognize that these secular cycles can be considerably longer than Ibn Khaldun's three or four generations. The cycles they find in Europe are comparable to Chinese dynasties, running typically

200–300 years. In these longer cycles, economic matters are more important, political ones less so.

Their driver is "demographic-structural" progression. From Ibn Khaldun they learned that cycles begin as powerful rulers reconstitute strong government after a dark age. Land is available, population increases, wealth increases, elites get rich. People are optimistic. This "expansion phase" is followed a generation or so later by a "stagflation phase," in which population rises more slowly and has become dense, cultivation has expanded about as much as it can, elites take more, and the land is stable and peaceful but with declining rural fortunes. Pessimistic ideologies begin to come into play. This is the time when artistic and scientific Golden Ages are most evident. The stagflation phase leads to a slow decline, as the land fills up and taxes rise. This leads to a period when elites truly gouge the poor, through more and more extraction of rents. Not only the absolute and relative numbers of elite individuals, but also their power as a class, is increasing steadily during this period, lagging after the increase in general population. So far, they remain close to Ibn Khaldun.

A newer concept is "elite overproduction." This is a basic concept for Turchin. It means the production of more and more elites, as elite families reproduce abundantly, and upward mobility brings some outsiders (few or many) into the elite ranks. Polygamous elites, like the old-time Muslim and Chinese ones, are particularly prone to overproduce. Soon the elites start fighting among themselves for the few ranks and privileges in the filling-up economy. It need not be literally the production of a huge number of elites. It can be simply a steady supply of elites in a declining economy. The point is that there are more elites than the system can absorb, leading to competition and even outright fighting for place. This was endemic in early Celtic and in Mongol and Turkic societies, notoriously. In these societies, a chief often had many children. Heirs were not usually designated, but selected by tanistry, i.e. agreement by a council or by the elite in general after a ruler's death. Polygamy led to many heirs for every slot, and they killed each other in the resulting political conflicts. Shakespeare in *Macbeth* described this as "bloody tanistry," a phrase now widely applied in Asian history.

Even without war or tanistry, overnumerous elites work harder and harder to crush the middle classes and poor. This involves two different struggles: elite vs. mass and richer elite vs. poorer elite—i.e., both dyadic and unstructured competition.

As a polity wanes, taxes become difficult to collect, most of the time, though there are periods when the state can extract high levels (Turchin and Nefedov 2009: 307), harming the workers still more. Pessimism dominates. People hoard wealth, making coin hoards a good indicator of troubled times; Turchin, seeking tangible signals of unrest, has studied the occurrence of these in European archaeology. This leads to "depression" or an "intercycle" dark age. Overpopulation, elite overproduction, and state loss of finances (as in Ibn Khaldun's theory) combine to produce decline, instability, fighting, rebellions, and finally collapse (see summary, Turchin and Nefedov 2009: 313–314). Somewhat downplayed, though not ignored, is Ibn Khaldun's stress on the psychological consequences of overpopulation and overproduction of elites: individualism at the expense of the state, corruption, alienation, intellectual stagnation. This leads to negative-sum gaming and each-for-himself politics. Societies fall, but cycle back (see studies of revival in Schwartz and Nichols 2006).

This is not always the case. Agricultural production can increase spectacularly in a very short time, when elites or even middle-class "improving squires" decide it is in their interest. This happened in Han China in the second century BC, Ummayad Andalusia in the 700s AD, Sicily in the 800 s, England in the 1500s and again in the period from the late eighteenth century on, and other times and places throughout history. Agricultural stagnation is a phenomenon that has to be explained. When it happens, the problem often turns out to be lazy or indifferent leaders and planners, or militarism that sees hope only in further conquests—not Malthusian limits. It may be, in other words, the result rather than a cause of cycling.

"Carrying capacity," likewise, is too vague and too easy to evade to be a real issue. There are times when considerable agricultural dynamism and expansion accompany decline and fall of an empire, as in China's Ming Dynasty. The New World food crops and the expansion of the southeastern Chinese economy helped Ming survive.

Elites may sharply increase or reduce rent extraction, depending on the pressures and incentives on them. Turchin and Nefedov present many cases, concluding that elites will try to extract what they can, especially when elite numbers are increasing relative to the poor. This, indeed, is one of their best-demonstrated findings. On the other hand, though, China's Han Dynasty coped with crisis by reducing taxes to incredibly low levels (around 3% of land income) in the 160s BC, and the Qing Dynasty repeated this measure in the 1700s. The Tokugawa in Japan also kept taxes relatively low, though not to Han levels.

Turchin and Nefedov conclude that there are indeed laws of history, causing such long cycles. The laws are mostly well-known ones following from logic of supply and demand (Turchin and Nefedov 2009: 313), but the whole picture, including the complicated but predictable dynamics of elites, is a new and exciting synthesis. This links back to Jack Goldstone (1991), who identified demographic drives to cyclic revolutions and periods of revolution. He also noted that major change often comes without political revolution.

Turchin (2016) has since applied secular cycles to the United States, where agricultural stagnation was clearly not an issue and Malthusian crises have not occurred. The cycles play out as in the older civilizations. Rise, stagnation, and decline in general well-being turn out to be as predictive as real Malthusian crises. Turchin uses a variety of measures: population growth, health, wealth, wealth concentration, differential between rich and working-class incomes, and, for recent decades, surveys of perceived social welfare, and even visits to national shrines (as a measure of social harmony). The measures are stunningly consistent, and equally dramatic is their consistence with Ibn Khaldun-Turchin hypotheses. The work uses mathematical models and relevant data in a creative synthesis. A China-watcher can only envy such databases; for most of Chinese history, we do not have even reliable population figures.

2.10 Turchin, Nefedov, and Longer Cycles

Particularly important are labor supply and elite overproduction. Labor supply fluctuates with immigration. On a rising cycle, labor is in demand; the gates of immigration are opened; that soon causes an oversupply, unemployment follows, the gates are closed again (as in the 1880s, 1924, and 2017), and labor becomes short as times improve. However, "oversupply" is in the eye of the employer. A vibrant economy will grow fast enough to absorb many workers, as in the United States in the 1880s and 2010s. More serious is elite overproduction and disunion. Everyone wants to cash in, so during a rising cycle people flood to law school (one of Turchin's measures), business school, and even the ministry, seeking to move up. Soon there are far too many of them, and they compete more and more fiercely.

Turchin finds what Ibn Khaldun would certainly have said: The United States had a cycle of rising harmony, prosperity and success until the 1820s, when things began to unravel. Political dissention rapidly escalated, reaching a fever pitch in the violent 1850s. Finally, the cycle crashed in the Civil War. There followed another cycle peaking in the 1870s and 1880s, with peace and prosperity, albeit the seeds of trouble in the escalating labor violence. That cycle declined into trouble, helped by WWI but heavily colored by continued escalation of labor vs employer violence and white-on-black crime (lynchings and the like). Political dissention was extreme from 1919 through the 1920s, with polarization, violence, and crackdowns. This cycle ended in the Great Depression. Recovery from that event unleashed another rise that climaxed in the 1960s and began to decline in the 1970s. Real wages began to slide along with union membership and other measures of well-being.

In the nineteenth century, and even into the twentieth, such dramatic measures as stature at ten years of age fluctuated with these cyclic changes. Ibn Khaldun would certainly predict breakdown around 2020. The increasingly pathological political dissention of the 2000s is almost eerily similar to that of the 1850s and 1920s. Particularly revealing is the return of extreme racist politics, as in those decades. The political center moved rapidly rightward; recall the Know-nothing Party of the 1850s, as well as the pro-slavery Democrats, and the thousands of lynchings and the brutal anti-labor-union violence that characterized the 1920s. The level of dissention and the extreme racism of the 2016 Republican election led to victory by a highly divisive neo-fascist government that initiated a classic omega-phase meltdown of the United States. Racist and neo-Nazi violence escalated out of control.

Individualism varies with these cycles. A study by Emily Bianchi (2016), unrelated to Turchin's research, shows that Americans get more individualistic on cyclic upswings, less so in disintegration phases. Apparently they feel that they and others can do fine by independence when times are improving, but that everyone will have to help everybody else as times get worse. This goes against some conventional wisdom and some other findings in other countries, where worsening times bring out hatreds; probably the main story is that people circle the wagons when times worsen, supporting their own but being more rivalrous and inimical toward other groups.

In comparing Turchin and Nefedov with Ibn Khaldun, the old sage has the better of it. Turchin and Nefedov rely too much on overpopulation, population pressure, carrying capacity, elite overproduction, and other abstractions. These always beg the question of why people allowed such things to occur and go unchecked. Why didn't they control fertility, find jobs for elites, increase farm effort? Ibn Khaldun tells us: because at first they were held together by charisma and available land and work options, but when the land filled up, society was slowly torn apart by group rivalry, distancing of elites from masses, increasing inequality in general, growth of layers of bureaucracy between social levels, and the psychological consequences of all the above. Ibn Khaldun bases his theory on known human psychological reactions to social-structural conditions, and the resulting individual decisions. Turchin is aware of Ibn Khaldun's points, but bases his theory on more biological features that may or may not be determinative, and may or may not play out in psychology. The fact that land-rich Russia and the food-and-land-rich United States went through the same cycles as food-poor Chinese and Arab dynasties should already give pause to simple Malthusian explanations.

A key variable in all these cyclic theories is the distance of elites from ordinary people. Whether elites are overproduced or not, whether they are fighting each other or not, once they are truly separate in lifestyle and self-interest from the masses, and separated from them by layers on layers of quasi-independent bureaucracy, the state declines.

These considerations guide us in seeking relevant markers in China's far less well documented history. We have no surveys of national opinion, no records of visitors to national monuments, not even good measures of stature, but we do have a great deal of data on elite dissention and labor impoverishment.

Hiroko Inoue and Christopher Chase-Dunn (2018) have further formalized Turchin's model. They look closely at his ideas on war and especially his ideas on population. They see the steady buildup of population in societies as a major driver. Population pressure—here defined, following Boserup, as increasing labor needed to produce the same returns from land and capital—must be dealt with, whether by migration, war, epidemics, intensification, innovation, or other means. This still does not adequately account for the tendency of people to plan their families, send excess children to monasteries and nunneries, and otherwise manage population growth. The data from East Asia that will be presented below indicates that, while population and climate are part of the back story, Ibn Khaldun still has the front story—and therefore the whole story—more adequately treated. Abstract forces play in the real world along with personal matters: mad kings, vengeful queens, warlike nobles, greedy and selfish landlords. Ibn Khaldun has shown us how immediate personal variables play in a world of greater backstage forces.

There are also path dependencies to consider. This is especially true in the case of East Asia, where civilizations have continued without major breaks for thousands of years. Japan has been under a single dynasty since ancient times. China is still self-consciously Confucian after 2500 years. Chinese systems of governance, agriculture, and urban planning have lasted since before Imperial times. Medicine, poetry, and religion have been slow to change. Buddhism, the great "new" influence on East Asia, became universal in the sixth and seventh centuries, and remained a major influence. Changes occurred, and indeed East Asia was far more dynamic and innovative than the older western history books said, but still the path-dependence of government and society must be recognized. he rhythms and needs of millet and rice agriculture, for one major example, are different from those of the wheat-barley-livestock world of Europe.

References

Allsen, T. (2001). *Culture and conquest in Mongol Eurasia*. Cambridge: Cambridge University Press (Cambridge Studies in Islamic Civilization.
Anderson, B. (1991). *Imagined communities* (2nd ed.). London: Verso.
Aristotle. (2013). *Aristotle's politics* (C. Lord, Trans.) (2nd ed.). Chicago: University of Chicago Press.
Bandura, A. (1982). Self-efficacy mechanism in human agency. *American Psychologist, 37*, 122–147.
Bandura, A. (1986). *Social foundations of thought and action: A social cognitive theory*. Englewood Cliffs, NJ: Prentice-Hall.
Barfield, T. J. (1989). *The perilous frontier: Nomadic empires and China, 221 BC to AD 1757*. Cambridge, MA: Blackwell.
Barfield, T. J. (1993). *The nomadic alternative*. Englewood Cliffs, NJ: Prentice-Hall.
Bianchi, E. C. (2016). American individualism rises and falls with the economy: Cross-temporal evidence that individualism declines when the economy falters. *Journal of Personality and Social Psychology, 111*, 567–584.
Chase-Dunn, C., & Anderson, E. N. (Eds.). (2005). *The historical evolution of world-systems*. New York: Palgrave Macmillan.
Chase-Dunn, C., Hall, T. D., & Turchin, P. (2007). World-systems in the biogeosphere: Urbanization, state formation and climate change since the Iron age. In A. Hornborg & C. Crumley (Eds.), *The world system and the earth system: Global Socioenvironmental change and sustainability since the Neolithic* (pp. 132–148). Walnut Creek, CA: Left Coast Press.
Collier, P., & Sambanis, N. (2005). *Understanding civil war: Evidence and analysis*. Washington, DC: World Bank.
Dale, S. F. (2015). *The orange trees of Marrakesh: Ibn Khaldun and the science of man*. Cambridge, MA: Harvard University Press.
Davis, B. C. (2017). *Imperial bandits: Outlaws and rebels in the China-Vietnam borderlands*. Seattle, WA: University of Washington Press.
Di Cosmo, N. (2009). In A. J. Frank & P. B. Golden (Eds.), *The Cambridge history of inner Asia: The Chinggisid age*. Cambridge: Cambridge University Press.
Durkheim, E. (1995). *The elementary forms of religious life* (K. E. Fields, Trans., French original 1912). New York: Free Press.
Goldstein, J. S. (1988). *Long cycles: Prosperity and war in the modern age*. New Haven, CT: Yale University Press.
Goldstone, J. (1991). *Revolution and rebellion in the early modern world*. Berkeley: University of California Press.
Grinin, L. E., Devezas, T. C., & Korotayev, A. V. (Eds.). (2014). *Kondratieff waves*. Volgograd: Uchitel.
Gunderson, L., & Holling, C. S. (2002). *Panarchy synopsis: Understanding transformations in human and natural systems*. Washington, DC: Island Press.
Hall, T., & Turchin, P. (2007). Lessons from population ecology for world-systems analyses of long-distance synchrony. In A. Hornborg & C. Crumley (Eds.), *The world system and the earth system: Global Socioenvironmental change and sustainability since the Neolithic* (pp. 174–190). Walnut Creek, CA: Left Coast Press.
Ibn Khaldun. (1958). *The Muqaddimah* (F. Rosenthal, Trans.). New York: Paragon.

Inoue, H., & Chase-Dunn, C. (2018). *A multilevel spiral model of sociocultural evolution: Polities and interpolity systems*. In Paper, International Studies Association, Annual Conference, San Francisco.

Kakkar, H., & Sivanathan, N. (2017). When the appeal of a dominant leader is greater than a prestige leader. *Proceedings of the National Academy of Sciences, 114*, 6734–6739.

Kiernan, B. (2007). *Blood and soil: A world history of genocide and extermination from Sparta to Darfur*. New Haven, CT: Yale University Press.

Latour, B. (2005). *Reassembling the social: An introduction to actor-network-theory*. Oxford: Oxford University Press.

Lieberman, V. (2003). *Strange parallels: Southeast Asia in global context, c. 800–1830. Vol. 1: Integration on the mainland*. Cambridge: Cambridge University Press.

Lieberman, V. (2009). *Strange parallels: Southeast Asia in global context, c. 800–1830. Vol 2: Mainland mirrors: Europe, Japan, China, South Asia, and the islands*. Cambridge: Cambridge University Press.

Modelski, G. (1987). *Long cycles in world politics*. Seattle, WA: University of Washington Press.

Mote, F. W. (1999). *Imperial China 900–1800*. Cambridge, MA: Harvard University Press.

Parker, G. (2013). *Global crisis: War, climate change and catastrophe in the seventeenth century*. New Haven, CT: Yale University Press.

Parsons, J. B. (1970). *The peasant rebellions of the late Ming dynasty*. Tucson: University of Arizona Press (for Association of Asian Studies).

Reardon, S. (2018). Model citizens. *Nature, 560*, 295–297.

Schwartz, G. M., & Nichols, J. J. (2006). *After collapse: The regeneration of complex societies*. Tucson: University of Arizona Press.

Scott, J. (2009). *The art of not being governed: An anarchist history of upland Southeast Asia*. New Haven, CT: Yale University Press.

Shi, T. (2010). *Sustainable ecological agriculture in China: Bridging the gap between theory and practice*. Amherst, NY: Cambria Press.

Smil, V. (1984). *The bad earth*. Armonk, NY: M. E. Sharpe.

Smil, V. (2004). *China's past, China's future: Energy, food, environment*. New York: RoutledgeCurzon.

Tainter, J. A. (1988). *The collapse of complex societies*. Cambridge: Cambridge University Press.

Te Brake, W. (2017). *Religious war and religious peace in early modern Europe*. Cambridge: Cambridge University Press.

Tong, J. W. (1991). *Disorder under heaven: Collective violence in the Ming dynasty*. Stanford: Stanford University Press.

Turchin, P. (2003). *Historical dynamics: Why states rise and fall*. Princeton: Princeton University Press.

Turchin, P. (2006). *War and peace and war: The life cycles of imperial nations*. New York: Pi Press.

Turchin, P. (2016). *Ages of discord*. Chaplin, CT: Beresta Books.

Turchin, P., & Nefedov, S. (2009). *Secular cycles*. Princeton: Princeton University Press.

Van de Mieroop, M. (2007). *A history of the ancient near east ca. 3000–323 B.C.* (2nd ed.). Oxford: Blackwell.

Walker, B., & Salt, D. (2006). *Resilience thinking: Sustaining ecosystems and people in a changing world*. Washington, DC: Island Press.

Warner, W. L. (1937). *A black civilization: A social study of an Australian tribe*. New York: Harper & Brothers.

Wei, Z., Rosen, A. M., Fang, X., Su, Y., & Zhang, X. (2015). Macro-economic cycles related to climate change in dynastic China. *Quaternary Research, 83*, 13–23.

Before Empire: State Formation in China and Proto-states Elsewhere

3.1 Language-Systems Before World-Systems

East Asia was settled from many directions, and its early settlement record is lost in the mists of time. By the dawn of history, the region was inhabited by a variety of peoples, speaking many unrelated languages. I have given details on prehistoric cultures and language spreads in a recent book (Anderson 2014). I will merely summarize here.

The dominant language phylum today in China is the Sino-Tibetan or Tibeto-Burman, probably the world's most populous and diverse language grouping. It includes the Chinese languages (there are approximately ten, sometimes miscalled "dialects"), Tibetan and Burman (of course), the Qiangic languages that became historically important in the medieval centuries, and thousands of diverse and often poorly-known languages in west and south China, northern Southeast Asia, and the Indian subcontinent. The phylum apparently developed in northwest China in the area from Xinjiang to Sichuan; we cannot pinpoint exactly where within that vast region.

A currently popular hypothesis holds that language spreads along with agriculture, since those who first develop and intensify agriculture are best placed to reproduce, expand, and occupy new lands. This hypothesis was developed by Peter Bellwood, describing the rise of the Austronesian phylum in southeast China and its spread to Taiwan and thence the entire tropical Pacific (Bellwood and Renfrew 2002). If this hypothesis generalizes beyond the Austronesians, there is a very good chance that the Tibeto-Burman languages spread along with millet agriculture (Anderson 2014 reviews the literature). It was developed at just about the right time and place: northwest China 8000 years ago. Millet agriculture certainly enabled the Chinese to form the earliest states in East Asia, and thus to be in a position to conquer what is now China, linguistically assimilating everyone in their path. Southeast China, for instance, was conquered early, and the speakers of the diverse "Yue" languages came to speak new languages based on fusing their own tongues with Chinese. Cantonese, for instance, can be described, with pardonable oversimplification, as Chinese spoken by Thai. It has the Thai tone system and many Thai words.

Striking northward, the Chinese in the centuries BCE progressively assimilated many groups, such as the Rong and Di. Nobody knows who the Rong and Di were. Possibly they spoke languages related to Mongolian or Tungus. They are often called "barbarians" in modern books, but the early Chinese sources—though showing prejudice against them as enemies—speak of them as reasonably civilized people, and archaeology shows complex cultures in the areas they inhabited. They are among the many people whose polities and languages have vanished without trace, absorbed by expanding core polities in East Asia.

The other great domestication event, the development of rice agriculture in the Yangzi Valley about the same time, fits perfectly with what we know of the radiation of the Thai-Kadai language phylum. The Thai languages remain a tight, closely related group, extending from far south China throughout Thailand and into neighboring countries; the few Kadai languages show that the phylum was once far more diverse and extended throughout southeast China.

The Hmong-Miao, Yao, and Austronesian language phyla have a very old presence in south China also, and probably spread along with the Thai, picking up rice agriculture very early, or perhaps even being present at the beginning thereof. The origins and spread of rice agriculture, with an implied link with the Thai-Kadai and Austronesian radiations, has recently been detailed by José Cobo and Collaborators (2019).

Later comers to historical visibility are the northern tier of languages that were formerly called "Altaic." The existence of an "Altaic phylum," always debatable, was substantially disproved by Alexander Vovin (2005; see also Shimunek 2017; Cecil Brown and his group have confirmed this; Brown, Personal Communication). The "Altaic" group consists of the Turkic phylum, originating in south Siberia and spreading early into China's far west; the Serbi-Mongolian phylum, including the Mongolian and Khitan languages; the Tungus languages, occupying a vast realm of boreal forest in north China and Siberia; and Korean. The Turkic, Tungus, and Mongolian languages have many obvious similarities today, but almost all are clearly due to borrowing over thousands of years. Turkic *gök* "sky-blue," for instance, gives us Mongolian *khökh*, the phonetic change $g > k > kh$ being historically well demonstrated as standard in Mongolian. Mongolian retains its own word for blue (and green), *tsenkher*. Mongolian *morin* "horse" was directly borrowed into Manchu (*morin*). This word, along with Korean *mol* (or *mor*; Korean *l* is pronounced *r* in this environment) and Chinese *ma*, may well be from the Indo-European root that gives us "mare" in English. The domesticated horse was likely introduced to eastern Asia by speakers of an Indo-European language, and the name may have spread. Even more interesting is Mongol *nom,* "law, teaching, dharma," which is the far-traveled Greek word *nomos,* "name, law," filtered through Syriac, Sogdian, and Uighur (Shimunek 2017: 22)!

Some links within the "Altaic" universe do hold up. Mongol is related to the Serbi languages, Serbi being the pronunciation (or at least one reconstructed pronunciation) of the word the Chinese wrote with characters now pronounced Xianbei. ("Shiwei" for another central Asian group may reflect a transcription of a slightly different form of the same word.) We now have a Serbi-Mongolic language family and phylum (Shimunek 2017). The Mongolian and Serbi languages may have separated as recently as 2500 years ago, though the lack of extensive Serbi materials makes timing hard to figure out. In addition to the Xianbei, the Khitan, Tuyuhun, and Awar spoke Serbi languages.

Relationship of Korean and Tungus has long been argued by Ki-Moon Lee (Lee and Ramsey 2011: 24–26), and this has now been supported by Cecil Brown and his group (Cecil Brown, Personal Communication 2017). Japanese appears to be a total isolate with no inferred history or record outside of Japan. Postulated links with Altaic never stood up, even in the days when "Altaic" was considered a valid phylum. The one intriguing but maddeningly ambiguous link with the mainland consists of a few words of the reconstructed Koguryŏ language of ancient northern Korea. When not similar to Korean or Tungus, they are similar enough to Japanese to suggest real relationship. Christopher Beckwith (2007) and Andrew Shimunek (2017) accept this, creating a Japanese-Koguryoic family. However, the reconstructions are tentative, and the whole issue is mysterious (Lee and Ramsey 2011: 43). It remains possible that Japanese is related to Korean and Tungus at some very distant level, but we have no solid proof. Russian linguists postulate a vast "Nostratic" superphylum that relates all these languages to Indo-European, Uralic, and other phyla; they have real reasons for this, but the resemblances are so few that they could be due to chance.

A pleasant feature of "Altaic" societies, as of many northern peoples, is mythic descent from noble animals: the Turkic peoples from a wolf, the Mongols from a wolf married to a fallow deer, the Koreans from a mountain god who humanized and married a bear. We are advised to see these as symbolic: the peoples have the fierce, noble qualities of the animals.

At the dawn of history, Indo-European languages dominated central Asia, including what is now Xinjiang, where the now-extinct Tokharian languages were probably the ones spoken by the blond, blue-eyed wearers of European-style clothes that are turning up in Bronze Age and Iron Age cemeteries there. Later, Iranic languages related to Persian dominated parts of central Asia up to China's modern frontiers.

Austronesian languages survive as the "aboriginal" languages of Taiwan, and evidently got there from the coast of Fujian and/or the Yangzi delta area, but no trace of them survives on the mainland north of the Cham of central Vietnam. They spread from Taiwan throughout Oceania.

A final player in the linguistic record is the Austro-Asiatic phylum, which apparently originated in eastern India, but was already established in south China and southeast Asia at the dawn of history. It probably spread along with rice agriculture from India, where rice agriculture was either independently invented (a likely scenario; Cobo et al. 2019) or borrowed at an extremely early time from China (less likely). (A recent claim that this phylum spread south from China is not credible.) The major Austro-Asiatic language we shall be considering here is

Vietnamese, but several other languages of this phylum are spoken in or near south China.

More than in Europe, language differences often went with social and cultural ones. European languages are almost all culturally related, most being Indo-European. It is quite easy, especially in western Europe with its Romance and Germanic families, to learn many languages and move easily from one to another. The situation is different in East Asia, where the phyla are very different indeed. It would be hard to imagine languages more different from each other than Chinese, Turkic, and Japanese. Chinese, Thai, and Vietnamese are an easier set to learn, because they have been influencing each other and borrowing from each other for 2000 years or more, but they too are quite different from each other.

East Asia developed pottery, rice agriculture, jade working, and many other arts of life. The earliest pottery in the world comes from Japan, Siberia, and northern China, with dates pushing back to 17,000 BCE (Barnes 2015: 93; Sato and Natsuki 2017; Shelach 2015). Millets and rice were exploited in China by 7000–8000 years ago, with domestication slowly emerging. The amelioration of climate and rise of sea levels after the Ice Age presumably stimulated expansion of plant use, including rice. Parts of the lower Yangzi became drier (Zuo et al. 2016), but parts were swamped by rising sea levels and only became available after 5000 BCE, then becoming a paradise of wetland and delta resources (Li 2018).

The earliest well-dated sequence of pottery is from Japan, from 15,000 BCE onward (Craig et al. 2013). All the early pots had fish and shellfish stew residues. Japan's love of sea food goes far back. Korea and neighboring Siberia are slightly later adopters, so far, but may prove to have had pottery as early as Japan.

3.2 East Asia: General Introduction

> The greatest achievement of the Chinese farmer has been the maintenance of the fertility of the soil for four thousand years under a constant burden of intensive production. When it is considered that strong, virgin farm lands in America have been exhausted in three or four generations and that too without the tremendous pressure of population on subsistence prevailing in China, the magnitude of this accomplishment can be more heartily appreciated. (Mallory 1926: 25)

Most of what we know about early East Asia is derived from archaeology, but by 500 BCE the Chinese historians become the main sources of knowledge, and after 500 CE the Korean, Japanese, and southeast Asian sources begin to appear. Westerners, and some Chinese, long wrote early histories off as too tradition-bound and Confucian to be trustworthy. More than a few westerners made the assumption (based on racism or colonialism) that no non-westerners could write accurate history, at least not in traditional times.

Archaeology has now confirmed much of Chinese historiography, and new documentary finds have confirmed much more, and we emerge with new respect for the historians. This extends to their moral judgments, often condemned but rarely on the basis of serious arguments. In particular, their judgments about the contribution to governmental immorality (corruption, irresponsibility) to dynastic decline are quite relevant. They were also aware of the military overextension that is basic to Ibn Khaldun's theory. They recognized that the semiperipheral marchers—the "barbarians"—would militarize in response, producing a feedback loop that was hard to avoid. On the other hand, their consistent biases against women, eunuchs, and "bad last" emperors (see below) do not invite confidence.

East Asia lacked the "slave societies," "feudal societies," and "capitalist societies" postulated for the west by Marx. There were never slave societies. Medieval Korea came close, with a huge servile labor force, but China and Japan had small and (at least in China) progressively declining numbers of slaves. Even in early times, society was based on free farmers, craftspeople, and traders. The state ran on taxation and tribute: the "tributary mode of production." This meant that there was always some balance between primary production and the trade-commerce-communication economy, a mix that encouraged scientific and economic progress. Like China's environmental attitudes, it was "ambiguous" (Thornber 2012)—never the anti-intellectualism and repression of Europe's Old Regime and its modern successors, never the riotous abandon of Enlightenment capitalism. Something much like feudalism did exist in the later Zhou Dynasty, but it was not true European feudalism. Contrary to Maoist usage, the imperial dynasties were not in the least feudal. They were bureaucratic tributary states. Japan in the medieval period had something close enough to European feudalism to deserve the label, though it showed differences from high medieval France or England. Korea and Vietnam were similar to China. Central Asia retained a rather tribal society, in which polities were vast and shifting aggregations of kindred or fictive kindred that could enlarge or contract rapidly by forming or breaking alliances. Over time, these became more and more statelike, evolving into full states when they settled down in oasis cities or in China.

3.3 Early China

Chinese agriculture dates back some 10,000 years. Complex societies with large towns and sophisticated crafts developed over the next several 1000 years (Anderson 2014; Barnes 2015; Li 2018). Just before Chinese urban civilization begins, the Hongshan, Shimao (Jaang et al. 2018), Liangzhu and Longshan societies reached notably high technological and social levels. Towns grew almost to urban size, and construction indicates labor mobilization on a large scale. Liangzhu flourished in the lower Yangzi, collapsing around 2300 BCE as colder and more flood-prone weather set in. The Longshan culture adapted to this and flourished in eastern China (Li et al. 2018: 82–86).

China's mythical history populates this time period with the legendary Yellow Emperor (Huang Di, sometimes translated "Yellow Thearch"—*di* means "divine emperor"—around 2800 BCE), the ox-headed Shen Nong ("Divine Farmer" at around 2700), and Yu the Great, who tamed the floods on the Yellow River.

Massive flooding did indeed take place along the Yellow River in the late third millennium BCE. Yu supposedly dyked and controlled the waters and drained them off. The myth tells that he created nine rivers, to distribute the Yellow River's water over the plain. If the idea was to break the Yellow River into nine distributaries fanning out, it was a good one. He passed his parents' door several times without stopping to visit, thus provoking centuries of debate over whether filial piety and politeness should have prevailed over public service.

This mythical period coincides with the decline of the warm, wet Altithermal period, which ended between 2200 and 2000 BCE. Cooler, drier weather followed, but the Yellow and Yangzi Rivers, at the time, endured several floods that climaxed in a great flood around 2000. It has inevitably inspired Chinese scholars to connect the Yellow River floods with Yu (Brooke 2014: 295). Perhaps the stresses of this period helped cause centralized, urban, state-level society to form. Yu may be a historians' fusion of several generations of local chieftains who built dykes and levees. The difficulties and fluctuations would probably have inspired people to group together for solutions. This could have given powerful chiefs an opportunity to consolidate control and institute states that could better control the rivers, or at least the populace. If so, this is a highly important case of climate change driving history.

Similar consolidation and rise of a centralized chiefdom occurred at Shimao and Taosi, in the dry northwest; Shimao conquered Taosi and would have threatened Erlitou if Shimao had not fallen—basically, to cold drought—before it could get there (Jaang et al. 2018). In Central Asia at this time, drought made people concentrate in river valleys and other moist areas, and apparently stimulated the rise of horse-riding and other arts (Chang 2018). Major events all over Eurasia are correlated; Near Eastern kingdoms suffered, local chiefdoms rose to power. There is thus a very good case to be made for cold and dry conditions forcing consolidation—often around still-wet areas—and thus empowering local rulers to organize for survival and then centralize for defense and conquest.

Unfortunately, we lack data other than the myths of Yu. A significant myth has him pacing all through the "nine lands," the nine regions that had proto-state and proto-civilization-level societies, roughly corresponding with the current central areas of China (Li 2018). The "paces of Yu" became, and remain today, part of Chinese ritual, enacted by religious officiants on a small scale. Chinese emperors in future, notably in the Song Dynasty (Zhang 2016: 118–124), used Yu as their model.

Yu supposedly founded the Xia Dynasty. It is solidly linked to the Erlitou culture that occupied the middle Yellow River in the period from 2000 to 1500 BCE (Allan 2002, 2007; Anderson 2014; Barnes 2015; Li 2018). It fits all the descriptions of the Xia state, and appears to have been the only truly state-level society in China at the time. (We cannot be sure, because writing is not known, so far, from Erlitou sites.) There was a huge primate city with monumental architecture and specialized craft quarters, subsidiary cities that were evidently tributary to the capital, and a consistent culture associated with that capital and spreading over a reasonably defined area—the "state" or proto-state—of several thousand square kilometers. No other candidate with these traits is known. A clincher is clear indication of Shang invasion of one of the cities near the Shang heartland just downriver, at about the time when Shang history claims that Shang invaded Xia (Barnes 2015). Agriculture was advanced. Cattle skeletons show stress consistent with plowing and hauling carts (Lin et al. 2018).

It now appears that the Liangzhu culture of the Yangzi Delta was advanced also, and it was even earlier, 3300–2300 BCE. By 3100 BCE it had huge waterworks involving dams and canals radiating from its capital, Liangzhu Ancient City, which covered 300 ha and may have had as many as 35,000 people (Liu et al. 2017; Pettit 2017). Thus, already, China had two poles: the central Yellow River area and the lower Yangzi. The Hongshan culture in northern China was also pushing the edge of civilization, though not quite there (Allan 2002; Anderson 2014). The Shimao culture in the northwest was also advanced, falling about the time Erlitou became major (Jaang et al. 2018). Thus China was polycentric. Probably these sites involved speakers of different languages. The cold weather and floods of the 2200–2000 BCE period seem to have stimulated centralization in several areas, but perhaps led to the fall of Liangzhu, which was in a flood-prone area.

Meanwhile, fires swept much of the eastern Tibetan uplands, indicating not only drier climate but also the rapid advance of human settlement there (Miao et al. 2017). This involved humans evolving tolerance for higher altitudes; Tibetans' blood holds and carries oxygen more efficiently than others' blood does.

Shang (ca. 1500–ca. 1050 BCE; Keightley 2000) was the first real dynasty and literate state, and consolidated power in central north China, the first "upward sweep" in power and area. History relates, and archaeology seems to confirm (Barnes 2015), that this was a case of conquest by a semiperipheral marcher state (Shang) of the core polity (Xia/Erlitou). China's ancient history records that the last emperor of Xia was a wastrel who had a liquor lake and a meat forest, explained as a pool that could be filled with ale and a forest hung with drying meat strips. This tale is less significant as history than as stereotype-maker. From then on, the myth of the "bad last emperor" was to dog Chinese historical writing into modern times. Often, as in the case of Shang's conquest of Xia, the "bad last" story was blatant propaganda by the following dynasty, using projections of wishful but sinful fantasizing.

Alas, such projection reduces the historical value of such tales. When we read of the fantastic ale consumption of the last Xia, or the incredible sexual athletics of the last ruler of Sui or the Empress Wu of Tang, we must take the stories as wish-fulfillment fantasies rather than as factual history. It makes Chinese historians' opinions on dynastic rise and fall somewhat dubious; the urge to glorify the founder and criticize the last few emperors—or de facto rulers, when the emperors were children and regents controlled the empire—was always hard to resist.

Shang history is known in outline (Loewe and Shaughnessy 1999), and excavations are rapidly adding to our knowledge (Cheung et al. 2017). Its capital was at first in Zhengzhou, a large city today, but later was at Yinxu, near Anyang. These were the first major cities in East Asia. They had writing, sophisticated art, bureaucracies, chariots, lavish burials involving sacrifice of many humans and animals, and all the other trappings of early states. They may have been demographically unsustainable, like most early cities, and may have had to draw on the countryside, in which case they may have done as early Mesopotamian cities did: use the unskilled labor to produce cheap imitations of imported goods, thus saving money for everyone [following Algaze (2008, 2018) on Mesopotamia]. On the other hand, their bronze industry, a world leader, involved high levels of expertise and artistic sophistication. They cast enormous vessels, showing they had the ability to lift and pour tons of molten bronze at a time.

Shang was conquered by Zhou. The founder of Zhou died soon after completing his conquest, and left the kingdom to a boy. The Duke of Zhou served as regent, and did such a good job that he became the model ruler for the rest of China's history. Zhou flourished in a time of climatic stability and relative bounty, rather like the twentieth century.

Zhou (ca. 1050-221 BC), was geographically a semiperipheral marcher state, though it shared the high culture of Shang. It was located in the Wei Valley, in Shaanxi, a mountain-ringed and very fertile valley fed by a large, reliable, not excessively flood-prone river—an ideal habitat. Being protected by mountains, the Wei Valley is almost impregnable from the more populous areas of China. It is seriously vulnerable only from the north, where a wide river valley gives an opening. Since the northern approaches beyond that valley are relatively infertile and dry, they did not usually support conquering states, but when they did the Wei Valley became a trap. It was also remote from the centers of wealth and power, which lie to the east. Thus there was always a temptation to move downriver to more central locations. This was convenient, but fatal, since such locations were easy to conquer. Only when the non-Han northern regimes developed Beijing did a capital arise that was fairly secure yet not isolated by near-impassable mountains from the networks of transportation and trade. (A long literature on these important geopolitical issues is summarized in Anderson 2014; Li 2006, 2008, 2013.)

Zhou started with an "upward sweep" of impressive size. Its twin capitals, Feng and Hao, lay on the Feng River (a southern tributary of the Wei). They reached enormous size and wealth, but were conquered from the relatively vulnerable north by the Quanrong in 771 BCE. (Quanrong is usually translated "Dog Barbarians," but could better be translated "Non-Chinese of the Noble Hound," since "quan" is a fancy literary borrowing rather than the usual word for a dog, and "rong" simply meant a non-Chinese-speaking northerner, not necessarily a barbarous one. *Quan* could simply be a transcription of a Rong word that had nothing to do with dogs.)

A list of all the wars of China up to this point has been compiled by Cioffi-Revilla and Lai (2001). They find dozens of wars of various sizes, many being purely mythical ones in pre-state times. The ones late enough to be believable—especially during Zhou before 770—are most often with the Rong, especially the Quan and West Rong.

The capital was moved downriver to Luoyi (more or less the modern Luoyang), in the old heartland. Wheat increased in the diet, though millet remained the staple in the north. Women fell in status, as shown by their bones; they got more of the prestigious wheat (which had to be ground to flour), but less meat (Dong et al. 2017).

Unfortunately, Luoyi was vulnerable to attacks. Zhou succumbed to regionalism. Feudal kingdoms and dukedoms became more and more independent until the emperor was a figurehead. The Spring and Autumns period, named from the historical annals that covered the period, lasted from 771 to 476. From it emerged several powerful local states, each one controlling a region, usually defined by river valleys and mountain borders. There followed the Warring States Period (476–221 BCE), from which important historical materials survive (Crump 1996, 1998; Hui 2005; Pines 2009, 2012). These states could not conquer each other, largely because every time one of them grew powerful, the others would form alliances to stop it.

The East Asian world-system during this period consisted largely of these conflicting states, and tribes living on their borders. At this time the term *zhongguo*, "central states," came into use—apparently applied not to one state alone, but to the major states of the North China Plain and central Yangzi Valley. The concept is very close to the idea of "core" states in world-systems theory. Only later did *Zhongguo* become the standard term for the Chinese empire. Since Chinese does not distinguish singular and plural nouns, the word could silently change meaning.

Finally, the strongest local state, Qin, in the old Zhou heartland, got powerful enough to roll up the rest, eliminating the otiose emperor as an afterthought.

At this time, the other regions that became integrated into the East Asian world-system were at a pre-state stage of development. The Rong, the Yue of southeast China, the Di in the west, and other little-known groups all had complex societies. None had anything close to writing, cities, or state societies. One important dynamic was the development of reflex states—polities that take on state administrative mechanisms, especially armed forces, so that they can defend themselves against the states that already exist. State formation spread like ripples on a pond from the central states. Xia presumably stimulated the rise of Shang, Shang led to the rise of Zhou and the North China Plain states, and these led to states such as Chu and Yue emerging in the Yangzi Valley. Here state-building came to a temporary end, while Qin conquered everyone. But with the rise of Han, state formation leaped the barriers around the China plains and valleys. The Xiongnu state and some smaller states arose in eastern Central Asia. States began to develop in Korea, as will be seen. Vietnam moved toward statehood. Thus began a process that continued into early modern times, with states finally stabilizing in Central Asia only in the past couple of centuries.

A typical evolution was for a peripheral tribal society to develop a state, develop until it could move up into the semiperiphery, get progressively incorporated into the core, and eventually be swallowed up by the expanding Chinese empire. This process seems to have reached an end with the recent incorporation of Tibet. Korea, Japan, Muslim Central Asia, and southeast Asia were all able to hold their own and avoid incorporation into China, though Vietnam was conquered and held by China in Han times and again in the medieval period.

3.4 The Rise of Philosophy

The decline of the Zhou state into warring polities led to revolutionary thinking about statecraft and governance. This was to have a fateful influence on the future, when the Han Dynasty carefully selected policies and ideology that, in a broad sense, set the course of empire in East Asia for the next couple of millennia. The long cycles (2–4 Ibn Khaldun cycles) of the great East Asian dynasties and military states depended on this policy set and its modifications.

Philosophy emerged from political advice during these troubled and rivalrous times. An interesting analysis of governmental failure was provided by Confucians and Legalists, notably Shen Buhai (Creel 1974). They discussed at length the concept of "rectification of names." Widely misunderstood to mean simply renaming offices (e.g. by Kidder et al. 2016), this meant—at least to Shen Buhai and his school—streamlining government such that titles go with offices that have named, designated responsibilities. Higher-ups were directed to do more general managing, and lower-down officials did more specific things. The Many Dogs Officer took care of the hunting packs. The *sima*—Officer of the Horse—managed the governmental steeds and the grooms and handlers. (This title gave us the common Chinese surname Sima.) People who interfered in lower-down officials' jobs, or took over their work, or tried to do more than their job description allowed, were taught to do otherwise. This common-sense policy, all too rare in Europe and the Near East at the time, probably had an enormous role in maintaining Chinese government for the next 2000 years. It was one of the main stabilizing factors that gave China such long cycles and such consistent governance.

The study of cycles must include attention to why stability is the norm—why East Asian dynasties often lasted for hundreds of years, making rulership in the western world look like a game of revolving chairs by comparison. The strongly centripetal geography of China explains some of the stability there, but less geographically centripetal Japan and Korea had even longer-lasting dynasties. Conversely, the great revolutions that took down or fatally weakened dynasties usually had ideologies that were millennarian and radical, advocating violent change rather than stability. Philosophy matters.

The thinkers cover a spectrum from near-fascism to total anarchism. Strong statist approaches were traditionally called *fa jia* by the Chinese; this literally means the Law School, and is usually translated Legalism. It covers a wide spectrum, from Hanfeizi (an authoritarian that Arthur Waley, in 1939, compared to Europe's fascists) to Shen Buhai.

The moderate middle ground was occupied by various forms of *Ru*. The Ru scholars are "Confucians" to the west, thanks to their honoring Kong Fuzi (551-479 BCE), "Confucius" in Latin transcription. Confucius left many sayings and brief passages, but was largely a selector of old texts. A significant human touch was his collection of folk and court songs, the *Book of Songs*. Few philosophical or political leaders have felt a major part of their mission to be compiling an anthology of romantic and charming popular poetry, but Confucius perceptively stressed the importance of song, music, and art in governance. It created a culture that could embed a healthy popular morality, which to the Ru was better than forced obedience.

Confucians ranged across a broad spectrum: from the near-Legalist Xunzi (who taught Hanfeizi) to the liberal Mencius (c. 371-c. 289 BCE), who defended the people's right to follow their consciences, even to the point of rebelling against a bad government. He may be the only ancient thinker to advocate the concept of rebellion, to a king's face, and get away with it. Confucians upheld tradition, ritual, and hallowed old ways. This sometimes made them hidebound (especially in later millennia), but was intended, at least in Mencius' form of Confucianism, as a device for resisting tyranny and arbitrary governance. One could always hark back to the ideally liberal, generous, charitable, and tolerant early rulers of Zhou, and then say one was merely following tradition when one resisted autocracy.

Ru philosophy focused on the family, and on the behaviors and virtues necessary to successful family life. An ideal was holding several generations together under one roof. The key virtue for doing this was *ren*, "humaneness"; like the English word, the Chinese word is simply a transform of the word for "person." It is written with the character for "two" next to the one for "person," indicating that it means "the way two [or more] people should behave toward each other." Early texts seem to have simply used the graph for "human" to cover both concepts. In the Tang Dynasty, one Zhang Gongyi supposedly kept his family together in a house already long used by the lineage: "When asked how his family managed to reside together for nine generations, Zhang Gongyi (fl. 665) wrote on a piece of paper the character *ren*...more than a hundred times" (Knapp 2003: 17, quoting from a story in the Older Tang History, by Liu Xu, 887–946).

The tension between tradition as dead hand and tradition as excuse for liberalism became, and remains today, a constant subtheme in Ruist thought. Confucians taught that the basis of all relationships lay in family and neighborhood, and that relations of blood kin, affinal kin, and friends were fundamental. These relationships undergirded and transcended relations of governance. The relation of political superior to inferior ("ruler to subject") was considered basic too, but family was never forgotten, and this was another way Mencius could resist the state.

China tended to combine Ruism with other philosophies, but Korea eventually adopted it wholeheartedly; the Chosŏn (Yi) Dynasty made it the official ideology of Korea for over five centuries. Governance was ostensibly based solely on Confucian principles, though—as elsewhere—Legalism covertly brought rule of law with an iron hand. Japan and Vietnam also embraced Ruism, but Buddhism and local traditions soon came to dominate religious and philosophic life in those countries. Still, perhaps the greatest modern statement of Ru morality is *The Tale of Kieu*, a Vietnamese epic poem of the eighteenth century; its author, Nguyen Du (1765–1820; see Kiernan 2017: 272), combined Ru with Buddhism. The work remains popular, combining a beautiful romantic surface narrative with a passionate, deeply moral and religious undercurrent. Similar poetic works in Japan and Korea kept the Ru philosophy foregrounded even as Buddhism dominated much of public life. Everywhere, however, the basic Legalist principles of strict uniform laws, enforced by some rewards and many punishments, was in fact the usual method of governance.

The anarchist end of the spectrum is now labeled Daoism. Dao simply means "way," in all senses: a way of doing anything, a way of being, or even just a street. All the schools claimed to have the Dao: the right Way of governing oneself and of governing the state. Moreover, a religion called Daoism arose later, in the Han Dynasty. The Warring States philosophers called "Daoists" were those who believed in quietism, meditation, stilling one's mind, and governing with a light touch. A leading book of this school was the *Dao De Jing* (lit. "The Classic of the Way and Its Virtue," but "virtue" in the old sense of "spiritual or moral power"). It contains the significant insight that governing the people is "like cooking small fish." Any cook will understand: it must be done with a *very* light hand, as quickly and skillfully as possible, to keep the fish from breaking up into mush.

Daoists too ranged over a spectrum. A near-Legalist end was anchored by the *De Dao Jing* (a version of the *Dao De Jing* with power emphasized; Henricks 1989). This text saw an autocratic imperial government as necessary, but counseled the ruler to do as little as possible, leaving the people some freedom and delegating considerable authority. At the other end, the soaring and uncompromising mysticism and anarchism of the philosopher Zhuangzi leaves no place at all for government or restraint. The current book of Zhuangzi includes several later works by or about other philosophers, giving us a better sense of the debates of the time (Chuang Tzu 1981).

These schools took place within a wider shell of religion: the imperial cult, with its sacrifices and rich and elaborate tombs, and the folk cult, with its offerings to a multiplicity of gods and spirits. These religious manifestations long predated the philosophic schools.

They all influenced statecraft—in fact, they created it—in the Warring States period, but their real flowering as government-makers will be tracked in the next chapter.

3.5 Summary

Little is known of these early dynasties. One thing, however, stands out: their rise and fall have nothing to do with climate, except for the likely influence of the unsettled conditions at 2200-2000 BCE on the rise of centralized authority, notably the state of Xia. The declines and collapses of the early dynasties and kingdoms did not coincide with significant climatic changes. A gradual rise in temperature and rainfall at the very end of the Warring States Period may have helped Qin, but it would have helped Qin's rivals too, so was probably a neutral factor.

The one clear factor in decline was the move of the Zhou capital to the site of Luoyi, impossible to defend against states with better geopolitical situations. Ironically, it was Zhou's old homeland in the Wei Valley that produced the strongest state, Qin—the one that ultimately conquered the rest. Other powerful states emerged in fertile-soil areas of the Yellow River Plain and the Yangzi Valley.

Chu, roughly today's Hunan, was the strongest state after Qin and the last to fall to Qin expansion. Apparently, geopolitics and the inevitably slow rise of the most militaristic state as an all-conquering polity were critical. Climate was not.

References

Algaze, G. (2008). *Ancient Mesopotamia at the dawn of civilization: The evolution of an urban landscape*. Chicago: University of Chicago Press.
Algaze, G. (2018). Entropic cities: The paradox of urbanism in ancient Mesopotamia. *Current Anthropology, 59*, 23–54.
Allan, S. (Ed.). (2002). *The formation of Chinese civilization: An archaeological perspective*. New Haven, CT: Yale University Press.
Allan, S. (2007). Erlitou and the formation of Chinese civilization: Toward a new paradigm. *Journal of Asian Studies, 66*, 461–496.
Anderson, E. N. (2014). *Food and environment in early and medieval China*. Philadelphia, PA: University of Pennsylvania Press.
Barnes, G. L. (2015). *Archaeology of East Asia: The rise of civilization in China, Korea and Japan* (3rd ed.). Oxford: Oxbow Books.
Beckwith, C. (2007). *Koguryŏ, the language of Japan's continental relatives: An introduction to the historical-comparative study of the Japanese-Koguryoic languages, with a preliminary description of archaic northeastern middle Chinese* (2nd ed.). Leiden: Brill.
Bellwood, P., & Renfrew, C. (Eds.). (2002). *The farming/language dispersal hypothesis*. Cambridge: Macdonald Institute, Cambridge University.
Brooke, J. L. (2014). *Climate change and the course of global history*. Cambridge: Cambridge University Press.
Chang, C. (2018). *Rethinking prehistoric Central Asia: Shepherds, farmers, and nomads*. Oxford and New York: Routledge.
Cheung, C., Jing, Z., Tang, J., Weston, D., & Richards, M. (2017). Diets, social roles, and geographical origins of sacrificial victims at the Royal Cemetery at Yinxu, Shang China. New evidence from stable carbon, nitrogen, and sulfur isotope analysis. *Journal of Anthropological Archaeology, 48*, 28–45.
Chuang, T. (1981). *Chuang Tzu: The inner chapters* (A. C. Graham, Trans.). London: George Allen and Unwin.
Cioffi-Revilla, C., & Lai, D. (2001). Chinese warfare and politics in the ancient East Asian international system, ca. 2700 B. C. to 722 B. C. *International Interactions, 26*, 1–32.
Cobo, J., Fort, J., & Isern, N. (2019). The spread of rice in eastern and southeastern Asia was mainly demic. *Journal of Archaeological Science, 101*, 123–130.
Craig, O. E., Saul, H., Lucquin, A., Nishida, Y., Taché, K., Clarke, L., Thompson, A., Altoft, D. T., Uchiyama, J., Ajimoto, M., Gibbs, K., Isaksoon, S., Heron, C. P., & Jordan, P. (2013). Earliest evidence for the use of pottery. *Nature, 496*, 351–354.
Creel, H. G. (1974). *Shen Pu-Hai: A Chinese political philosopher of the fourth century B.C.* Chicago: University of Chicago Press.
Crump, J. I. (1996). *Chan-kuo Ts'e*. Ann Arbor, MI: University of Michigan, Center for Chinese Studies.
Crump, J. I. (1998). *Legends of the warring states*. Ann Arbor: University of Michigan, Center for Chinese Studies.
Dong, Y., Morgan, C., Chenenov, Y., Zhou, L., Fang, W., Ma, X., & Pechenkina, K. (2017). Shifting diets and the rise of male-based inequality on the Central Plains of China during central Zhou. *Proceedings of the National Academy of Sciences, 114*, 932–937.
Henricks, R. G. (1989). *Lao-Tzu Te-Tao Ching: A new translation based on the recently discovered Ma-Wang-Tui texts*. New York: Ballantine Books.

Hui, V. T.-b. (2005). *War and state formation in ancient China and early modern Europe.* Cambridge: Cambridge University Press.

Jaang, L., Sun, Z., Shao, J., & Li, M. (2018). When peripheries were centres: A preliminary study of the Shimao-centred polity in the loess highlands, China. *Antiquity, 92*(364), 1008–1022.

Keightley, D. (2000). *The ancestral landscape: Time, space, and community in Late Shang China (ca. 1200–1045 B.C.).* Berkeley: University of California Press.

Kidder, T. R., Haiwang, L., Storozum, M. J., & Zhen, Q. (2016). New perspectives on the collapse and regeneration of the Han dynasty. In R. K. Faulseit (Ed.), *Beyond collapse: Archaeological perspectives on resilience, revitalization, and transformation in complex societies* (pp. 70–98). Carbondale, IL: Southern Illinois University Press.

Kiernan, B. (2017). *Viet Nam: A history from earliest times to the present.* New York: Oxford University Press.

Knapp, K. (2003). *Selfless offspring: Filial children and social order in medieval China.* Honolulu: University of Hawai'i Press.

Lee, K.-M., & Ramsey, S. R. (2011). *A history of the Korean language.* Cambridge: Cambridge University Press.

Li, F. (2006). *Landscape and power in early China: The crisis and fall of the Western Zhou 1045–771 BC.* Cambridge: Cambridge University Press.

Li, F. (2008). *Bureaucracy and the state in early China: Governing the Western Zhou.* Cambridge: Cambridge University Press.

Li, F. (2013). *Early China: A social and cultural history.* Cambridge: Cambridge University Press.

Li, M. (2018). *Social memory and state formation in early China.* Cambridge: Cambridge University Press.

Li, L., Zhu, C., Qin, Z., Storozum, M., & Kidder, T. (2018). Relative sea level rise, site distributions, and Neolithic settlement in the early to middle Holocene, Jiangsu Province, China. *The Holocene, 28,* 354–362.

Lin, M., Luan, F., Fang, H., Xu, H., Zhao, H., & Barker, G. (2018). Pathological evidence reveals cattle traction in North China by the early 2nd millennium BC. *The Holocene, 28*(3), 095968361877148. https://doi.org/10.1177/0959683618771483.

Liu, B., Wang, N., Chen, M., Wu, X., Mo, D., Liu, J., Xu, S., & Zhuang, Y. (2017). Earliest hydraulic Enterprise in China, 5100 years ago. *Proceedings of the National Academy of Sciences, 114,* 13637–13642.

Loewe, M., & Shaughnessy, E. (Eds.). (1999). *The Cambridge history of ancient China.* Cambridge: Cambridge University Press.

Mallory, W. H. (1926). *China: Land of famine.* New York: American Geographic Society.

Miao, Y., Zhang, D., Cai, X., Li, F., Jin, H., Wang, Y., & Liu, B. (2017). Holocene fire on the northeast Tibetan plateau in relation to climate change and human activity. *Quaternary International, 443B,* 124–131.

Pettit, H. (2017). Oldest known waterway system that took a mystery Neolithic civilisation 3000 people and nearly a decade to build is discovered in China. *Kaogu, 5.* http://kaogu.net.cn/en/News/New_discoveries/2017/1206/60319.html.

Pines, Y. (2009). *Envisioning eternal empire: Chinese political thought of the warring states era.* Honolulu: University of Hawai'i Press.

Pines, Y. (2012). *The everlasting empire: The political culture of ancient China and its imperial legacy.* Princeton: Princeton University Press.

Sato, H., & Natsuki, D. (2017). Human behavioral responses to environmental condition and the emergence of the world's oldest pottery in east and Northeast Asia: An overview. *Quaternary International, 441,* 12–28.

Shelach-Lavi, G. (2015). *The archaeology of early China from prehistory to the Han dynasty.* Cambridge: Cambridge University Press.

Shimunek, A. (2017). *Languages of ancient southern Mongolia and North China: A historical-comparative study of the Serbi or Xianbei branch of the Serbi-Mongolic language family with an analysis of northeastern frontier Chinese and old Tibetan phonology*. Wiesbaden: Otto Harrassowitz.

Thornber, K. L. (2012). *Ecoambiguity: Environmental crises and east Asian literatures*. Ann Arbor: University of Michigan Press.

Vovin, A. (2005). The end of the Altaic controversy: In memory of Gerhard Doerfer. *Central Asiatic Journal, 49*, 71–132.

Zhang, L. (2016). *The river, the plain, and the state: An environmental drama in northern Song China, 1048–1128*. New York: Cambridge University Press.

Zuo, X., Houyuan, L., Li, Z., Song, B., Deke, X., Zou, Y., Wang, C., Huang, X., & He, K. (2016). Phytolith and diatom evidence for rice exploitation and environmental changes during the early mid-Holocene in the Yangtze Delta. *Quaternary Research, 86*, 304–315. http://www.sciencedirect.com/science/article/pii/S0033589416300424.

The Creation of Stable Dynastic Empires in East and Southeast Asia

4.1 Qin and Its Fall

Climate, however, was clearly involved in the rise and success of the Qin and Han dynasties, as it was in the rise of Rome (Brooke 2014; Campbell 2016). The formation of major empires in much of eastern Asia coincided with a warm, moist period ca. 250 BCE-220 CE, well known also from Roman history, since it coincided with the glory days of the Roman Republic and Empire (Brooke 2014: 322–323). It often called the Roman Maximum. Little is known of fluctuations during this time. It seems to have been rather uniformly good for all of East Asia. Records of local sharp downturns are found, however, and one of these coincides roughly with a palace coup that led to a brief interregnum in China.

It was during this long favorable period that Korea slowly came onto the historic scene, with the development of three kingdoms from a group of chiefdom-like societies that the Chinese regarded as their "commanderies." Japan began a rise from local nonstate polities to dynastic states. Southeast Asia developed its first city-states.

After 200-220 CE, the climate deteriorated. This seems clearly linked to the fall of Han, which occurred at precisely the time that colder, drier weather sharply broke the Roman Maximum (Brooke 2014: 326–327). The period of disunion that followed was rather similar in climate to the twentieth century, but with many sharp fluctuations. Korea, Japan, and even Central Asia (which faced much worse conditions) do not seem to have been much affected by the climatic deterioration.

China's great historic dynasties begin with Qin and Han. Qin lasted a mere 15 years, 221 to 207 BCE, while Han was China's longest dynasty, with a 426-year run. Qin state was regarded by the other states as somewhat peripheral and even uncouth. Thus, its takeover of all China in the third century BCE represents one of the most classic cases in history of a semiperipheral marcher state conquering an empire.

It created the Qin Dynasty (221-207 BC; Loewe and Shaughnessy 1999; Pines et al. 2014) in a classic upward sweep. The founder, Qin Shi Huang Di, invoked tough Legalist policies. These had proved extremely successful in Qin state, but on an imperial scale they proved unpopular, especially with Confucians, who were coming into prominence as administrators. His powerful but tyrannical government succeeded extremely well during his lifetime, pushing Qin's frontiers to encompass the core of modern China. He thus created China more or less as we know it, and the name "China" derives from Qin. He standardized weights and measures, built roads of standard width, developed irrigation works, passed strict and comprehensive laws, and indeed did everything to make China a great empire with a solid legal, mercantile, and physical infrastructure.

A particularly significant example of Qin success was the control of the Min River in Sichuan by the Li family of engineers, starting around 250 BCE, before Qin conquered eastern China (Anderson 2014). They tamed the fierce, flood-prone river by the Daoist strategy of going with its natural tendency: splitting it into distributaries, keeping them deep and clear, and keeping levees low. They left a huge rock inscription by their successful waterworks: DYKES LOW, CHANNELS DEEP. For more than 2000 years, at least when governments were acting responsibly, the dykes have indeed been kept low and the channels deepened—the river being diverted into two of its three main distributaries so that the other can be cleared out, in rotation. The system still works today, and preserves the city of Chengdu (over five million people) from destructive flooding, as well as feeding the irrigation of the fertile Chengdu plain. The contrast with the constant overdyking of the Yellow River, causing it to aggrade its bed and eventually flood the countryside, has been noted for centuries. The Lis have been honored with a beautiful old temple, and certainly deserve it more than almost any others in history so honored. (In the west, we honor saints that did nothing but pray, and forget water engineers and city planners—an interesting comparative note.)

But tyranny makes enemies. Qin Shi Huang Di killed a large but uncertain number of scholars and others. He supposedly burned all books not concerned with practical matters, though in fact many books survived. The brilliant but outspoken scholar Jia Yi (201-169 BCE), not long after Qin's fall, ascribed the dynasty's collapse to Qin Shi Huang Di's refusal to listen to others, delegate authority, or follow tradition in spreading rule (Bodde 1986); there is surely something to this. The important historian Derk Bodde (1986: 89–90), however, stressed the overextension that the incredibly ambitious imperial project involved.

Tyranny also makes for bloody family relationships. When he died at 49, his son Huhai and a eunuch, Zhao Gao, moved to eliminate the crown prince Fusu. Huhai took over at 21. Further intrigue by Zhao Gao led to the death of Huhai and the succession of Ziying, who turned the tables and eliminated the eunuch (Bodde 1986). By this time, rebellions had broken out, and chaos reigned. A general assigned to govern Yue—southeastern China—broke off and made the region a separate country for almost a century (Brindley 2011, 2015).

Another general, Liu Bang, took the capital and captured Ziying. His superior Xiang Yu then entered the city, sacked and burned it, killed Ziying, and prepared to run the state. Liu Bang refused to accede, and the two fought for 4 years. Liu Bang won in 206, founding the Han Dynasty (206-BC-221; on Han history, see Lewis 2007; Twitchett and Loewe 1986). He is thus known to posterity as the "High Ancestor" (Gaozu) of Han. (Chinese emperors are rarely known by their personal names, unless, like Liu Bang, they founded a dynasty.) Han did not subdue the powerful Chu state till 202 (hence the dynasty is sometimes said to start only from that date). The Chu area was one of the last to fall to Qin, had broken from Qin as soon as Qin weakened, and was to remain troublesome.

History records that Xiang Yu was autocratic and cold, Liu Bang a man of the people, but of course it was the victors who did the writing. Liu deliberately fostered a cult of personality around himself, a sensible thing to do, especially given the unpopularity of Qin Shi Huang Di. Even discounting his sponsored history substantially, he seems to have been a charismatic person. As a farmer's son, he apparently had the common touch.

4.2 The Rise of Han and the Formation of Chinese Society

Han's rule was broken by palace coups and countercoups. The empire proved resilient when Liu Bang died in 195. His widow installed her son, aged 15, as successor. She then dominated the boy thoroughly, and when he died in 188, she took over herself, reigning as Empress Lu (r. 188-180 BCE). When she died in 180, the Liu family recaptured the throne without difficulty. This was the first, but not the last, case of a woman ruling China. It was also an early case of a major crisis coming at the beginning of a dynasty, but failing to bring the vigorous new dynasty down.

Liu Heng was enthroned, later to be titled Wen Di, the "Cultured Emperor" (r. 180-157 BCE). He apparently reigned with as much decency and clemency as he could reasonably do, but given the coup and countercoup, he was faced with a violent situation that necessitated the extermination of whole rival families. (In almost all ancient imperial societies, the whole family was obligated to avenge the death of a member, so an emperor, king, or khan could not usually stop by executing the one person who was his direct rival.) Land taxes sank in 168 to 1/30 the value of product, a low sometimes equaled but rarely excelled in China or anywhere else, and then were abolished totally in 167. His successor, Emperor Jing (r. 157-140 BCE), restored the taxes in 156, but only to the 1/30 level (Loewe 1986: 150). This created the classically "Good Times of Wen and Jing." The state promoted the arts of peace, especially agriculture.

Somewhat before this time, Jia Yi warned the dynasty: "He who does not forget the past is master of the future." As the great historian Derk Bodde (1986: 87) remarked, this anticipates by two millennia George Santayana's famous line that "those who will not study history condemn themselves to repeating it." Jia Yi's outspokenness later cost him his life; he committed suicide at 33 (Loewe 1986: 148), partly or entirely because of imperial disapproval.

Under Han, the harsh Legalist policies of Qin were softened somewhat, but more importantly they were supplemented with more benevolent and tolerant theories from Confucianism and Daoism. At least under early emperors, this was taken seriously. There is a claim that Han was simply Legalist, using Confucianism and Daoism only to conceal the iron hand in a velvet glove.

Evidence for a real liberalizing includes the dramatic progress of agriculture, the open debates recorded in historical sources, the very existence of these detailed and open historical sources, and the rise of civil service examinations for the bureaucracy. These exams, sporadically tried by Qin and probably other Warring States, became established in early Han. They were to grow in importance in later dynasties. They created an increasing coterie of scholar-bureaucrats dependent on the imperial dynasty for recognition and usually for pay. These often came from well-to-do families, but many started poor, and none were drawing only on family privilege. True nobility in the European sense began to wane. Great families came back in Later Han and dominated in the troubled times that followed, but China never again had anything quite like the independent, titled nobilities of Europe.

At this time began an evolution that was to continue and prove important throughout Chinese history. Slaves and serfs were freed in increasing numbers. The role of free farmers increased. China was never a "slave society," but the ratio of slaves and serfs to free persons in Former Han was high. Throughout all Chinese history, the rate dropped, until by Qing slavery was uncommon and serfdom virtually abolished, though low-status households and debt "slaves" were common. Still, chattel (as opposed to debt) slavery lasted until the Communist period (and, illicitly, even then). It was often a punishment of convicts, or sexual enslavement of prostitutes. Less often, it was full plantation or household bondage.

China never had a peasant population in the European sense—that is, a special separate class of small farmers, with fewer rights and privileges than other classes (Mote 1999: 365). Chinese society held farmers in very high respect, above craftsmen or merchants. A farmer, if not a slave or serf, was a full citizen. In the terms of Europe's Old Regime, he or she was a "yeoman," not a "peasant." The Marxists, when they conquered China, misapplied Marx' European concept of peasants to Chinese farmers, with devastating results (Day 2013). Marx shared the most reactionary bourgeois stereotypes of peasants, comparing them to potatoes in a sack and otherwise slighting them (Mitrany 1951). Mao Zedong, in spite of his own farming origin, picked up the full dose of this bigotry (see quote, Day 2013: 20). In fact China's farmers were a progressive group, higher in status than their wealth would predict, and generally careful and forward-looking. The ranks of fully free farmers increased, both absolutely and relatively, from early times. The Chinese rulers knew that slaves were inefficient and unmotivated, and a source of independent and local power for great lords and landowners. A free yeoman farmer population was easier to tax, easier to control, and easier to set against the independent local lords if there was a conflict.

This fondness for yeoman farmers—based on experience, which showed they were the most productive class of rural labor—spread to the other East Asian agrarian societies, but faded rapidly with distance from China; Japan eventually developed a more genuinely feudal society, with limited rights to farmers. Japanese and Korean farmers became peasants in the European sense, changing only in modern times.

"The good times of Wen and Jing" were not to last. Further troubles occurred when a partial internal coup in 140 BCE led to an irregular takeover by a prince later designated Wu Di, "Martial Emperor" (r. 140-87 BCE). This came some 66 years after the start of the dynasty, almost the right time for a short Ibn Khaldun cycle.

The East Asian world-system was expanding from China and its immediate border states to include seriously large territories and seriously dangerous enemies. Reaction to the rise of the Chinese state inevitably produced reflex states that arose partly in self-defense, partly to organize raids and strikes against the now-giant core polity.

First to challenge China from outside the classic "central states" area was the Xiongnu Empire (first visible in 209 BCE), a vast steppe and desert realm west and northwest of China, ruled by a group whose language is still essentially unknown. It arose in what is now Mongolia, helped by the warmer, moister weather. It pushed more and more boldly at the Chinese borders. Growing along with the Han Dynasty, it conquered and took over the good farmland in what is now northern and central Shaanxi. With this new food source, it could mount a formidable threat to Han. Xiongnu armies supposedly numbered up to 200,000 men. Presumably it rose partly through trade with China and the west, but we have surprisingly little information on this.

That farmland was, at the time, wetter and warmer than now, and its hills and mountains not yet deforested. It was thus considerably more productive and valuable than in later, drier times. Thus, though they began as nomads, the Xiongnu found their wealth and power deriving increasingly from settled agriculture. The strength and success of the Xiongnu would be hard to imagine in a less fortunate climatic time. The Xiongnu could strike down the valleys of tributary rivers to the Wei River, near which Han's capital, Chang'an, was situated. (Chang'an, the modern Xian, was very close to the old Feng-Hao dual capital of early Zhou.) Chang'an is otherwise ringed by mountains and substantially impregnable.

Wu Di raised taxes to sky-high levels, mobilized a huge army, and attacked the Xiongnu with crushing might. He won a Pyrrhic victory, weakening the Xiongnu and ending their threat to the capital but almost ruining China and losing hundreds of thousands of men in the process. This left China with high taxes and a still-serious Xiongnu threat, a situation that increasingly weakened Han. The Xiongnu took over small states in northwest China after Han fell, but eventually moved west, where their descendants and followers became the Huns who so troubled Rome (Vaissière 2005).

During the reign of the Emperor Wu, the ideology of the Chinese empire was hammered out in a series of debates and discussions. It was to remain foundational for 2000 years. Confucianism supplied the individual morality, the fondness for tradition, the devotion to literature, and some degree of openness in advocacy. Confucian morality, with its focus on immediate human relations and on humaneness, became both official and popular. In 136 the Confucian classics—histories, rules, conservation ideas, and folk poetry—were formally established as bases of education (Loewe 1986: 154). Confucianism also taught the importance of ritual in engaging the people and getting them to accept the state and its rules. Some early Confucian and Legalist writings anticipate Emile Durkheim quite precisely on the role of ritual and ceremony in maintaining social order.

Legalism, however, remained the basis of actual working polity. Harsh punishments were modified by Confucian mercy, but only to a small extent. Legalist impositions of strict standards and carefully collected statistics show that the Legalists understood "seeing like a state" long before James Scott (1998). Legalism gave China a government of laws that supposedly transcended the government of men; this concept, much refracted, was enshrined by America's Founding Fathers, who got the idea from Jesuit writings on China (among other sources). The effect of this were to be far-reaching. The Confucian-Legalist mix, along with court ritual and ceremony and the religious underpinning behind that, remained the foundational structure and ideology of the Chinese state. It was all-important in maintaining that state as a dynamic, evolving, but continuous entity through succeeding dynasties. Other East Asian polities copied the basic ideas, and were strong in proportion.

One of Emperor Wu's harsh moments involved the elimination of his relative Liu An, king of Huainan, in 122 BCE. Liu had been driven to rebellion, or perhaps he was framed. In any case, he had run an independent court that was a center of science and learning. His great collection the *Huainanzi* (Liu 2010) summarizes much of the knowledge of the time. Even more impressive is the fact that the *Guanzi* (Rickett 1965, 1985, 1998)—the teachings of an ancient economist, with added commentaries—was probably compiled in Huainan. It summarizes a great deal of the agricultural science of the time. It is also likely that the *Zhuangzi* was put together in its present form by the same court.

Closer to the central capital at Chang'an in northwest China, one Fan Shengzhi compiled the world's first known agricultural extension manual around 120 BCE, and it records government-sponsored agricultural research that includes the world's first known case/control experiments (Anderson 1988; Sadao 1986; Shih 1973). It gave excellent, practical advice on farming in that colder, drier realm. Works like the Guanzi and Fan Shengzhi paralleled the great agricultural manuals of the Roman Empire. They were less comprehensive and systematic, but they show a high level of agricultural sophistication.

It is hard to avoid some conjectural history at this point. If Liu An had taken over, would he have enthusiastically backed Fan Shengzhi and others? Would he have started serious agricultural science in China, 2000 years ahead of the west? Surely, Liu would have backed a mix of Daoist ideology and practical, scientific, economically sophisticated governance, rather than the Legalist-Confucian fusion created by Wu Di. China would have been a very different civilization indeed.

Han also began a process of liberating imperial hunting parks and other open lands to free farmers (Anderson 2014). The great poet Sima Xiangru immortalized this in poetry (Sima 1987). Sima praised lavishly the park with its trees and animals, but then praised the emperor for releasing it to the people, an early example of what Karen Thornber calls "ecoambiguity" (Thornber 2012). Again, this set a precedent that was followed until modern times. China preserved little and constantly expanded the frontiers of intensive agriculture.

Buddhism came to China in Han, largely in Later Han but presumably with some prior contacts. The effect of this quietist religion on Chinese society was complex. It preserved its own teachings and monastic system, but, in the public mind, it broadly merged with existing religious traditions. To these it contributed a peaceful, charitable, other-oriented morality that reinforced existing tendencies in those directions.

4.3 The Fall of Former Han

Former Han came to a sudden end. Palace politics were the immediate cause. Emperor Cheng (r. 33-7 BCE) fell under the spell of Zhao Feiyan ("Zhao the Flying Swallow"), a singer and dancer who became proverbial throughout Chinese history for her bewitching beauty, like Helen in Greek myth. Supposedly, the emperor so lost his head over her that she and her sister worked evil intrigues right and left, but this is another "bad last" (or near-last) story, and thus of dubious value.

More serious were the intrigues of Wang Zhangjun, the mother of Emperor Cheng. She not only ran his life but installed the next two emperors, Ai (r. 7-1 BCE) and Ping (r. 1 BC-6 CE). All three of them died without heirs, in one of those demographic disasters that sometimes struck Chinese royal families. The latter two were young and short-lived.

In Ping's time, the world's oldest surviving census was taken, showing China with about 60 million people, heavily concentrated in the North China Plain, Wei Valley, and Sichuan Basin. China at that time had recently conquered the Red River delta in what is now Vietnam, and that was also a population focus, though a thoroughly non-Chinese one. A later census in 140 showed rapid colonization in the Yangzi drainage.

Ping's early death left Wang Zhangjun's nephew Wang Mang in power He took over as emperor (r. 9-23; Lewis 2007: 23–24; Loewe 1986: 215). One interesting bit of success was his (more accurately, his mother's) promulgation of a strong code of conservation and environmental protection throughout the empire, as shown by Charles Sanft (2009, 2010). However, Wang attracted challenge, and increasing dissatisfaction led to revolt. Farmers painted their foreheads red, and rose against taxes and labor requirements. The chaos of this "Red Eyebrows" outbreak allowed the imperial Liu family to rise and retake the empire. Liu Xuan became the Gengshi Emperor in 23 CE. History has not been kind to him or to Wang Mang (Bielenstein 1986). The Red Eyebrows continued, taking and sacking Chang'an and eliminating the pathetic Liu Xuan in 25 CE. Liu Xiu became the leader of the Liu restoration movement, taking power as Guangwudi. He had a terrible fight on his hands, first against the weakening Red Eyebrows and then against many partisans of various leaders. China was not reunified until 36 CE (Bielenstein 1986: 255).

The decline of Former Han and rule of Wang Mang has recently been treated as an archetypal story of decline and fall in an important paper by Tristram Kidder and others (2016). They exaggerate the degree to which this was a bloody collapse. It was a coup and countercoup, like those of the old days in Latin America. It occasioned a good deal of violence, especially at its end. Wang fell to a peasant rebellion in which more and more elite families joined. On the other hand, it was not a true dynastic collapse.

Kidder's group reports that there were more floods and droughts than usual in the years leading up to 9 CE, then a major one in 11 CE, and then a huge flood in which the Yellow River broke its dykes and flooded a vast area in central China in 14-17 CE. The Yellow shifted its mouth to the north from a more southerly route. These floods followed on decades of expanding cultivation, involving deforestation and plowing. The light, fragile loess soils of the north and northwest were extremely erosion-prone; in fact, they are the reason for the river's name, since the enormous amounts of silt they produce colors its water. Thus, as they ably put it, politics was "playing out against a backdrop of climatic and especially anthropogenically induced environmental change that was constantly shifting and reconfiguring the resource base on which so many relied" (p. 85).

They are not naïve climatic determinists. They point out that the wars with the Xiongnu Empire to the west had exhausted the country and its treasury. They conclude that the collapse of a dynasty occurs when "the disjunction between rules and resources reaches a threshold so stark that agents at all social levels stood to gain more by challenging the status quo than they did by conforming to it" (p. 86). This is the story of dynastic fall in a nutshell—never better put. They point out that "...the collapse of Western Han actually spans multiple temporal cycles from minutes to millennia, which is, we suspect the case for most collapse events" (p. 87). Indeed it is.

Later Han never rose as high as Former Han. The capital was moved for safety from Chang'an in the west to Luoyang in the center of China. Hence the two halves of Han are often called Western and Eastern Han. Eastern Han was surprisingly tranquil and prosperous. Medicine and agriculture flourished. China grew more populous, especially in the south. The whole period lacked the spectacular highs and lows of other dynastic centuries.

Contact with the west became important. Former Han had already sent envoys westward, returning with Central Asian products including alfalfa and wine grapes, but Later Han was far more involved, holding all of eastern Central Asia and settling the oases there. These oases began as kingdoms, but many ended up incorporated into Han. These oases were very well known, as were the routes and trading opportunities, and detailed summaries of this knowledge were recorded in the *Hou Han Shu* (History of Later Han). A major translation and study of relevant parts of this work, by John Hill (2009), has added what we now know from archaeology and other sources. The East Asian world-system had definitely expanded to incorporate the eastern half of Central Asia. The importance of that region to China and points eastward, and of the Silk Road trade, was firmly established.

Trade across Central Asia truly developed during Han. The Chinese knew of the Roman Empire and the Romans knew of China as a source of silk. No direct contact is known. Roman coins abound as far east as the west coasts of India, but they do not occur in China, nor do many Chinese items show up in the west. Trade was through middlemen on the Silk Road—actually a cluster of caravan routes—of inner Asia and the "maritime silk road" around southeast Asia via India and the Arabian Sea.

The oases of Central Asia hosted a number of small states to take advantage of the rapidly rising trade. They became cultural, religious, and ethnic melting pots. Direct Han trade was extensive throughout what is now Xinjiang, and at its widest extent Han held and occupied most of that vast area. Trade with the farther states of western Central Asia was notably less. Silk was evidently a prime good, but agricultural commodities, books, slaves, spices, and a range of other commodities traveled the routes. At this point, trade was not enormous. It was probably an exciting event when a caravan rolled into Rouran or Kashgar.

China and Rome did not contact each other directly, but Rome was known from Central Asian accounts, and Chinese silks began their travels thither—to be duly denounced as idle luxuries by Roman Empire moralists. The enormous extent of the silk, spice, and general Asian trade to Rome has only recently become known, with the excavation of Berenike and other ports (Chew 2018; McLaughlin 2010; Sidebotham 2010). The "maritime silk route," across the Indian Ocean from southeast Asia to east Africa, the Persian Gulf, and the Red Sea, was already a major trade route, well known to mariners (*Periplus of the Erythraean Sea*, 1912). Rome did not reach China, but did reach India, at least through traders. Like the accounts of western regions in the *Hou Han Shu*, the *Periplus* gives detailed accounts of trading cities and peoples, covering the western Indian Ocean.

Merchants prospered. In Later Han they could dominate their regions and prevent monopolies and other controls by government. The small farmers—90% of the population—tended to lose ground to all these demographic elements (Nishijima 1986). The gap between the rich and the poor grew steadily. Writers complained about this in terms strikingly similar to those used today to denounce the "1%" (see Ebrey 1986; Nishijima 1986, for examples). All these blocs fell more and more into competition with each other instead of cooperation to help the realm. "After 140 the government gradually lost its ability to provide relief; then to maintain order in particular areas; and finally to maintain order at all" after 184 (Ebrey 1986: 628).

Droughts and floods mattered as background action in the decline of Former Han and climate deterioration in the fall of Later Han. It appears that the growing abstraction of palace politics from the real world, and the growing bureaucratization and remoteness of the government, were more decisive. The histories of the time—the great historical works by Sima Qian and Ban Gu that really started Chinese historical scholarship—tended to blame individual morality and huge military expenses. Especially noteworthy are the historians' charges that monarchs and bureaucrats were not doing their jobs.

A long breakdown period began in 168 CE. Emperor Huan Di (r. 146-168) died childless. The Empress Dowager had the real power to name a successor, and she named Liu Hong, "a boy of eleven or twelve" (Beck 1986: 318). Eunuchs then captured and killed the Empress Dowager and her allies. As Emperor Ling Di, Liu Hong reigned from 168 to 189, after which a massacre of eunuchs weakened the court relative to great clans and generals. Mansfeld Beck's definitive study of the fall of Han makes little mention of climate, and much of imperial weakness and failure to control the realm (Beck 1986). The real deterioration of climate came after 200 CE, by which time Han was fatally weak.

The empire was devastated by the rebellion of the Yellow Scarves in 184. (They are usually called the "Yellow Turbans," but the Chinese word means "scarves," and that is what the rebels wore.) Hundreds of thousands of farmers and other workers joined a millennarian religious movement. Other rebellions, warlord risings, and palace intrigues occurred. A particularly murderous outbreak by Dong Zhuo and his troops destroyed Luoyang and set up the young boy Liu Xie as the new emperor Xian Di.

The dynasty was shattered. It never controlled the whole empire again. It survived through military might, but power consolidated in the hands of the great general Cao (155-220). While he lived, he maintained Xian Di, but when Cao died in 220, his son Cao Bi (186-226) immediately "accepted"—i.e., forced—the abdication of Xian Di and the passage of imperial power to the Caos. The Liu family continued to hold out in remote parts of China, and were a historical force for another 1000 years.

4.4 China Disunited

Cao Bi dropped the Han fiction and set up his own state of Wei, ruling north China. The warlord Liu Bei (161-223) had already seized control of his own domain, in the west, under the name of Shu, the old name for Sichuan. He promptly renamed his realm Han and claimed (largely on the thin excuse of his surname being Liu) to be continuing the dynasty. A third kingdom, Wu, lasted a short 36 years in the southeast under the brilliant warlord Sun Quan (182-252).

This "Three Kingdoms" period was China's heroic age. Its story had been told in countless forms since the fourth century. China is sometimes said to be without an epic, but the manifold versions of the tale of the Three Kingdoms actually fill the function. Stories of the Three Kingdoms were retold in countless histories, poems, plays, and storyteller versions. The novel version is the *Romance of the Three Kingdoms*, written Luo Guanzhong in the fourteenth century on the basis of earlier works (there are many translations).

Liu Bei swore with Guan Yu (Guan Gong, 160-219) and Zhang Fei the Oath of the Peach Orchard, in which they took a blood oath to fight to the death for each other. Guan served so loyally and heroically that he became the God of War and the God of Loyalty and Trust, and as such is now represented and respected in countless Chinese places of business, from stores in Singapore to restaurants in Los Angeles. There is certainly no question that they, Sun, and Cao all commanded 'asabiyah. If any one of them had lived longer, he would very likely have united the empire. As it was, each had enough brilliance and charisma to balance the others out. No one state could overcome the other two.

The Battle of Red Cliff, in which Sun Quan's troops destroyed Cao Cao's in 208 (long before Han officially fell), became the archetypal horrific battle in Chinese history and legend. Of countless poems and essays about it, the most popular are probably Su Shi's from the Song Dynasty. His personal responses to visiting the

Cliff were expressed in powerful, direct style that has made the tragedy—and by extension the horrors of war—real for generations of readers.

In 249 the Sima family took over from the Caos, and in 265–266 they formed a new dynasty called Jin. They eclipsed Shu, and then in 280 Jin conquered Wu (Beck 1986: 373). The empire was thus reunited in a dramatic and rapid conquest. This unity did not last. In 316 Jin was conquered in the north by a new Han state, which rose in 304 under Liu Yuan, a Xiongnu who claimed Han imperial descent (Beck 1986: 369–370; Lewis 2009: 51). Jin survived, with a shrunken territory, reconstructing itself as "Eastern Jin" in 317. The emperor was so poor that neck bones were the great luxury in his court, reserved for him (Stephen Owen's footnote to Du Fu, Du 2016, vol. 5, p. 391). Pork neck bones are a classic poverty food in the United States. Jin provides a fascinating mystery. How did they succeed so well? Why did they fall so fast? Records of that period were largely lost in successive wars, leaving us with thin understanding.

Notably important here is the triumph of the Xiongnu, or one line descended from them and possibly from Han also, at a time when climate was dry and cool, in sharp contrast to the golden days of the Xiongnu. This provides an unusual case of the old story of drought sending the nomads on their campaigns of invasion (Chen 2015). On the other hand, Liu Yuan had been ruling a settled state in north-central China, not a nomadic tribe on the frontier.

At this time—after 260 and increasingly until 600—colder weather made north China and central Asia turn away from millets and toward the more cold-tolerant wheat and barley (D'Alpoim Guedes and Bocinski 2018). Flour milling became ever more important, with noodles and dumplings becoming staple foods. People began to move south; perhaps as many as one in eight farmers deserted the cold north for the warmer southern valleys (D'Alpoim Guedes and Bocinski 2018). Trade operated again: the flour milling and dumpling-making technology was all Central or West Asian, and came along the Silk Road, where we have many finds of flour foods from about this time.

A succession of "Northern and Southern Dynasties" after 316 kept China broken into many feuding states. (On this troubled period, see Lewis 2009.) In 420 Jin lost the south, and disappeared. "For the next two centuries, the southern dynasties (Liu Song, Liang, and Chen) were ruled by military men of humble origins, who increasingly asserted their authority at the expense of the great families. These dynasts could not challenge the magnates in their localities.... But the dynasts stripped the great lineages of the three bases of their political power: military force, administrative authority, and wealth" (Lewis 2009: 70; see also 72–73). All this activity, plus the needs of feeding the military in continual wars, prevented any leader from getting enough traction to take the empire from the south.

Meanwhile, the north was struck by successive waves of Central Asian peoples, who built up cavalry power and who could draw on resources from oases, deserts, steppes, and northwestern Chinese farmlands. Non-Chinese ("barbarian") rulers took over. The north was divided for a century into the so-called "Sixteen Kingdoms." At first, many were Serbi (Xianbei). Then the Turks became important, conquering eastward from the vast Gők Turk empire that had formed in central Asia.

The Tabghach (correctly Taghbach, but a mistaken inversion of letters has become established in the literature) were the most important Xianbei group, conquering the north and appearing as the Tuoba of Chinese history. Generally considered Turkic in the earlier literature, they spoke a language now known to be related to Mongol and Khitan, not to Turkic (Shimunek 2017). Arising in the late 300 s, they slowly but surely expanded to control all north China by 440. They instituted the Wei Dynasty, establishing a court at Luoyang in 493 (Lewis 2009: 80–81).

The climate background is complex. Overall, the climate was much like that of the twentieth century. Eastern Central Asia was cold and dry, but western Central Asia—west of what is now the China border—was probably wetter and less unbearably hot than it had been in Han or Three Kingdoms times, so groups there could build up wealth and manpower. A "negative" phase of the North Atlantic Oscillation was responsible for this.

A major collapse in 536 changed these rather easy times. Violent volcanic activity darkened the sun, producing a Year without a Summer. In a separate cooling episode, the sun's radiation drastically decreased, producing the deepest solar minimum in millennia, considerably lower than the Maunder Minimum of the seventeenth century (Harper 2017: 255). The result was a cold, harsh time, probably hastening the fall of Wei, and certainly involved—but in very mysterious and unclear ways—in the violence of the next century that ended with Tang dynastic control of China.

Wei had the distinction of being ruled for a long time by an Empress Dowager, during a period of child-emperors (Von Glahn 2015: 172–173). Wei maintained varying degrees of control till 550, but never held all of China and rarely even most of it. As a Central Asian dynasty, Wei had many close ties with the west, and introduced a great deal of Near Eastern and Indian culture, via Central Asia. The Buddhist religion flourished, and Wei Buddhist sculpture remains famous and widely admired. Possibly more important were introductions of food, medicine, and technology. Records are spotty or nonexistent, but it is now clear that many of the Central Asian introductions usually credited to Tang, and made famous by Edward Schafer (1963), came during Wei (see e.g. Dien 2007, including a fascinating study of a Central Asian leader buried in China; Lewis 2009). Trade flourished through Central Asia, but not so much oceanward.

Wheat, which had become widespread but uncommon in Zhou times, and then commoner in Han with improved milling machinery, now came into its own. Mills propagated widely. Flour and flour foods were now staples (von Glahn 2015: 194). This seems due in part to the Turkic peoples. They apparently introduced the familiar and varied dumplings now so familiar in China, Korea and Japan. The Chinese word *mantou* and Korean *mantu* for steamed bread rolls is clearly the Turkic word *mantu* or *manty*. Now unstuffed, *mantou* in those days were stuffed dumplings, like *jiaozi*. Some survived, dried out, in an early Tang tomb in far west China, to be studied by modern experts.

One fateful change that was under way by this time is the demographic shift from north to south. This was to accelerate in the next few centuries. It affected the ways that climate influenced history: north China is much drier and colder, and thus devastated by cooling periods, especially by short, sharp cold spells. The south is much more stable and reliable. Moreover, agriculture in the north is largely rainfed; in the south, the more productive agriculture is irrigated, largely in rice paddies. Thus, *the progressive demographic shift made China less and less exposed to the worst climatic problems*, an important point (Fan 2010). The corresponding events were notoriously devastating in the west, where the Romans were trapped by the Mediterranean and the deserts (Harper 2017).

Another important shift was the rise of great families. Beginning in Han, powerful regional families arose, developing local centers of control. With the constant unrest of the period from 200 to 500, they became the real power in much of China.

All the warfare and division did not prevent a great cultural flowering. Some of China's greatest and most foundational poetry was written in Jin and Wei times by men like Ruan Ji and Tao Yuanming. The great and unique encyclopedia of domestic arts, the *Qimin Yaoshu* ("Knowledge Needed by Ordinary People"), by Jia Sixie, appeared under Wei. Tao Hongjing, synthesizer of China's medical and herbal traditions as well as leading scholar of Daoism, lived at about the same time, but in the south. Even sin became refined and arcane; an official named Wang Sengda (423-458) got in trouble because he loved falconry, ran wild, and even "feigned illness to watch a duck fight" (Wallace 2017: 126).

Directly connected with world-system dynamics was the rapid westernization of China. A vast flood of western learning—everything from Buddhist scriptures to Iranian horse gear—came to China. Buddhist monks from India and Central Asia introduced western foods, medicines, and other knowledge.

Even so, the period of disunion was not pleasant. Someone saw fit to record a folk song that sums it all up, and was intended to:

Man—pitiful insect,
Out the gate with fears of death in his breast,
A corpse fallen in narrow valleys,
White bones that no one gathers up.
(Translated by Burton Watson 1984: 196).

4.5 Southeast Asia Enters History

Until mid-Han, China had been far ahead of the rest of eastern Eurasia. Local chiefdoms and tribal societies prevailed. The unsettled climatic period in the early first century CE, and the rapid rise of Han to wealth and power, seem to have had a dramatic effect on nearby areas.

First to rise was northern Southeast Asia, affected by southward migration from China. Thai-speaking peoples were probably the main groups involved in Thailand, but in Vietnam the expanding Qin and Han Dynasties met northward-expanding ethnic Vietnamese on what had probably been Thai ground. Today, Chinese, Thai, and Vietnamese sound alike and share many words, but this is due to borrowing in historic times. (On Vietnamese history, see Kiernan 2017; Lieberman 2003, 2009; Tarling 1992.)

By 1500 BCE, some groups, very possibly ancestral Thai, were developing sizable settlements in north Thailand (Higham 2015, gives a particularly good recent summary). These climaxed in spectacular Bronze Age chiefdoms by 500 BCE. At this time, similar societies, with striking and distinctive art and world-class bronze-casting abilities, occupied Yunnan; they probably spoke Thai and Tibeto-Burman languages.

Many of the Yunnan sites share with northern Vietnam the Dongson Culture, a spectacularly brilliant Bronze Age culture or multiethnic tradition that created not only superb bronzes but also fine pottery, jade work, and early urban sites. By 200-300 BCE, Co Loa, north of Hanoi, was developing into a true city, with multiple walls, moats, an irrigation system, and other appurtenances of civilization (Kim 2015).

The Qin Dynasty sent an army to conquer southeast China, at that time a mosaic of Thai, Austronesian, and other ethnic groups. The name Yue, originally applied to a Yangzi Delta state, extended to the whole region. The Qin emperor sent General Tu Ju to conquer it, which he did, from 221 BCE onward. The southern Yue proved refractory, and eventually the Qin sent another general, Zhao Tuo. By the time he finished conquering the entire region, Qin had fallen and Zhao's immediate commander was dead, so Zhao naturally set up his own kingdom. He lived long and subdued several minor polities in what is now Vietnam (Kiernan 2017: 67–70). This lasted until his son lost it to Han Wu Di in 112 BCE. The tomb of a Yue king survived, amazingly unlooted, until today, and can be seen with all its magnificent contents in the modern city of Guangzhou (Linna 2007). Some of the decorated objects are Dongson-like or even Oceania-like in artistic style.

The lands south of the core region of Yue—south coastal China—were naturally if unimaginatively designated "South of Yue," which the inhabitants thereof pronounce Vietnam (it is Yuenan in modern Mandarin). By the time of the famous census in 2 CE, the Red River delta was the most populous, and presumably the richest, area south of the Yangzi (Kiernan 2017: 74).

Han conquest led to a widespread revolt in 39-43 CE (Kiernan 2017: 78–82), led by the Trung sisters, who became and remain great national figures in the Vietnamese pantheon. They are often shown riding heroically on war elephants. Whether they rode these or not, they commanded plenty of them, to the discomfiture of Han troops. However, sheer force of numbers prevailed, and the Trung sisters died martyrs' deaths in 43 CE.

The area reasserted independence soon after Han fell. Small kingdoms ruled northern Vietnam for brief periods, while independent polities to the south, notably Linyi, raided the weakly-held domains (Kiernan 2017: 99). Then and subsequently, Vietnam struck for independence whenever China weakened—a point not lost on the many experts who warned the United States not to involve itself in the Vietnam War.

By this time, civilization, in the sense of urban and literate society, was not only spreading south from China; it was also spreading north from beachheads in Cambodia and the Mekong Delta region. Influence from India, including some outright immigration, led to the rise of city-states such as Oc Eo in these regions, between 1 and 200 CE. This was succeeded by or expanded to become Funan, in the Mekong Delta and points south; "Funan" is the modern Chinese pronunciation of what was probably a Chinese transcription of Phnom, as in Phnom Penh. They could not have been more different from China in culture, being Indianized Austro-Asiatic societies, but they quickly became throughput sites on the developing China-India trade route (Abu-Lughod 1989; Chaudhuri 1985). Local states arose, with strikingly original cultures and no deep political debts to the larger neighbors (Chew 2018). An independent and powerful state, Nan Yue, was declared in what is now Vietnam in 544, and lasted until Sui conquest in 602. Long-distance trade, connecting China with India and the west, was largely in fabrics, aromatics, and spices. A delightful modern Malay proverb says: "Even though ten ships come, the dogs will have no loincloths but their tails." The ships, of course, are bringing rich foreign fabrics. Such was trade in the old South Sea.

4.6 Korean Civilization

Meanwhile, civilization came also to Korea. Urbanized and highly developed states had flourished in northeast China for some time. Han defeated an early state or chiefdom called Chosŏn—later to be revived as Korea's name in the early modern period—and created commanderies by 108 BC. The most famous one was Lelang (Lolang), a focus for the spread of Han civilization and power. These commanderies brought full civilization to the peninsula.

Later, society crystallized into three kingdoms: powerful and very Han-influenced Koguryŏ (or Goguryeo, 37 BCE-668 CE) in the north, Paekche (18 BCE-660 CE) in the southwest, and Silla (or Shilla; 57 BCE-935) in the southeast. The starting dates are traditional, but are semi-legendary and too early to be believable. The real dates are unknown, but it seems that China's disunion, especially after 311, left Korea and Japan free to form strong independent states, which they promptly did (Seth 2011: 26–31).

Paekche is mentioned in Chinese annals by 372 (for an event in 371; Best 2007: 22–25), and apparently had taken southeast Korea from the "Mahan," a loose grouping of tribal proto-states of which it was one member. It probably became a state between 290 and that point (Best 2007). Koguryŏ, known earlier, and incorporating what is now northeast China as well as the northern half of the Korean peninsula, was particularly important. It too started later than claimed, but apparently before Paekche, since it was important in Chinese annals from at least that early. In 313 it conquered the Han commandery of Lelang, the main outpost of civilization in northern Korea. In 370 it defeated the Xianbei (Pratt and Rutt 1999: 226). It became tributary to the Chinese states. The name was later revived and shortened to Koryŏ to serve as the name for a medieval dynasty; this name was pronounced "Korai" in Japanese, which gives us the modern name "Korea."

In the 400 s, Korea found itself wedged between an increasingly feisty Japan and the great power of China. Even in division, north China was the gorilla in the East

Asian room. Japan rapidly increased its military power, and used it to raid the peninsula, harrying Silla and reducing Paekche to tributary status. Even powerful and remote Koguryŏ suffered raids. Harassment ranged from piracy to outright war. China was much more prone to ally itself with Korean powers, sending missions and allying itself with Korean states against the tribes of what is now northeast China.

The effects of the 536-650 cold period on Korea are unclear. Food production must have been hurt, but the kingdoms proved resilient. Korea is protected by geography from Chinese conquest. Han and, later, other dynasties failed to conquer it.

However, Korea was always eager to learn from China, and to trade. Koguryŏ's elite copied Chinese styles faithfully, but their art already shows some of the zany humor that always defined it against Chinese norms. A painted and lacquered box from a Korean tomb, just before the rise of Koguryŏ, is decorated with highly humorous scenes that would do credit to a modern cartoonist (McCune 1962: 57–59). More impressive is the Anak tomb, from the Koguryŏ period, which revealed a vast range of painted scenes; their portrayals of warriors indicate the source of many influences on Japanese military equipment (Takeshi 1993: 302). Trade no doubt ran heavily to furs, ginseng, and other Manchurian products, exchanged for the metal and ceramic goods we know from burials. Korea, walled by mountains from China and protected by the sea from the worst climatic effects of the downturn after 200, continued to be a center of high civilization during China's time of disunion.

4.7 Japan Appears in the Records

The source of the Japanese language remains totally obscure. It was probably the language of the Yayoi, and was probably related to the language of Koguryŏ. It may have been a language spoken by some of the Jomon. Ainu, a language of equally obscure links, may (also?) have been a Jomon language. Just after Han times, the Japanese-speaking Yamato culture, ancestral to dynastic Japanese civilization, emerged in the south and central parts of the archipelago.

Japan had only recently picked up intensive rice agriculture and chiefdom-level social organization with the Yayoi immigrants from Korea after 400 BCE (Barnes 2015; Brown 1993a, b, c; Kidder 1993). They were apparently not directly ancestral to modern Koreans; they were ancestral genetically and culturally to the modern Japanese, as were the Jomon people who previously inhabited the islands—the two mixed, especially after 1 CE. The Jomon tribes became known to early Japanese as the Emishi, and the expanding Japanese state was to fight them for centuries.

At this point, envoys from the "Wo" ("dwarfs," the Chinese term from then throughout Imperial times for the Japanese) began reaching Han and then its successor Three Kingdoms. A Queen Himiko was especially notable in contacting the state of Wei, and she and her country are described in detail in the Wei annals, which refer to a few earlier records (Arikiyo 2018; Batten 2003: 147; Kidder 1993; Walker 2015). Japan had no history of its own at the time, so she remains little known. The Wei annals speak of many small chiefdoms or kingdoms, one of which was called "Yamatai" (or something like that; the Three Kingdoms pronunciation is not well known). It probably represents Yamato, the central region that progressively conquered the rest of Japan from that time onward. Japan consolidated a state under the Yamato dynasty around the third century CE. According to Japan's own histories, the Sun Goddess was mother to the dynasty, whose first clearly human ruler was Jimmu Tenno (*tennō* being the word of the time for "emperor").

However, Japan is barely noted otherwise in Chinese records, and has almost no written records of its own, until it bursts suddenly into full light in the Tang Dynasty. One earlier written record in Japan itself is "an inscription carved on a fifth-century sword.... It includes the names of six clan (*uji*) chieftains who served Yamato kings" (Brown 1993a, b, c: 2). A gold seal supposedly from about that period is more than a bit suspicious as to authenticity. More interesting is a seven-pronged sword preserved in a shrine storehouse; it was made in Paekche, around 369, and shows important diplomatic and military relationships between Japan and Korea; diplomatic and military activity increased sharply in the 500 s (Brown 1993b: 121, 154–156). By this time Japan was militarily involved in Korea, a point more fully treated in the next chapter. The cold period of the sixth–seventh centuries did not greatly affect the Yamato state, with its ocean-protected, subtropical central regions. This enabled it to extend control over northern parts of Honshu, which must have suffered terribly.

4.8 Verdict

This was a period in East Asian history when climate does appear to have affected history. Han's success was certainly helped by the Roman optimum, but, unfortunately for Han, the northwest frontiers were helped relatively more. The Xiongnu became the first of the long succession of formidable steppe empires to attack China. The rise of Vietnam, Korea, and Japan from chiefdoms to civilized states may also have been assisted by climate, but we have no real evidence; these areas were less affected than north China and the northwest steppes.

The decline of the climatic optimum coincides with the fall of Han, and may be part of the explanation for the fall, but Han had entered a long decline well before the downturn. Rival states were building fast in areas increasingly hard to control. Han's final fall was not a full collapse, but a coup, and the result was war with rising rival states.

However, colder and drier weather does appear to have motivated the Xiongnu and Serbi peoples to strike directly for the heart of China, and they were able to establish powerful states during a long period of disunion.

The effect of this cooler, drier time on the rest of East Asia is not clear. It seems to have been minor. It may have led to more consolidation of power in Korea, where the classic "three kingdoms" emerged. The dynamics of statecraft in rapidly rising polities seems to have been much more a matter of human enterprise and creativity than of climatic forcing.

References

Abu-Lughod, J. (1989). *Before European hegemony: The world-system A.D. 1250–1350*. New York: Oxford University Press.

Anderson, E. N. (1988). *The food of China*. New Haven, CT: Yale University Press.

Anderson, E. N. (2014). *Food and environment in early and medieval China*. Philadelphia, PA: University of Pennsylvania Press.

Arikiyo, S. (2018). *Treatise on the people of Wa in the chronicle of the Kingdom of Wei: The world's earliest written text on Japan* (J. A. Fogel, Trans.). Portland, ME: MerwinAsia.

Barnes, G. L. (2015). *Archaeology of East Asia: The rise of civilization in China, Korea and Japan* (3rd ed.). Oxford: Oxbow Books.

Batten, B. (2003). *The end of Japan: Premodern frontiers, boundaries, and interactions*. Honolulu: University of Hawai'i Press.

Beck, B. J. M. (1986). The fall of Han. In D. Twitchett & M. Loewe (Eds.), *The Cambridge history of China. Vol. 1. The Ch'in and Han empires, 221 B.C.-A.D. 220* (pp. 317–376). Cambridge: Cambridge University Press.

Best, J. W. (2007). *A history of the early Korean kingdom of Paekche, together with an annotated translation of the Paekche annals of the Samguk Sagi*. Cambridge, MA: Harvard University Press.

Bielenstein, H. (1986). Wang Mang, the restoration of the Han dynasty, and later Han. In D. Twitchett & M. Loewe (Eds.), *The Cambridge history of China. Vol. 1. The Ch'in and Han empires, 221 B.C.-A.D. 220* (pp. 223–290). Cambridge: Cambridge University Press.

Bodde, D. (1986). The state and empire of Ch'in. In D. Twitchett & M. Loewe (Eds.), *The Cambridge history of China. Vol. 1. The Ch'in and Han empires, 221 B.C.-A.D. 220* (pp. 20–102). Cambridge: Cambridge University Press.

Brindley, E. F. (2011). Representations and uses of Yue identity along the southern Frontrier of the Han, ca. 200-111 BCE. *Early China, 33–34*(2010–2011), 1–35.

Brindley, E. F. (2015). *Ancient China and the Yue: Perceptions and identities on the southern frontier, c. 400 BCE-50 CE*. Cambridge: Cambridge University Press.

Brooke, J. L. (2014). *Climate change and the course of global history*. Cambridge: Cambridge University Press.

Brown, D. (1993a). Introduction. In D. Brown (Ed.), *The Cambridge history of Japan. Vol. 1: Ancient Japan* (pp. 1–47). Cambridge: Cambridge University Press.

Brown, D. (1993b). The Yamato kingdom. In D. Brown (Ed.), *The Cambridge history of Japan. Vol. 1: Ancient Japan* (pp. 108–162). Cambridge: Cambridge University Press.

Brown, D. (Ed.). (1993c). *The Cambridge history of Japan. Vol. 1: Ancient Japan*. Cambridge: Cambridge University Press.

Campbell, B. M. S. (2016). *The great transition: Climate, disease and Society in the Late-Medieval World*. Cambridge: Cambridge University Press.

Chaudhuri, K. N. (1985). *Trade and civilisation in the Indian Ocean: An economic history from the rise of Islam to 1750*. Cambridge: Cambridge University Press.

Chen, Q. (2015). Climate shocks, dynastic cycles and nomadic conquests: Evidence from historical China. *Oxford Economic Papers, 67*, 185–2024.

Chew, S. (2018). *The Southeast Asia connection: Trade and polities in the Eurasian world economy, 500 BC-AD 500*. New York: Berghahn.

D'Alpoim Guedes, J., & Bocinsky, R. K. (2018). Climate change stimulated agricultural innovation and exchange across Asia. *Science Advances, 4*(10), eaar 4491. http://advances.sciencemag.org/content/4/10/eaar4491/?fbclid=IwAR1JRsDxkdZ6w13-kXl4auW9Hyr1dUCpMfyPlimgmGszK6ES6q6_9gEtCr0.

Day, A. F. (2013). *The peasant in postsocialist China: History, politics, and capitalism*. New York: Cambridge University Press.

Dien, A. (2007). *Six dynasties civilization*. New Haven, CT: Yale University Press.

Du, F. (2016). *The poetry of Du Fu. Tr. Stephen Owen. 6 v*. Berlin: Walter De Gruyter.

References

Ebrey, P. (1986). The economic and social history of later Han. In D. Twitchett & M. Loewe (Eds.), *The Cambridge history of China. Vol. 1. The Ch'in and Han empires, 221 B.C.-A.D. 220* (pp. 608–648). Cambridge: Cambridge University Press.

Fan, K.-w. (2010). Climatic change and dynastic cycles in Chinese history: A review essay. *Climatic Change, 101*, 565–573.

Harper, K. (2017). *The fate of Rome: Climate, disease, and the end of an empire*. Princeton, NJ: Princeton University Press.

Higham, C. F. W. (2015). Debating a great site: Ban non Wat and the wider prehistory of Southeast Asia. *Antiquity, 89*, 1211–1220.

Hill, J. E. (2009). *Through the jade gate to Rome: A study of the silk routes during the later Han dynasty, 1st to 2nd centuries CE*. Cooktown: John E. Hill.

Kidder, J. E., Jr. (1993). The earliest societies in Japan. In D. M. Brown (Ed.), *The Cambridge History of Japan. Vol. 1. Ancient Japan* (pp. 50–107). Cambridge: Cambridge University Press.

Kidder, T. R., Haiwang, L., Storozum, M. J., & Zhen, Q. (2016). New perspectives on the collapse and regeneration of the Han dynasty. In R. K. Faulseit (Ed.), *Beyond collapse: Archaeological perspectives on resilience, revitalization, and transformation in complex societies* (pp. 70–98). Carbondale, IL: Southern Illinois University Press.

Kiernan, B. (2017). *Viet Nam: A history from earliest times to the present*. New York: Oxford University Press.

Kim, N. C. (2015). *The origins of ancient Vietnam*. New York: Oxford University Press.

Lewis, M. E. (2007). *The early Chinese empires: Qin and Han*. Cambridge, MA: Harvard University Press.

Lewis, M. E. (2009). *China between empires: The northern and southern dynasties*. Cambridge, MA: Harvard University Press.

Linna, L. (Ed.). (2007). *Treasures from the Museum of the Nanyue King*. Beijing: Museum of the Nanyue King and Cultural Relics Press.

Lieberman, V. (2003). *Strange parallels: Southeast Asia in global context, c. 800–1830. Vol. 1: Integration on the mainland*. Cambridge: Cambridge University Press.

Lieberman, V. (2009). *Strange parallels: Southeast Asia in global context, c. 800–1830. Vol 2: Mainland mirrors: Europe, Japan, China, South Asia, and the islands*. Cambridge: Cambridge University Press.

Liu, A. (2010). *Huainanzi* (J.S. Major, S. A. Queen, A. S. Mayer, & H. D. Roth, Edited and Trans.). New York: Columbia University Press.

Loewe, M. (1986). The former Han dynasty. In D. Twitchett & M. Loewe (Eds.), *The Cambridge history of China. Vol. 1. The Ch'in and Han empires, 221 B.C.-A.D. 220* (pp. 103–222). Cambridge: Cambridge University Press.

Loewe, M., & Shaughnessy, E. (Eds.). (1999). *The Cambridge history of ancient China*. Cambridge: Cambridge University Press.

McCune, E. (1962). *The arts of Korea: An illustrated history*. Japan: Charles E. Tuttle.

McLaughlin, R. (2010). *Rome and the distant east*. London: Bloomsbury Academic.

Mitrany, D. (1951). *Marx against the peasant: A study in social dogmatism*. Chapel Hill: University of North Carolina Press.

Mote, F. W. (1999). *Imperial China 900–1800*. Cambridge, MA: Harvard University Press.

Sadao, N. (1986). The economic and social history of former Han. In D. Twitchett & M. Loewe (Eds.), *The Cambridge history of China. Vol. 1. The Ch'in and Han empires, 221 B.C.-A.D* (Vol. 220, pp. 551–607). Cambridge: Cambridge University Press.

Periplus of the Erythraean Sea. (1912). *Tr. Wilfred H. Schoff*. New York: Longmans, Green, and Co.

Pratt, K., & Rutt, R. (1999). *Korea: A historical and cultural dictionary*. Richmond, Surrey: Curzon.

Pines, Y., von Falkenhausen, L., Shelach, G., & Yates, R. D. S. (Eds.). (2014). *Birth of an empire: The state of Qin revisited*. Berkeley: University of California Press.

Rickett, W. A. (1965). *Kuan-Tzu* (Vol. 1). Hong Kong: Hong Kong University Press.

Rickett, W. A. (1985). *Guanzi: Political, economic, and philosophical essays from early China* (Vol. 1). Princeton: Princeton University Press.

Rickett, W. A. (1998). *Guanzi* (Vol. 2). Princeton: Princeton University Press.

Nishijima, S. (1986). The economic and social history of former Han. In D. Twitchett & M. Loewe (Eds.), *The Cambridge history of China. Vol. 1. The Ch'in and Han empires, 221 B.C.-A.D. 220* (pp. 545–607). Cambridge: Cambridge University Press.

Sanft, C. (2009). Edict of monthly ordinances for the four seasons in fifty articles from 5 C.E.: Introduction to the wall inscription discovered at Xuanquanzhi, with annotated translation. *Early China, 32*, 125–208.

Sanft, C. (2010). Environment and law in early Imperial China (third century BCE-first century CE): Qin and Han statues concerning natural resources. *Environmental History, 15*, 701–721.

Schafer, E. (1963). *The Golden peaches of Samarkand*. Berkeley: University of California Press.

Scott, J. C. (1998). *Seeing like a state*. New Haven: Yale University Press.

Seth, M. J. (2011). *A history of Korea, from antiquity to the present*. Lanham, MD: Rowman and Littlefield.

Shih, S.-H. (1973). *On "Fan Sheng-chih Shu," an agriculturist book of China written in the first century B.C.* Peking: Science Press.

Shimunek, A. (2017). *Languages of ancient southern Mongolia and North China: A historical-comparative study of the Serbi or Xianbei branch of the Serbi-Mongolic language family with an analysis of northeastern frontier Chinese and old Tibetan phonology*. Wiesbaden: Otto Harrassowitz.

Sidebotham, S. E. (2010). *Berenike and the ancient maritime spice route*. Berkeley: University of California Press.

Sima Zhangqing [Sima Xiangru]. (1987). Rhapsody on the Imperial Park (Chinese orig. ca. 139 BCE.). In *Wen Xuan, by David Knechtges* (pp. 73–114). Princeton: Princeton University Press.

Takeshi, M. (1993). Early Kami worship (Janet Goodwin, Trans.). In D. Brown (Ed.), *The Cambridge history of Japan. Vol. 1: Ancient Japan* (pp. 317–359). Cambridge: Cambridge University Press.

Tarling, N. (Ed.). (1992). *The Cambridge history of Southeast Asia. Vol. 1. From early times to c. 1800*. Cambridge: Cambridge University Press.

Thornber, K. L. (2012). *Ecoambiguity: Environmental crises and east Asian literatures*. Ann Arbor, MI: University of Michigan Press.

Twitchett, D., & Loewe, M. (Eds.). (1986). *Cambridge history of China. Vol. 1. The Ch'in and Han empires, 221 B.C.-A.D. 220*. Cambridge: Cambridge University Press.

de la Vaissière, É. (2005). Huns et Xiongnu. *Central Asiatic Journal, 49*, 3–26.

Von Glahn, R. (2015). *The economic history of China*. Cambridge: Cambridge University Press.

Walker, B. L. (2015). *A concise history of Japan*. Cambridge: Cambridge University Press.

Wallace, L. V. (2017). Wild youths and fallen officials: Falconry and moral opprobrium in early medieval China. In N. Harry Rothschild & L. V. Wallace (Eds.), *Behaving badly in early and medieval China* (pp. 122–134). Honolulu: University of Hawai'i Press.

Watson, B. (1984). *The Columbia book of Chinese poetry: From early times to the thirteenth century*. New York: Columbia University Press.

High Empire: The Glory Days of Early Medieval Eastern Asia 5

5.1 The Development of the East Asian World-system

"The works of the Tang masters are piled on my desk:
Half of them are poems about chaos and separation."
Chu Yunming, ca 1500 (tr. Jonathan Chaves; Chaves 1986: 189).

The words of this Ming Dynasty poet alert us to some hard times. The savagely cold, dry trough in 536–650 CE devastated Eurasia (Campbell 2016; Harper 2017) and much of the world. Several volcanoes erupted, some spewing great amounts of dust and sulphuric acid into the air.

This was followed by amelioration, and most of the period to 900 was much like today, or pleasanter, with a rather average number of disaster events. Climate was not a major problem for East Asia from 650 to the fall of Tang in the 900s. Philippe Beaujard (2017: 20) sees the good climate after 650 as helping Tang agriculture.

The period after 600 was in some ways the most important formative time in East Asian history. China was finally reunited by the Sui Dynasty (589–618). The Tang dynasty arose, consolidated control by 620, flourished, and fell (in 907). Under Tang, south China came increasingly to be "Chinese" rather than a thinly populated region of largely non-Han ethnicity. The East Asian world-system was thus changing: China was expanding spatially, and the Chinese-speaking world was expanding through assimilation of groups that had formerly been independent and peripheral. Korea developed and united; Japan rose to become a spectacular civilization.

In Central Asia, a Turkic empire had formed. Turkic power was to dominate that region throughout much of subsequent history. The Turks eclipsed the earlier peoples, assimilating them or driving them out; some had gone to Europe, where the Xiongnu transformed slowly into the Huns, the Rouran into the Avars. Turkic power in the north and Iranic power in the south divided Central Asia between them. The dominant Iranic group in much of the region was the Sogdians, a large mercantile population speaking an East Iranic language (a descendant of it survives

© Springer Nature Switzerland AG 2019
E. N. Anderson, *The East Asian World-System*, World-Systems Evolution and Global Futures, https://doi.org/10.1007/978-3-030-16870-4_5

in one valley today). These groups were critical to the expanding world-system, since they dominated long-distance trade, largely in prestige goods, and thereby also dominated information flows at the time.

However, cold in 626–632 hit the Eastern Gök Turk empire hard, allowing Tang to conquer them in 630, and move on to conquer the Western Gök Turks in 659. The Tang history explicitly mentions cold as part of the reason for Turkic collapse. Nicola Di Cosmo et al. (2017) find a more nuanced story, with the rise of Tang and slow erosion of Turkic rule contributing. Later, a major drought in 783–850, shown by tree rings in Mongolia (Crow 2018), contributed—along with internal rivalry and conflicts with Tang—to the fall of the Uighur empire (formed in 744) in 840, with both cold and Central Asian enemies to blame. This drought seriously impacted the Silk Road and probably contributed to the decline of Tang. Its scope and extent remain unclear; it does not seem to have affected western central Asia significantly.

Northern Vietnam was still under Chinese control. It had a thinner population than in Han times (Kiernan 2017: 106). A new player in the East Asian world-system was the kingdom of Nanzhao, centering in what is now Yunnan. It attacked Vietnam and other southeast Asian states, and was a major regional player until the Mongols destroyed it much later. Unlike the other states, whose languages survive and are well known, Nanzhao remains ethnically obscure; it probably had a Tibeto-Burman dynasty, but Thai was widely spoken, and the fantastic diversity of languages spoken in Yunnan today was presumably as great then. Nanzhao engaged in extensive trade with China and southeast Asia.

Tibet also became a major kingdom and center of power and political and cultural sophistication during this time. It served as a cultural bridge between India and China. Tibet warred against China along China's western frontiers, forcing borderland developments that affected history thereafter. Trade and commerce brought Indian learning via Tibet to west China.

All these polities except Japan were fully integrated into one economic and political system, either under Chinese control, or allied to China, or fighting it along its western borders. Japan was rapidly becoming integrated into the Chinese orbit, becoming theoretically "tributary" and receiving gifts as well as trade and embassy missions. The Tang court sent lavish presents to the court; many of these were stored with other imperial valuables in the Shōsō-in treasury, built in 756 at Nara, and they are still there, giving us a unique look at Tang high culture. The treasury is rebuilt at regular intervals, preventing decay. It preserves medicines, art works, and other important cultural items from China, showing that these were extensively imported by Japan at the time.

The region was tied together by military activities, but much more by trade and by cultural contact. Not only was Chinese culture widely spread; Buddhism from India was now thoroughly established as a major religion throughout East Asia. It became the definitive religion of Korea, Japan, and Vietnam, existing along with local folk religions (heavily shamanistic, especially in Korea) but winning the elite. Monks and scholars crisscrossed lands and seas, spreading the word. Several monks made the appallingly difficult journey from China to India, to study Buddhism in its home, and to return with authentic sutras. A whole industry of translation arose, ably chronicled by Pierce Salguero (2014). Tibet, meanwhile, developed its own extremely distinctive and original forms of Buddhism, largely tantric.

The Turkic and Iranic realms of Central Asia became centers of throughput trade. They were religiously tolerant. Buddhism was pervasive, but Manichaeanism, Christianity (including the Eastern or "Nestorian" form), Zoroastrianism, and local shamanistic faiths were all found. The Sogdians became famous as traders, businessmen, and caravaneers. They tied Central Asia and China together in webs of trade and kinship.

In the period of disunion after Han, and still more strikingly in Tang, west and Central Asian influences on China became profound. Literature reflected new norms. The ballad of Mulan, now made famous by Disney studios, is a Central Asian epic form (Chen 2012). Much of the arts of daily life, from food and medicine to chairs and beds, were Central Asianized (Chen 2012; Laufer 1919; Schafer 1963). North China developed a true fusion cuisine and fusion culture, reflected in the incredibly eclectic art and religious life of the Dunhuang oasis, where a nominally Buddhist stronghold witnessed Zoroastrian, Manichaean, Nestorian, and other travelers and scriptures. Agriculture, food, and medicine were all transformed by the new foods and medicinal plants brought from the west (Anderson 2014; Laufer 1919; Schafer 1963). The East Asian world-system came close to joining the west in a pan-Eurasian system.

Confucianism, at this time more philosophy than religion, spread also, influencing especially Korea, but also Japan and Vietnam. Islam entered with Arab and Persian armies in the 700s, and slowly became dominant. There has generally been a mutually exclusive relationship between the Confucian and Islamic worlds. The exception is the Hui nationality (culturally Han Chinese, but religiously Muslim), a dispersed community within China.

Buddhism gained imperial patronage in most countries of East Asia. The epic journey of the most famous Chinese traveler to India in search of scriptures, Xuanzang, was faithfully recorded by himself and his followers (Xuanzang 1996). It later accumulated a vast overlay of folktales. These became immortalized in the Chinese novel *Xi Yu Ji* (*The Journey to the West*), by Wu Cheng'en (Eighteenth century).

One major result of the spread of Buddhism was the spread of Buddhist literature and art (Heine 2018), first from India to China, then from China to the rest of East Asia. It came from Korea to Japan in the mid-500s (the traditional date is 552, but it was known earlier than that). China had developed spectacular Buddhist art under the Wei Dynasty. China and the other polities produced more in the Tang era, with Japan rapidly rising to first rank in this regard. The integration of East Asia into a single artistic area not only shows the penetrance of Buddhism, but also the extent to which the East Asian world-system had formed. We have only a tiny idea of how splendid it was, from the few surviving examples at Dunhuang in west China and in Japan and Korea. Dunhuang, a tiny and remote town, is now regarded as a wonder of the world; if it was an example of the bottom tier, one cannot even imagine what the great cities were like.

A final part of the world-system story in Tang times is the progressive assimilation of non-Han peoples in the Chinese empire. The Cantonese, today, call themselves *Tong yan* (Mandarin *Tang ren),* "people of Tang," because they were Sinicized during that dynasty. This contrasts with the usual self-label Han, from the Han dynasty. (Today, Cantonese are more and more often calling themselves Han.) Linguistic assimilation—but not cultural borrowing—reached a sudden, sharp stop at the Korean and Vietnamese borders, however. The Korean and Vietnamese languages accepted countless loanwords, but kept their extremely distinct core vocabularies and grammars. More remote realms like Tibet were even less influenced.

5.2   Events in China After the Mid-Sixth Century

The Sui dynasty (581–618) was, like Qin, a case of a single powerful ruler whose son could not hold the conquests. It was precisely at the worst climatic time that Yang Jian, later titled Wen Di, rode out from the cold north, conquered all before him, and reunited the country. He had been the maternal grandfather of the ruler of the feeble Northern Zhou state—another child-emperor done in by his maternal family. A dark, brooding, remote man, he did not make himself popular with the great families. He reigned by "stern authoritarian and Legalist principles" (Wright 1979: 63). He rebuilt the capital (Chang'an, now Xi'an), reformed the government, and began the process of taking down the great aristocratic families that had arisen in the period of disunion. This was a long process that was to continue for centuries before being completed.

He had a few wives, but only one true love. She was the empress Wenxian, who "was born into a powerful and long sinicized Hsiung-nu [Xiongnu] clan which had intermarried for centuries with the great families of Northern Wei" (Wright 1979: 63). Her importance in his life was "consistent with her upbringing as a northern woman with their strong feelings about monogamy and marked managerial propensities. She was...a literate and cultivated woman with strong political instincts. Yang Chien and his wife were very close" (Wright 1979: 64). He made her his co-regent, refusing to marry the daughters of the great lords (Wright 1978, 1979). This outraged powerful families that wanted to contract alliance by marriage. Those Chinese imperial harems were there for a reason. Marrying a daughter to the Emperor was a way for a great family to succeed. Marrying such a girl was important for holding that family's loyalty.

Thus, the Sui base began to erode. Decline set in as the emperor aged and became more and more suspicious (a typical progress of China's constantly-beset rulers). His son Yang Di (r. 604–617) could not control the state. He did his best, beginning a huge canal construction program, including the Grand Canal, that was to benefit China in future (Wright 1979: 134–138). This was done to bring more grain from the south. Many farmers had deserted the north for the south as weather got steadily worse in the north (D'Alpoim Guedes and Bocinsky 2018). He also tried to conquer Korea, and failed in a series of disastrous wars, causing massive disaffection with his rule. He was deposed, hunted, finally murdered by an enemy. He came to be portrayed as a weak, erratic, women-addicted "last emperor." In truth he seems to have been inadequate to the task, but not as bad as stereotypes (let alone salacious fiction) painted him. He lost the empire: not because of climate, though it was bad enough, but because of the disaffection and disunion of the great elites.

As so often at the end of Chinese dynasties, a leading general took over, founding Tang (620–906). Li Yuan, the Emperor Gaozu (r. 618–626, d. 635), conquered China from the northwest during the cold, dry period. The double conquest from the north—first by Sui, then by Tang—goes against the "common sense" climatic explanation of events. However, it fits perfectly with the general rule that whoever holds the Wei River valley and nearby Yellow River has a grasp on all China. Yang Jian had started without it, but quickly moved to take it.

It is possible that the cold dry years weakened the central provinces more than they weakened the powerful military machine, dominated by Turko-Chinese elites, that had developed as a quasi-independent world in the northwest. Both Yang Qian and Li Yuan were part Turkic, and they were used to harsh winters and harsh fighting. But such speculation does not change the fact that they won despite climate. Qiang Chen (2015) mistakenly cites them as evidence for the old idea that "nomads" conquered China in drought times, but both were generals from settled agricultural states. Their nomad ancestors were generations back. More reasonably, Chen cites the unification as facilitated by the cold. This might be seen as special pleading—he usually attributes dynastic rises to warm, favorable conditions—but we have noted above that cold weather can stimulate centralization.

War and weather combined to cut drastically the population of north China (Twitchett 1979), where most Chinese lived at the time. Tang consolidation, and cold weather, lasted till around 623 (the cold until 650). China's migration to the south compensated for decline in the north.

Li Yuan's family had been a military one for generations, and he had served as a general in Taiyuan for years. He took the capital of Sui in 617, declared the new dynasty in 620, and moved on to conquer essentially all China by 628 (Lewis 2009: 31). History records that he was charismatic, but, as usual, it was the victors who wrote the history.

Li Yuan's sons were an unruly and violent lot. One of them, Li Shimin, led a dramatically successful expedition against the Eastern Turkic empire. He then murdered his brothers, took over the position of heir apparent, and forced his father into retirement (Wechsler 1979a: 182–187). As Emperor, later titled Taizong (Great Ancestor), he ruled from 627 to 649—a glorious age of power, conquest, and wealth. However, he did not have a smooth ride. Guilt for his way of winning the throne supposedly possessed him. According to one folktale, Shimin was troubled with nightmares. His two murdered brothers and his displaced father appeared as wronged ghosts. He posted his generals (real ones) at the doors, and the nightmares stopped. Since then, Chinese have used paintings of these three magnificent generals to keep ghosts out of the house—two generals on the two leaves of the front door, sometimes one on the back door. They still protect countless Chinese doors today.

Li Shimin failed to register all the households, raise enough taxes, and otherwise regulate the empire fully (Lewis 2009; Wechsler 1979b). He continued the struggle to bring the great aristocratic clans to heel; four major ones from the northeast proved especially difficult. He fought the rising power of the Turks, who were rapidly emerging as the real power in central Asia—there to remain since.

The Turkic state had emerged during China's period of disunion; the Turks had been followers of the Xiongnu and then the Xianbei. The Turks were supposedly skilled in metal work. With the fall of the Xianbei the Turks formed the Gök empire. *Gök* is a significant word in Turkic languages, referring to the deep blue of Heaven. The religion of the Turks was devoted to *tengri,* Heaven. Sky-blue was a sacred color, as it still is throughout central Asia, from the domes of the mosques of Samarkand to the ritual scarves tied around sacred objects in Mongolia. The empire broke into two in 627, and the fiercely cold weather devastated their flocks (Lewis 2009: 35), allowing Li Shimin to conquer them: another case of direct climatic influence on history.

A war with Korea proved as unsuccessful as usual. China briefly conquered Korea in 668–70, or most of it, but it broke away in 676. However, Tang turned defeat into victory: their pressure caused Paekche to fall in 660, and the war destroyed Koguryŏ in 668, leaving Tang's ally Silla in control of all Korea, as Great Silla (676–935). Japan intervened to try to save Paekche, at Paekche's urgent request, but was defeated by Tang forces. This was a classic case of a semiperipheral state becoming a battleground in a proxy war between two more powerful states—a presage of the United States and Communist China battling in Korea 1300 years later. Tang-Silla relations naturally became close. Silla rose with Tang to an eighth-century peak, then declined, and fell shortly after Tang (Best 2007; Seth 2011). The loss of Koguryŏ and Paekche, then, was due to military pressure, not to internal or climatic factors. The classic Chinese myth of the "bad last emperor" attached itself to the unfortunate last king of Paekche (Best 2007: 185), but he was probably no more fond of wine, women and song than other kings.

Tang pressure led to greatly increased contact and influence. The kingdoms had already converted to Buddhism—a tolerant religion that did not greatly interfere with folk shamanism or other faiths—and now Silla became a world center of Buddhist scholarship. Korean scholars traveled to China, apparently in some numbers, to study Buddhism. Chinese influenced the Korean language. Integration of Korea in the East Asian world-system was complete.

Silla also became a maritime power of note, dominating trade and commerce in the seas around it. It exported "silver and gold bullion, textiles, and ginseng," and imported Chinese luxury goods (Seth 2011: 65), showing that it was still in a rather peripheral status vis-à-vis its huge neighbor. However, it was sophisticated enough to export made goods to Japan. It seems to have been a major source of technically advanced wares for the Japanese. Silla maintained a very active sea trade, as well as sending large missions to China (Seth 2011: 66).

The state of Parhae (Chinese Behai) developed to the north, and Japan allied with it against Silla, seeking to neutralize the China alliance, but Parhae was destroyed by the Khitan in 926, just after Tang fell. The Khitans appeared during Parhae rule of what is now far northeastern China, in the seventh century.

Toward the very end of Tang, Silla declined. After the fall of its great supporter, China, Silla could not hold out; it fell in 935. This seems to have been purely a knock-on effect of Tang decline, though unrest late in the dynasty was evident. Wang Kŏn established the Koryŏ (or Goryeo, 935–1392) Dynasty. China's disunion gave it a window to rise and establish power. It was to be a splendid, unified dynasty, like Tang, but largely contemporary with Song and Yuan.

5.3 High Tang

A short cycle within Tang ended dramatically with seizure of power by the Empress Wu in 683 (Guisso 1979; Lewis 2009; Twitchett and Wechsler 1979). She emerged as the consort of Li Shimin's successor, the (future) Gaozong (r. 649–683). After he died in 683, she established her own "Zhou Dynasty" (691–705)—another tough, no-nonsense woman shaking up the "patriarchy."

She reinstated the bureaucratic examination system. Future generations were to tell lurid stories about Empress Wu's wild sex life, but the stories surely owe more to later Tang propaganda than to reality. (For a gloriously racy summary of this literature, see Fitzgerald 1956; for a curt dismissal, based on modern scholarship, Lewis 2009: 36.) She moved the capital to the more central but more vulnerable Luoyang. It was to move back to Chang'an after her fall. She was an ardent Buddhist, favoring that religion over Daoism.

Meanwhile, the Tibetans reached perhaps the all-time summit of their power, consolidating control over all of what is now Tibet and conquering far into China. Their power declined slowly. Tang was to drive them from most of China by 801 (Guisso 1979; Lewis 2009: 64–66). The Khitan also surfaced as a major power on the north (Guisso 1979: 317–319); they were to rise to glory after Tang fell.

Tang conquered far into Asia. "By the 670s T'ang protectorates had been established up to the borders of Persia, the Chinese had occupied the Tarim and Zungharia, and destroyed Koguryŏ in Korea" (Twitchett 1979: 13). Arabs stopped China by defeating the Chinese army at Talas River, almost in the exact center of Asia, in 751. This set China's frontier more or less permanently; the border is still just east of that small stream. The boundaries of the East Asian world-system were set in the process of integrating it more tightly with the western part of Eurasia.

In the meantime, Empress Wu was finally displaced in a coup in 705, and soon died. She was succeeded in quick succession by the hapless emperors Zhongzong and Ruizong (r. 705–710, 710–712). Woman power was not dead: the empire was in fact run by Zhongzong's Empress Wei. She initially took a lover, Wu Sansi, nephew of Empress Wu. He was murdered in 707, Wei herself in 710. The whole period was memorably described as "the calamity of the empresses Wu and Wei" by Song Dynasty historian Yuan Shu (Guisso 1979: 329).

Yet another woman, the princess Taiping, then took the real power, her brother Ruizong being hopelessly weak and sickly. He retired in hopeless illness in 712. The most functional heir of the Li family, Li Longji, took over in 713, blocking a coup by Taiping and her forces. She committed suicide (Twitchett 1979).

Li Longji is known to history as Xuanzong, the Dark or Mysterious Ancestor. A larger-than-life figure, he has attracted a mass of legends. The glory of his reign is still known to every educated Chinese person as the Golden Age of the civilization. At his court were the greatest poets in Chinese history, men such as Li Bai and Du Fu. Supposedly the painters and musicians were equally skilled, but we do not have their works. Wu Daozi was said to be the greatest painter in China; according to another of those stories that I used to hear in my young travels in Asia, he climaxed his career by painting a vast scene of Paradise in the imperial palace; when all was done, he painted a door in the middle of the fresco, opened it, walked through, and was never seen again.

Xuanzong loved women, and had 59 children (Twitchett 1979: 373); Turchin's "elite overproduction" exemplified. Partly because of his lavish lifestyle, the period was financially strapped, a source of major weakness. Chang'an continued to be difficult to provision (Twitchett 1979: 355–360, 419). Factions and court politics became more vicious.

The last independent king of Persia, displaced by Arab conquest, had earlier taken refuge at the Tang court. Everything from Persian pottery to Persian bread became popular in Chinese cities. *Naan* is ancestral to many modern Chinese breads, including *shaobing*. Central Asian medicine was influential, introducing ideas from as far as Greece. In Tibet, one doctor of the time even called himself Galen (Garrett 2007). Edward Schafer's classic *The Golden Peaches of Samarkand* (1963) describes this age in exquisite detail. Back-and-forth problems with Turkic and Khitan peoples were partially resolved by military strength (Twitchett 1979: 364–365, 426).

This was the height of the cyclic Golden Age of Tang, but it was also time for another Khaldunian crisis. Xuanzong became excessively devoted to the beautiful Courtesan Yang. Their romance led to neglect of the state and to alienation of the families of neglected palace women. An Lushan, a central Asian and a leading general in the Tang armies, became a friend of Courtesan Yang, who adopted him as her son in 751. He was rumored to be her secret lover. He launched a huge rebellion in 754. This brought down the government. Since Empress Wu took power in 683, 71 years had passed. Xuanzong fled; Yang was killed by mutinous troops. Incompetent generalship and the inevitable pernicious factioneering almost led to Tang downfall, but the Uighurs stayed loyal—for a high price—and turned the tide. The dynasty was restored and Chang'an retaken in 757, but the rebellion sputtered on in remote areas till 763 (Twitchett 1979: 453–463).

Tang was never the same. Brooke (2014: 358–359) credits vaguely-coincident bad weather with the An Lushan Rebellion of 754–757, and with the later fall of the dynasty in 906, but the case is not clear. The bad weather around An Lushan's time was real, but brief and probably not adequate to affect the rebellion significantly. Between An Lushan and Huang Chao, the dynasty sputtered and faced problems, but endured. The country never centralized well after 763 (Peterson 1979). Much of northeast China became semi-independent.

The romance of Xuanzong and Courtesan Yang became the great tragic love story of Chinese tales. Bai Juyi wrote a long and stirring poem about it, and later met with ordinary people in inns and streets who could recite the poem. One would have to work to find any active Chinese writer from Tang to Qing who did not refer to the story at some point. Tourists to modern Xi'an, the Chang'an of old, are guided to countless places where "the Emperor Xuanzong dallied with Courtesan Yang."

Further rebellions occurred. Even the non-Chinese of the southern frontier managed to put together an army of 200,000, which ravaged the south in 756 (Benn 2002: 11). They are called "barbarians" and "aborigines" in the sources, but an army of 200,000 requires a state, or a couple of allied states. Probably these were Tibeto-Burman and Thai speakers, like those of Nanzhao.

The provinces of the northeast broke away after 781. Tang never regained real control of several large parts of the empire. It survived by maintaining control over key areas around the capital, in the northwest (the old power source), the rich east, and the central river and canal routes (Peterson 1979: 494). Further Tibetan incursions also troubled China.

All these episodes, plus strong efforts by the emperors to consolidate and centralize power, led to a steady decline of the old aristocratic families. Slowly they lost out to imperial reach (Lewis 2009; Mote 1999; Twitchett 1979). This was the end of dispersed oligarchic power in China; steady centralization proceeded in subsequent dynasties. State and noble control of land and economy also began to give way to free trade and small farming. The latter restored Han momentum in that direction.

Warm but not overly hot or dry times also helped Central Asia, and thus enabled the rise and glory days of the Silk Road. It had come into its own since the end of Han, flourishing especially from 500 through Tang. It flourished again under the Mongols. As Valerie Hansen (2017) reminds us, we are not to see the Silk Road in terms of a modern freeway; it was a horribly difficult, demanding, lonely set of trails (never a single road) through deserts and mountains. Yet it carried significant trade, both local and international (Anderson 2014). This trade led to a steady westernization of China. From Arab and Persian recipes and spices to chairs and watering bottles, western forms were popular, spreading like wildfire in China and soon moving on to Korea, Japan, and Vietnam. As usual, China's great export was textiles: silks, brocades, ornamental weavings of all kinds. Some medicines accompanied these. Relatively few coins were circulated; silk bales were the functional currency. The price of everything from wine to slaves was listed in standardized bales of silk. It is hard to estimate the extent and value of this trade, but it was enormous, even without freeway-level traffic.

Chinese food became more like Central Asia's. Herbals and medical books list more and more west Asian items; accounts of food mention more and more west and Central Asian items (Anderson 2014; Schafer 1963). Items as various as spinach and *naan* bread, both from Iran, became common. The favor was not returned over the Silk Road, since most of the distinctive Chinese foods did not grow well in cold dry climates. Instead, oranges, tangerines, and other Chinese foods spread via the Maritime Silk Road.

Silk Road trade could not thrive during cold periods. When China is wet, western Asia is usually dry. Warm periods bring northward the Chinese monsoon rains, but also the hot desert climate of subtropical western Eurasia. Cool periods bring drought to China, rain to Central Asia, though if times are cold Central Asian winters are exceedingly harsh.

The Silk Road was critically important in transmission of religion during this time (see e.g. Foltz 2010). Zoroastrianism and its descendant Manichaeanism were particularly notable, especially since they slowly declined into near-vanishing in later centuries, leaving modern scholars often unaware of their enormous influence in the early medieval period. They spread non-Chinese ideas of absolute Good and absolute Evil, and of the Millennium to come, very widely. Manichaeanism is generally credited for much of the millennarian aspect of Chinese folk culture, seen in many rebellions. However, Chinese millennarianism antedated its arrival, as in the Yellow Scarves rebellion in Han.

Crises at Khaldunian intervals destabilized the dynasty. Empress Wu's coup when the dynasty was 63 years old; An Lushan's rebellion 70 years later. Other, smaller rebellions came in and after 860 (Fan 2010). The Huang Chao rebellion which began the dynasty's end came in 880, 123 years after the restoration.

5.4 Tang's Decline and Fall

The Emperor Dezong (r. 779–805) did much to save Tang, shoring it up with a mix of shrewd policy and no-nonsense intolerance of dissent. His successor Xianzong (r. 805–820) tried reform, which led to his murder by eunuch courtiers (Peterson 1979: 537), though others say he died of overdosing on—ironically—longevity drugs (Dalby 1979: 654–655). Tibetans harried the borders, but the Uighurs—a relatively Sinicized Turkic group—were generally allied, though demanding heavy rewards.

As in Han, a succession of weak emperors, and perhaps climate, led to collapse and disunion (Lewis 2009). Warlords arose. Money was scarce. Eunuch courtiers, always troublesome because of their lack of direct ties to the public sphere, proliferated and seized power. Muzong (r. 820–824) succeeded at 24; his son Jingzong (r. 824–827) also succeeded and died young; Wenzong (r. 827–840) died at 30.

In 835, under Wenzong, the court scholars tried to strike back against the eunuchs, arranging a faked fall of "sweet dew" (an ancient auspicious omen) on a pomegranate tree in the court garden, hoping to lure the leading eunuchs out. The scholars and their allies would fall on the eunuchs and kill them. The eunuchs saw through the plot. It was the scholars and courtiers who died. This ended all hope of stopping eunuch rule. It dispirited the young emperor Wenzong, who took ill (or took those drugs) and died (Dalby 1979: 656–658). By this time everyone realized that the dynasty was in serious trouble.

5.4 Tang's Decline and Fall

Wuzong (r. 840–846) launched an enormous and cruel persecution of Buddhists, apparently triggered less by religious intolerance than by desire to appropriate their vast wealth for the failing empire (Benn 2002: 16; Dalby 1979; Peterson 1979). It may even have been related to mania induced by the "immortality" drugs (Dalby 1979: 663). Xuanzong (r. 846–859; a different word from the earlier Xuan) did better and consolidated rule, propping up the failing dynasty, but could not save it. Succession crises followed, and Yizong (r. 859–873), Xizong (r. 873–888), Zhaozong (888–904) and Jingzong (r. 904–907) did nothing to stem the decline. Of these, Xianzong, Wenzong, Wuzong, Xuanzong, and others very possibly died of drug poisoning, either deliberate or taken in a mistaken attempt to prolong life. It was easy for eunuchs to substitute fatal doses for recommended ones, and no one now knows how often that happened. Meanwhile, court factions worked so hard to destroy each other that the court became a negative-sum game.

Rebellions and banditry increased as court politics—factions, eunuch power plays, and imperial poisoning—increased out of control. A bandit code of *renxia*, mutual loyalty (Somers 1979: 724), proved more successful than court rule. Even the small, remote Nanzhao state provided major military setbacks to Tang in the 850s. Vietnam and other southern polities were troublesome (Somers 1979: 694).

The Huang Chao Rebellion, a violent bandit outbreak with millennarian aspects, lasted from 878 to 884, sacking and burning the capital, Chang'an. Huang attempted to set up his own dynasty, but it collapsed in appalling violence. Lurid tales of cannibalism and other horrors surround this extremely violent revolution (Kuhn 2009: 17 retails some obviously exaggerated stories). Other rebellions such as Chian Neng's broke out. A Shatuo Turk, Li Keyong, was instrumental in quelling the rebellion, and became a major warlord thereafter. Final brief climatic deterioration in the early 900s (rather like the 550–650 period but less intense) had some role in bringing down the dynasty; crop failures led to increasing rebelliousness.

At the end: "Hatred and intrigue grew to grotesque proportions between the ministers...and the eunuchs, with each side willing to pay any price to damage the other" (Somers 1979: 780). This, with appropriate changes in the cast of characters, could be said of the very last phase of almost any IK cycle.

A warlord, Zhu Wen, took control and forced the teenage emperor to abdicate in 907. He moved the capital to Luoyang, abandoning the ruin of Chang'an. Unfortunately, Zhu Wen and his Liang state could not control the empire, which fell apart into many warring states. Another period of disunion had begun. It was much shorter than the Warring States period or the interval between Han and Sui, but it was, if anything, even more bloody and unsettled. For a while, over fifty local polities contended, possibly an all-time record for China.

Unsettled, erratic climate at the start of the Medieval Warm Period hurt China at this time. Philippe Beaujard (2017: 20) notes that the decline of the Abbasid Caliphate in Mesopotamia occurred at the same time, with erratic weather occurring there also. This may have hastened the decline and kept China from re-uniting, but Tang was already lost long before, having never recovered from An Lushan's rebellion.

5.5 Meanwhile in the South...

Sui reconquered northern Vietnam in 602. Linyi fell in 605, but was not fully conquered, and remained a tributary state till it disappears from view in 793, succeeded by Huanwang. Tang gave the region its long-standing name of Annam, "pacified south," in 679. (The name had previously applied to the once-fully-independent lands southward; Kiernan 2017: 105). Funan morphed into Chenla, based upriver from the old Funan.

A major new state then entered the picture: Champa. Chinese records first note it in 657 (Kiernan 2017). It expanded at the expense of the other polities, becoming a major state with a capital near Danang. The modern Cham speak a Malayo-Polynesian language, influenced by Mon-Khmer, and presumably Champa was at least partly Malayo-Polynesian-speaking. Its culture was strongly Indian-influenced.

Rebellions against China and pressure by Champa on Chinese territory followed, and when Tang fell, its weak successor states lost control of Vietnam. Its regional states maintained independence. The Nanzhao kingdom, rising around 738, became a major southeast Asian power. It collapsed in 937, to be succeeded by the Dali kingdom, which fell to the Mongols in 1253. Little can be said of dynastic cycles and climatic effects in these southern polities. Even sorting out their names is difficult enough. They were largely reflex states to Chinese power, and had little opportunity to develop beyond peripheral status.

5.6 Japan, 550–950

Japan was not part of the East Asian world-system until Tang, and then only marginally. It was a tiny world-system of its own, comprising the Nihon kingdom and the Ainu and other local hunter-gatherers and simple agricultural people of the islands.

Japan developed full-scale civilization (referring, as above, to literate urban-centered society) between 400 and 600 CE. Kings and queens were buried in enormous keyhole-shaped mounds known as *kofun*. Japan enters detailed history with the rise of the Wa state by 600 and its renaming as Nihon in 669. Nihon, or Nippon, "sun root" or "land of the rising sun" (*Riben* in Mandarin), is the name corrupted into "Japan." The Yamato imperial family of Nihon is, at least in ideology, still the imperial family of that country. The record of the next 400 years is one of a rapid national rise, consolidation, and dynamic Golden Age.

Japan entered recorded history in Sui. Never a country to minimize itself, practically its first entry into the record was a letter from Empress Suiko in 607 to Sui Yangdi, beginning: "The Child of Heaven of the land where the sun rises sends a message to the Child of Heaven of the land where the sun sets" (McCullough 1999b: 83). She thus equated her power with the venerable empire of China.

The Sui example led to a rapid and dramatic Sinicization, which historian Inoue Mitsusada (1993: 163) compares to the period after 1868 as the two great periods of learning from foreign nations in Japan's history. As he puts it, Japan developed "a strong and persistent urge to build a powerful Chinese-style state" and showed "increasing openness to diverse expressions of Chinese art and learning" (Mitsusada 1993: 163). This led to a rapid rise in the power of the Yamato regime. Writing, in Chinese characters, became common. In 553 there "arrived in Yamato from Paekche...one diviner, one calendar specialist, one physician, two herbalists, and four musicians" (Mitsusada 1993: 171); thereafter the immigration of Chinese and Korean specialists steadily increased.

Buddhism, a great integrator of the East Asian world-system, became entrenched and dominant. By 624 there were 46 temple compounds with 1384 clerics, many Korean (Kōyŭ 1993; Mitsusada 1993: 176). Buddhism soon became a dominant part of Japanese life, involving enormous effort and wealth. Edward Gibbon would surely have extended from Rome to Japan his sour comments on the waste of money and human talent seen in monasticism and religious "enthusiasm," had he read of these efforts, but the Japanese believed Buddhism was the key to security and prosperity. The elites found it congenial (Kōyŭ 1993). Buddhist stress on hierarchy, kingship, and the results of karma were not lost on these elites. Japanese animism and shamanism (ancestral to Shinto) were relatively egalitarian. Shinto had developed from folk animism into a state system with its own hierarchy and validation of power (Takeshi 1993). Still, the elites needed a system that naturalized disparities of wealth and social place.

Confucianism is the other great ideological integrator of East Asia, and it appears in the famous Seventeen Injunctions of the early 600s. They are "expressions of a Confucian-oriented, emperor-centered state ideology" (Mitsusada 1993: 181). They were supposedly passed by Prince Shōtoku in 604, but are found in the *Nihon shoki* and suspect of being heavily modified there. Numbers 3 and 10 are ostensibly Buddhist, but immediately lapse into Confucian ideas and rhetoric (Kōyŭ 1993). Number sixteen orders the officials "not to trouble farmers at planting and harvesting times" (Mitsusada 1993: 181). This famous Confucian instruction has had real environmental effect, at least in China and probably in Japan. By this time, involvement in Korea, often using it as a foil or battle ground to confront China, was intense and continual.

In 645, a group of reformers assassinated the de facto ruler, Soga no Iruka, and installed a new emperor (Mitsusada 1993: 192–193). Notable among the conspirators was Nakatomi no Kamatari. He later took the name Fujiwara no Kamatari, and instituted the Fujiwara line that dominated Japan for centuries. Though this was a major power transfer, it shows none of the characteristics of a dynastic fall; it was a fairly straightforward coup. The new rulers immediately instituted new tax, census, and governance policies, based on Chinese tradition.

In the late 600s "Japan's first Chinese-style capital was built at Fujiwara" (Brown 1993: 35). It was heavily derivative on Tang China but had distinctive characteristics: Shinto religion involving worship of *kami* (gods, spirits), the *waka* poetry form, the refined aesthetic of elegant simplicity and love for evanescent beauty, and other familiar parts of Japan's heritage. The Fujiwara-early Nara decades are a classic "r" phase portion of a cycle; wealth and culture blossomed, the economy thrived, and Yamato expanded by military conquest of the Emishi and other rivals.

Korean immigrants, including refugees from the Tang-Silla conquest of Paekche and Koguryŏ, contributed much to the rise of civilization; they included architects, designers, politicians, teachers, Buddhist scholars, artists, and other skilled workers.

Trade with Korean and China developed rapidly, beginning with missions bearing gifts and establishing somewhat fictive "tributary" relations (Verschuer 2006). The first missions were humble gifts of things like goats and donkeys, but by late Tang Japan was exporting precious raw materials and importing craft work. By the 800s, they were exporting high-quality paper—Japan's first case of learning a skill (from China) and quickly surpassing their mentors (Verschuer 2006: 15). A huge Chinese mission came to Japan in 671, causing fear of Tang invasion, already a lively worry. It merely sought aid for further adventuring in Korea, but Japan took steps to improve defenses (Kōjirō 1993: 222).

A major civil war occurred in 672, when the emperor appeared to shift his approval from one son to his half-brother, and then died without the issue fully settled. Inevitably, the brothers fought it out, the crown prince proving victorious (Mitsusada 1993: 216–220). He moved the capital to a new location, picking Nara (then called Heijō), construction of which was finished in 710.

This initiated the brilliant Nara period, when Japanese civilization took shape (Mitsusada 1993: 229–230). The capital was modeled on Chinese capitals. Chinese-style administration was developed, with some local modifications, creating the *ritsuryō* penal, administrative, and regulatory regime. Chinese art, literature, music, religion, and philosophy prevailed. By the Nara period, educated men and the best-educated women were expected to be able to produce a Tang-style poem, in proper formal Chinese, as occasion demanded (H. McCullough 1999). Since Tang poetry involved exceedingly complex composition rules as to prosody, subject matter, and allusions to earlier literature, this was no mean achievement. Japan imported all manner of art objects, medicines, and manufactures from China, exporting largely raw and semifinished materials in return—a classic semiperipheral pattern. Fear of Tang conquest may have fueled much of the Nara effort. Tang's reduction of Korea to vassal status was uncomfortably close (Brown 1993: 38–40).

The *Kojiki* (written originally around 712) and *Nihon Shoki* (ca. 720), Chinese-style histories, were the first major writings in Japan. A great deal of exchange with China took place, with major missions flowing back and forth, while Buddhist missionaries and students flowed in a steady stream.

Further faction fighting between branches of the Imperial and Fujiwara families led to another move of the capital, to Nagaoka, ending the Nara period in 784 (Kōjirō 1993). This change of capital went without major conflict, and the Heian period was one of stability and prosperity. The Japanese developed a new capital at Heian-kyo, now Kyoto, in 794, and the period from then until the collapse of the Japanese imperial system in 1185 is known as the Heian period (see Shively and McCullough 1999). Heian-kyo was modeled after Chang'an, though indirectly, via earlier Japanese capitals. Japanese governance and court behavior were extremely influenced by Tang.

Kyoto was to remain the capital throughout the centuries of imperial Japan, though it lost real power to Kamakura in the medieval period and to Edo (Tokyo) in the 16th-17th century and after. Like Koreans, Japanese came to China to study Buddhism. The monk Ennin, who traveled in Tang China in 838–847, left a fascinating record of Tang life (Reischauer 1955). Japanese also raided Silla, but Silla fought back, deterring Japan from further activity (Seth 2011: 68).

Over the early medieval period, Japan moved from a system in which all land was owned by the state to a more feudal situation. The *ritsuryō* system of administration had introduced something like China's well-field system, in which public land is parceled out by the state on a grid pattern. Slowly, the great families of Japan received grants that resembled the medieval manors and estates of Europe, with villages, castle keeps, and large areas of productive land. These developed from simple land grants; by the 1000s they were full-scale *shōen*, the famous estates that were the backbone of Japan's economy for the next few centuries (Toshiya 1993).

Meanwhile, Japan was expanding its own little world-system, conquering the Emishi and spreading Japanese civilization into their realms. The Ryukyu Islands also entered history, being first mentioned in the history of the Sui Dynasty. They then vanish from the record for centuries.

Japan's changes of capital and of dominant family were less dramatic than the dynastic rises and falls of China. They were, basically, conflicts within elite lineages. They show elite overproduction, but few other evidences of decline. They also fail to show the correlations with climate changes. They lack the correlations with China's dynamics so evident in Korean dynastic cycling. They were, however, to some extent the result of a problem that was to dog Japan for more than a millennium: the tension between the Emperor and his family and the great military families that increasingly dominated Japan.

5.7   China Divided After Tang

In China, disunion and brief local kingdoms followed from 907 to 960. This was the time of "five dynasties and ten kingdoms," local power bases with their own royal families, usually noted for murder, amoral coups, and warlordism. Not long after its end, the great scholar Ouyang Xiu (1007–1072) wrote a history of it, highlighting the peaks of loyalty and decency that stood out above a truly dismal landscape (Ouyang 2004). The book is a sobering read for those who do not believe individual moral agency matters in history. Ouyang is sometimes dismissed as a mere moralist, but his solid data and close argument should convince critics otherwise.

He could find few peaks. Unlike the period of unrest between Han and Sui, the period from 907 to 960 is well known. The dynasts seem usually to have been too busy killing rival family members to consolidate strong positions. The levels of murder and treachery did not allow a charismatic leader to emerge, until China was so leveled down that a strong general with some steadfast values could consolidate control. There were surely other factors—economic, social, perhaps even climatic— but the personal problems of the leadership in the period are hard to ignore.

The period was a true alpha phase in Holling's terms. The small and warring kingdoms hammered out a more streamlined, organized way to doing business, slowly getting away from the bureaucracy and religiosity of late Tang. Trade even flourished locally. A shipwreck discovered off Java revealed vast amounts of fine ceramics as well as copper, tin, and gold items and "nearly 190 kg of silver ingots…presumably disbursed to purchase aromatics, spices, and other exotic goods" (Von Glahn 2015: 227). Like other medieval Chinese shipwrecks, this revealed widespread trade with high levels of diverse merchandise. Few had expected such commercial activity from the period.

5.8 Verdict

Tang recapitulated to a striking degree the history of Han: a major early crisis involving female rule that did not dislodge the royal family; a later, more serious coup that truly shook the dynasty (Wang Mang's Xin Dynasty and An Lushan's rebellion); and a final long, slow decline that turned into collapse after a major popular rebellion. None of these except the very end of Tang involved or correlated with major climate shocks. The difference in outcomes of the events, roughly one short Khaldunian cycle apart, was due to their timing within the overall long-cycle histories of these dynasties.

The Tang Dynasty, and the Silla in Korea, fell during a dry period, the same, in fact, that was instrumental in bringing down Maya civilization and perhaps some other New World cultures. Drought intensified in the 900s. Gergana Yancheva and her colleagues (2007) suggested these droughts brought down Tang. However, in China they were local and not prolonged. Tang lasted about as long as any other successful Chinese dynasty. Its decline and fall took 150 years, starting with the traumatic rebellion of An Lushan, during a time of good climate. China's vast latitudinal spread, once again, protects it from the worst effects of climate change.

Accounts of its end stress politics, and do not usually mention climate as a major factor (Twitchett 1979). Climate may have led to productive stagnation that hurt tax revenues. This *might* have contributed to a general hopelessness which accompanied corruption and negative-sum competition during the decline of Tang, but evidence is essentially nonexistent. Certainly, the poetry and other writings of Tang after An's rebellion display a great deal of alienation and protest.

Even so, one wonders. There is a coincidence in time of the fall of Tang, the fall of Silla, and the troubles in Japan where the Fujiwara lineage's many branches were competing with each other and with a succession of retired emperors. The early tenth century hit China with a short, sharp reversal in the warming trend. The monsoon declined in the late ninth century and bottomed out in 910–930, bringing cool, dry conditions and widespread hardship. This may well have contributed to the failure of anyone to reunite China (Brooke 2014: 358; but see Fan 2010). The usual reports of floods, droughts and locusts may not mean much by themselves. However, it is hard to discount the role of the unsettled and fast-changing climate of the period.

The Medieval Warm Period slowly gathered strength after 930, easing life after Tang fell. It helped Vietnam by increasing rainfall (Kiernan 2017: 13), especially in the drier seasons, and would have done the same for China. It would have been particularly beneficial to China's dry north, still the heartland in Tang times. It continued, with fluctuations, until about 1300; the 1200s were the height of reliable warmth and good rain. The same warming brought drought to Cambodia, possibly destroying the Angkor civilization, and farther afield it brought devastating drought to Mexico, ending the Classic Maya period (Gill 2000).

If climate fits well, Ibn Khaldun's model fits even better:

Rise. Initial triumph under a popular military leader.
Expansion. Tang's conquests far into central Asia in the early decades are impressive, and the economic and cultural expansion of Tang are famous.
Widening gap. Narratives and rebellions make this clear.
Elite overproduction. Family expansion and conflict, as in other cases.
Increasing corruption. Alleged; poisoning of emperors by "immortality" drugs is a related problem.
Decline. Clear after the middle eighth century.
Widespread unrest.
Military overextension. Maintaining the central Asian conquests proved impossible.
Collapse.
Fall. Again, an in-house general rather than a semiperipheral state actually ended the dynasty.

Fan Ka-wai (2010), in an important article, reviewed the evidence for Tang's fall depending on climate, and finds it wanting. He notes that the end of Tang was not associated with exceptional problems; the droughts noted above were neither massive nor long-lasting. Internal dynamics were enough to reduce climate to a minor contributory cause. The consensus of historians is that Tang never recovered from the An Lushan rebellion, and the Huang Chao rebellion was a fatal blow, especially given the weak, young, short-lived emperors that both preceded and followed it. "The final dissolution of T'ang authority, it is clear, admits of no simple explanation" (Peterson 1979: 560).

References

Anderson, E. N. (2014). *Food and environment in early and medieval China*. Philadelphia: University of Pennsylvania Press.
Beaujard, P. (2017). *Histoire et voyages des plantes cultivées à Madagascar avant le XVIe siècle*. Paris: Karthala.
Benn, C. (2002). *China's golden age: Daily life in the Tang Dynasty*. New York: Oxford University Press.
Best, J. W. (2007). *A history of the early Korean Kingdom of Paekche, together with an annotated translation of the Paekche Annals of the Samguk Sagi*. Cambridge, MA: Harvard University Press.
Brooke, J. L. (2014). *Climate change and the course of global history*. Cambridge: Cambridge University Press.
Brown, D. (1993). The Yamato Kingdom. In D. Brown (Ed.), *The Cambridge history of Japan. Vol 1: Ancient Japan* (pp. 108–162). Cambridge: Cambridge University Press.

Campbell, B. M. S. (2016). *The great transition: Climate, disease and society in the late-medieval world*. Cambridge: Cambridge University Press.
Chaves, J. (1986). *The Columbia book of later Chinese poetry: Yuan, Ming and Ch'ing Dynasties (1279–1911)*. New York: Columbia University Press.
Chen, S. (2012). *Multicultural China in the early middle ages*. Philadelphia: University of Pennsylvania Press.
Chen, Q. (2015). Climate shocks, dynastic cycles and nomadic conquests: Evidence from historical China. *Oxford Economic Papers, 67*, 185–2024.
Crow, D. (2018). "How a Eurasian Steppe Empire Coped with Decades of Drought." *Sapiens,* May 2, https://www.sapiens.org/archaeology/uyghur-empire-drought/
D'Alpoim Guedes, J., & Bocinsky, R. K. (2018). Climate change stimulated agricultural innovation and exchange across Asia. *Science Advances, 4*(10), eaar 4491, 31 Oct., http://advances.sciencemag.org/content/4/10/eaar4491/?fbclid=IwAR1JRsDxkdZ6w13-kXl4auW9Hyr1dUCpMfyPlimgmGszK6ES6q6_9gEtCr0.
Dalby, M. T. (1979). Court politics in late T'ang times. In D. Twitchett (Ed.), *The Cambridge history of China. Vol. 3, Sui and T'ang China, 589–906, Part 1* (pp. 561–681). Cambridge: Cambridge University Press.
Di Cosmo, N., Oppenheimer, C., & Büntgen, U. (2017). Interplay of environmental and socio-political factors in the downfall of the Eastern Türk Empire in 630 C.E. *Climatic Change, 145*, 383–395.
Fan, K.-w. (2010). Climatic change and dynastic cycles in Chinese history: A review essay. *Climatic Change, 101*, 565–573.
Foltz, R. (2010). *Religions of the silk road* (2nd ed.). New York: Palgrave MacMillan.
Garrett, F. (2007). Critical methods in Tibetan medical histories. *Journal of Asian Studies, 66*, 363–387.
Gill, R. (2000). *The Great Maya Droughts*. Albuquerque: University of New Mexico Press.
Guisso, R. (1979). The reigns of the empress wu, Chung-tsung and Jui-tsung. In D. Twitchett (Ed.), *The Cambridge history of China. Vol. 3, Sui and T'ang China, 589–906, Part 1* (pp. 290–332). Cambridge: Cambridge University Press.
Hansen, V. (2017). *The silk road: A new history with documents*. New York: Oxford University Press.
Harper, K. (2017). *The fate of Rome: Climate, disease, and the end of an empire*. Princeton: Princeton University Press.
Heine, S. (2018). *From Chinese Chan to Japanese Zen: A remarkable case of transmission and transformation*. New York: Oxford University Press.
Kiernan, B. (2017). *Viet Nam: A history from earliest times to the present*. New York: Oxford University Press.
Kōjirō, N. (1993). The Nara State. In D. Brown (Ed.), *The Cambridge history of Japan. Vol 1: Ancient Japan* (pp. 221–267). Cambridge: Cambridge University Press.
Kōyū, S., with Brown, D. M. (1993). Early Buddha worship. In D. Brown (Ed.) *The Cambridge history of Japan. Vol 1: Ancient Japan* (pp. 163–220). Cambridge: Cambridge University Press.
Kuhn, D. (2009). *The age of confucian rule: The song transformation of China*. Cambridge, MA: Harvard University Press.
Laufer, B. (1919). *Sino-Iranica*. Chicago: Field Museum of Natural History.
Lewis, M. E. (2009). *China's cosmopolitan empire: The Tang dynasty*. Cambridge, MA: Harvard University Press.
McCullough, H. C. (1999). Aristocratic culture. In D. Shively & W. H. McCullough (Eds.), *The Cambridge history of Japan. Vol. 2, Heian Japan* (pp. 390–448). Cambridge: Cambridge University Press.
McCullough, W. H. (1999b). The Heian Court, 794–1070. In D. Shively & W. H. McCullough (Eds.), *The Cambridge history of Japan. Vol. 2, Heian Japan* (pp. 20–96). Cambridge: Cambridge University Press.

References

Mitsusada, I. (1993). The Century of Reform. In D. Brown (Ed.), *The Cambridge History of Japan. Vol 1: Ancient Japan* (pp. 163–220). Cambridge: Cambridge University Press.

Mote, F. W. (1999). *Imperial China 900–1800*. Cambridge, MA: Harvard University Press.

Ouyang, X. 2004. *Historical records of the five dynasties*. R. L. Davis (Trans.), New York: Columbia University Press.

Peterson, C. A. (1979). Court and Province in Mid- and Late T'ang. In D. Twitchett (Ed.), *The Cambridge history of China. Vol. 3, Sui and T'ang China, 589-906, Part 1* (pp. 464–560). Cambridge: Cambridge University Press.

Reischauer, E. O. (1955). *Ennin's Travels in T'ang China*. New York: Ronald Press.

Salguero, C. P. (2014). *Translating Buddhist medicine in Medieval China*. Philadelphia: University of Pennsylvania Press.

Schafer, E. (1963). *The golden peaches of Samarkand*. Berkeley: University of California Press.

Seth, M. J. (2011). *A history of Korea, from antiquity to the present*. Lanham, MD: Rowman and Littlefield.

Shively, D., & McCullough, W. (Eds.). (1999). *The Cambridge history of Japan. Vol 2: Heian Japan*. Cambridge: Cambridge University Press.

Somers, R. M. (1979). The end of the T'ang. In D. Twitchett (Ed.), *The Cambridge history of China. Vol. 3, Sui and T'ang China, 589-906, Part 1* (pp. 682–789). Cambridge: Cambridge University Press.

Takeshi, M. (1993). Early Kami Worship. J. Goodwin (Trans.). In D. Brown (Ed.), *The Cambridge history of Japan. Vol 1: Ancient Japan* (pp. 317–359). Cambridge: Cambridge University Press.

Toshiya, T. (1993). Nara economic and social institutions. In D. Brown (Ed.), *The Cambridge history of Japan. Vol 1: Ancient Japan* (pp. 415–452). Cambridge: Cambridge University Press.

Twitchett, D. (1979). Hsüan-tsung (reign 712–756). In D. Twitchett (Ed.), *The Cambridge History of China. Vol. 3, Sui and T'ang China, 589-906, Part 1* (pp. 333–463). Cambridge: Cambridge University Press.

Twitchett, D., & Wechsler, H. J. (1979). Kao-tsung (reign 649-83) and the Empress wu: The Inheritor and the Usurper. In D. Twitchett (Ed.), *The Cambridge History of China. Vol. 3, Sui and T'ang China, 589-906, Part 1* (pp. 241–289). Cambridge: Cambridge University Press.

Von Glahn, R. (2015). *The Economic History of China*. Cambridge: Cambridge University Press.

von Verschuer, C. (2006). *Across the Perilous Sea: Japanese trade with China and Korea from the seventh to the sixteenth centuries*. K. L. Hunter (Trans.), Ithaca, NY: East Asia Program, Cornell University.

Wechsler, H. (1979a). The Founding of the T'ang Dynasty: Kao-tsu (reign 618-626). In D. Twitchett (Ed.), *The Cambridge History of China. Vol. 3, Sui and T'ang China, 589-906, Part 1* (pp. 150–187). Cambridge: Cambridge University Press.

Wechsler, H. (1979b). T'ai-tsung (reign 626-649), the Consolidator. In D. Twitchett (Ed.), *The Cambridge History of China. Vol. 3, Sui and T'ang China, 589-906, Part 1* (pp. 188–241). Cambridge: Cambridge University Press.

Wright, A. F. (1978). *The Sui Dynasty: The unification of China, A.D. 581-617*. New York: Alfred A. Knopf.

Wright, A. F. (1979). The Sui Dynasty (581-617). In D. Twitchett (Ed.), *The Cambridge History of China. Vol. 3, Sui and T'ang China, 589-906, Part 1* (pp. 48–149). Cambridge: Cambridge University Press.

Xuanzang. (1996). *The Great Tang Dynasty Record of the Western Regions. Translated by the Tripitaka-Master Xuanzang under Imperial Order. Composed by Sramana Bianji of the Great Zongchi Monastery*. Tr. Li Rongxi. Berkeley: Numata Center for Buddhist Translation and Research.

Yancheva, G., Nowaczyk, N., Mingram, J., & Haug, G. H. (2007). The influence of the intertropical convergence zone on the East Asian monsoon. *Nature, 445*, 74–77.

The Rise of Central Asia: Coastal Golden Ages Increasingly Threatened by Conquest Dynasties from the Deep Interior

6.1 Medieval Warmth

The rise of Song (960–1279) in China and Koryŏ (918–1392) in Korea came along with the increasing warmth of the Medieval Warm Period (MWP, aka Medieval Climatic Anomaly), which made East Asia warmer and wetter. Based in the mid-north with a capital at Kaifeng on the middle Yellow River, Sung benefited from this. Whether that benefit was enough to be critical in the fight to subdue the populous south is difficult to decide. In any case, the south was subdued, and good conditions prevailed for most of the rest of Song. The expansion of the southern economy, particularly of the lower Yangzi region, continued and accelerated.

Following a brief lag in the Oort Minimum (1010–1050), when solar radiation was reduced, solar radiation strengthened, making the 1100s and especially the 1200s among the warmest years since the last interglacial. This strengthened the Asian monsoons, producing more rain for the eastern coast and interior, but drying up Central Asia. Particularly critical, as will appear, was the increased warmth and rainfall in Mongolia, especially after 1200.

Koryŏ benefited from warmer and moister times. Japan, already warm and moist, profited little. Southeast Asia's rain center moved north and left Cambodia and southern Vietnam increasingly dry. Northern Vietnam profited greatly; the increased rain was called "rice from the sky" (Kiernan 2017: 158). The result was a major power shift: the northerly kingdoms of southeast Asia—Thailand, Vietnam, Burma—expanded at the expense of Khmer and Mon civilization. The great Khmer city of Angkor was devastated by drought. This expanded China's sphere of influence at the expense of India's. The dominance of Indian culture in southeast Asia waned drastically, especially in the later MWP. Apparently, climate change drove history in this particular situation.

The golden age of good monsoons and warm, moist north was punctuated by some cold, wet periods, especially in the early twelfth century, pursuant on volcanic eruptions that introduced vast amounts of sun-blocking dust and sulphuric acid droplets into the atmosphere (Campbell 2016: 55–58). These had devastating effects on northern China.

6.2 World-system Dynamics

During this period, the integration of the East Asian world-system increased. Enormous ships constantly shuttled back and forth to southeast Asia, Japan, and Korea, carrying cargoes of coin, metal, pottery, books, cloth, and food. Bulk goods trade was firmly established. Trade with Malaya had already begun before Tang; it now became extensive and well developed (Heng 2009). It was to continue in the future. Japan and Korea were now trading as semiperipheral states, sending fine craft and manufactured goods as well as raw materials, and doing well in the trade (see e.g. Hurst 1999: 636). The Japanese were trading lacquerwork and swords as well as minerals, but still importing "hi-tech" goods from China, to the point of abandoning coinmaking to import Chinese coins as daily tender.

If one looks at a map of East Asia, one sees that Japan, Korea, and China almost meet. The southern islands of Japan swing close to the mainland. The tripoint, where they are all about equally close and a few tens of miles apart, is near Cheju Island, off the mouth of the Yangzi River—China's center of wealth and economic progress then as now. In the medieval period, seas connected, lands divided. Seafaring in the China seas was far ahead of the rest of the world technologically. Land transportation was slow. China's canals and Japan's inland sea made access much easier within those nations. All this guaranteed great importance to three-way trade between China, Korea, and Japan.

Soon the Ryukyus were incorporated in the network; they were a small kingdom, but strategically placed in regard to sea-trade routes. Trade often went via these islands, which thus finally became integrated in the East Asian world-system. Okinawa at this time was divided into rival local chiefdoms or petty kingdoms. Local kings competed for control over the next few centuries, while Chinese and Japanese shipping passed through the islands. Oddly, Taiwan remained a wild stronghold for Austronesian tribal people, many of them head-hunters; they were not integrated into literate urban society, let alone the world-system, for centuries to come. They maintained a splendid independence. Why China and Japan neglected this huge, resource-rich island is a mystery.

During the Song Dynasty (960–1279), Central Asia was the source of enemies, Japan and Korea were the source of traders and students. The Silk Road lost somewhat in relative importance. The sea lanes saw glory days, establishing the "Maritime Silk Road" as the major channel for Chinese goods.

With this went full-scale cultural exchange. Song was a golden age for Chinese science, technology, art, literature, and philosophy, and East Asia rushed to take advantage of this. Koryŏ (Seth 2011) and Japan proved willing learners. Korean celadon pottery, for one example, rapidly equaled Song's in quality; many consider these two celadon traditions among the most beautiful ceramics ever made. Buddhist schools, Neo-Confucian philosophy, new traditions in art, new poetic and prose forms, and even new foodways and recipes spread. Considerable numbers of travelers went back and forth to learn the most refined and arcane matters at the feet of masters. Everywhere, the rule was creative integration rather than exclusive, mindless following. China's greatest writers followed late Tang examples in integrating Daoism, Buddhism and Confucianism in "Three Teachings" syntheses, producing in men and women like Mei Yaochen, Su Shi, and Wang Anshi some of the greatest moral thinkers (as well as some of the greatest poets) in the medieval world. Chinese Daoist-Buddhist Confucian, Korean Buddhist-Confucian and Japanese Buddhist-Confucian-Shinto cross-fertilization produced cultural brilliance. Thus did progressive integration of the world-system, in a favorable climate, produce a Golden Age.

An excellent case in point is the spread of Chan Buddhism, recently chronicled by Steven Heine (2018; see also Collcutt 1990). This school emerged in China in late Tang. It was influenced by Chinese Daoism. It spread rapidly to Japan, as Zen, and Korea, as Sŏn (all the names are derived from Sanskrit *dhyāna*, "meditation"). Populist cults within Zen not only held ceremonies for rural people, but helped with farming and agricultural infrastructure (Collcutt 1990: 631–2). Zen developed its great influence on Japanese aesthetics, but previous Japanese art and musical forms influenced Zen in turn.

This gave it a wholly different character from that it had in China. In Song, it was an arcane meditative and philosophic discipline, for the most recondite sages. In Japan, it became a hugely important ideological and organizational environment for emerging militarism, conflict, and politics. In China, monks on mountaintops eclectically learned from each other and from other traditions; in Japan, increasingly rigid chains of transmission, passing on of the ritual whisk, and canonical bodies of teachings defined strict, set lineages and monastic orders. The Gempei War of 1183–85 brought out the most martial elements. Supposedly nonviolent Buddhism led to cyclic decline by contributing to warfare. That strife and the Ōnin War of 1464–76 set Zen back sharply, destroying much of its temple infrastructure, but it survived and flourished.

The spread of Zen and other Buddhist movements came at the expense of the highly ritualized, ceremonialized, and expensive Buddhism of previous centuries. Zen, Pure Land, and other movements served as pietist movements opposed to lavish ritualism. An inevitable counter-reaction from the Ru (Confucians) produced Neo-Confucianism, first in China, but spreading rapidly to Korea and Japan. It incorporated many Buddhist meditative techniques and a rather idealistic philosophy, especially through the work of Zhu Xi (1130–1200; see below).

In sum, the meeting of these three empires and their intensive trade, communication, and mutual learning produced a cultural brilliance hard to match anywhere in the world at the time. It was, however, contemporary with a brilliant period in central Asia and nearby west Eurasia. This was by no means an independent development. The two centers were connected closely by the Silk Road.

6.3 Japanese Crisis

After 900, Japan came increasingly under the control of the Fujiwara clan, an enormous descent group with many branches that often competed for dominance. The Heian period (794–1185) declined, with Fujiwara dominance overwhelming after 850; much of their dominance involved manipulating emperors such that adults abdicated and the throne was held by children (Hurst 1999; W. McCullough 1999a, b). The retired emperors, however, continued to hold great power. Shirakawa (1053–1129, r. 1073–1087; not to be confused with Go-Shirakawa, 1127–1192) controlled the country well enough to claim, according to the *Heike Monogatari*: "Three things refuse to obey my will: the waters of the Kamo River, the fall of backgammon dice, and the monks of Enryakuji Temple" (Rizō 1999: 683). He once was kept from presenting a festival by rain, and after four days of it he lost his temper, collected a vessel full of it and threw it into prison (Hurst 1999: 603). Such was the power of retired emperors.

Eventually, other clans or families rose to power. The Minamoto (or Genji) in the east and the Taira (or Heike) in the west, rose to power and attacked each other. Both had imperial blood in their veins. The retired emperor Go-Shirakawa and the fighting monks of the Kyoto region added enormous complexities to the story, as did the inevitable tendency of some family members to side with the opposite family. ("Genji" is the Japanese pronunciation of the Chinese words for "Minamoto family"; Heishi or Heike is similarly the equivalent of "Taira family." On this and the war's complex background, see Hurst 1999; Rizō 1999.)

Japan fell apart, disintegrating in war. The Gempei War (1180–1185; the name is a fusion of "Gen" and "Hei") was Japan's defining war, producing the great and mournful *Tale of the Heike*, Japan's true national epic (*Heike Monogatari*; beautifully translated by Helen McCullough, 1990). The Taira abducted the young emperor Antoku, but Go-Shirakawa set an even younger figure, a grandson, on the throne as the emperor Go-Toba. The Taira and Minamoto were fairly evenly matched, but famine struck the Taira base in the west, possibly deciding the war (Frédéric 1972). It was probably a result of one of the local fluctuations within the Medieval Warm Period, so this is one case where climate or weather may have had a small but real role in determining history. The Minamoto brothers Yorimoto and Yoshitsune managed the war and led their side to victory, after which Yorimoto betrayed and eliminated his brother, in the classic fashion of central and northeast Asian politics. Yoshitsune's brilliant career and unhappy end made him one of the great tragic heroes in the Japanese pantheon. Japanese culture thrives on tragedy. Like Homer's *Iliad*, the Heike epic is named after the losers, not the winners.

Minamoto victory was sealed at the famous naval battle of Dannoura in 1185. The Taira fleet was destroyed. In a sad presage of the end of Song, the emperor Antoku—about 5 years old at the time—was held by his grandmother as she leaped overboard to end their lives. The crabs of Dannoura have an odd pattern on their backs that is taken to resemble a death's head, the pattern having been imprinted after the Taira's catastrophe.

The 1180s were one of the optimal times climatically in the Medieval Warm Period. One can say this gave opportunities to the Minamoto, but good times floated all boats; the famine that may have brought down Taira was local. Much more critical to the Taira's fall was the complex politics of the time, especially the rivalry between the Retired Emperor and the Taira. Neither climate nor Ibn Khaldun's model fits this case well enough to score as a hit for either hypothesis.

The Minamoto conquest instituted the Kamakura Period, with Japan ruled from the Minamoto base in Kamakura near modern Tokyo (Mass 1990). The emperor in Kyoto became a figurehead, despite occasional attempts by individual emperors to regain power. The real power lay with the *bakufu*, the government headed by the *shogun* ("generalissimo"). The period from then to 1603, when the Tokugawa shoguns reunited the country, is Japan's medieval age, complete with pirates and robber barons. Life during this time included burning of cities and castles every few years, which put an enormous drain on Japan's wood supply. The economy focused on counterproductive ends: war, recovery from war, and preparation for more war. (Material life in this period has been chronicled by Frédéric 1972 and Verschuer 2016.)

It is hard to escape the conclusion that Japan's descent into militarism and then into chaos was influenced by the collapse of Tang and then the fall of Northern Song. The dates can be seen to track those events. China had no direct influence on Japan except through trade, but the period did not encourage stability anywhere.

Korea suffered similar troubles. In 1170, a coup by Chŏng Chung-bu (1106–1179) led not to a new dynasty but to the subjugation of the king (Seth 2011: 104), as in Japan. The king was exiled to an isolated island. The dynasty had become old, with the usual widening gap between elite and mass, and the usual corruption, palace intrigues, eunuch politics, and economic malaise.

Military factions then vied for control, until the Mongol invasion in 1217 forced Korea to unite against a common enemy. The Korean resistance was fierce and unrelenting—a mistake in dealing with the Mongols. The result was utter devastation, though the Korean government held out until 1270, fleeing to the tiny and remote island of Kanghwa (Seth 2011: 110–113). They finally broke free of the Mongol yoke when the Ming Dynasty eliminated Mongol power from the coastal regions. Freedom, however, merely returned Korea to governmental strife, until the fall of Koryŏ and rise of Yi in 1392.

The coincidence of the beginning of Mongol rise to power, the similar initiation of Jin strength in the northeast, and the military takeovers in Japan and China coincides perfectly with an optimal time in the Medieval Warm Period. It is hard to escape the conclusion that these rising powers benefited from warmer years. On the other hand, there is no reason why warmth should have helped the military triumph over the civil bureaucracy in Korea and Japan. The Korean coup of 1170 and the Gempei War seem logical extensions of forces that had been building for centuries, not responses to sudden climate change.

Through crises, however, trade with China and Korea continued to flourish, and Japan made a major contribution to the arts of life by inventing the folding fan around 1100 (Verschuer 2006: 72). They also perfected swordmaking, for which

Japan is still famous—using today techniques based on the superior technology of the 1100s and later. It is doubtful if anyone has ever made better swords. These became major exports, not only for use but for prestige (Verschuer 2006: 74–75).

6.4 Southern Approaches

Vietnam by this time had broken free of Chinese control. "From the tenth century, Vietnamese history comes into its own" (Kiernan 2017: 131). Independent Nam Viet was established in 939, but it later became tributary to Song. Tributary kingdoms to China ranged from dependent semiperipheral statelets to completely independent major polities (including Burma and Thailand in later centuries) that merely sent trading missions that the Chinese proudly but overambitiously called "tribute missions." Nam Viet, later Dai Viet ("Great Viet"), was more of the latter sort. Attempts by Song to bring it under better control ended in humiliating defeats in 982 and 1076—another pair in the long record of failures to control the Vietnamese. Dai Viet returned to tribute-paying independence, and promptly invaded and devastated Champa (Kiernan 2017: 145–146). Times were good in the north, as the early phase of the Medieval Warm Period produced generally good weather, with reliable heavy rainfall through the 1200s.

In southeast Asia at the same time, serious and prolonged droughts devastated the economies of Burma, southern Vietnam, and the Khmer Empire, contributing heavily to their downfall (Iannone 2016). The worldwide warming drove the inter-tropical convergence zone far enough north to starve the tropics of rain while increasing rain in the monsoonal temperate zone. Droughts may have extended north from these areas into Yunnan and other southern frontiers.

The Ly dynasty controlled Dai Viet from 1009 to 1225 (Kiernan 2017; Lieberman 2003: 352–358), when the imperial succession passed to a 7-year-old girl. She had suffered child marriage to one Tran Thai Tong (r. 1225–1258), whose uncle took over actual rule and named the boy as emperor, producing a new dynasty without conflict or trouble—a most unusual event in East Asia, and far from any Ibn Khaldun dynamics. The Tran dynasty was to rule from 1225 to 1400 (Lieberman 2003: 358–362).

6.5 China's Warm Period

A general powerful and canny enough to found a dynasty arose in the small state of Later Zhou, take it over, and use its large army to subdue the rest of China one bite at a time. Zhao Kuangyin was a brilliant strategist, able to calculate how to take on the weaker realms and build up to the stronger, instead of moving too fast and uniting all against him. He also seems to have combined wisdom and statecraft well enough to be a true Khaldunian conqueror.

The Khitan might well have taken all China if Zhao had not done so. They had risen against their Parhae overlords in 926 and established their own Liao dynasty in northeast China. In 938 Liao took 16 commanderies in northern China. Song tried repeatedly and unsuccessfully to take these back, exhausting many military resources in the effort. Liao conquered the Jurchen between 983 and 985 and took northern Korea in 1005 (Pratt and Rutt 1999: 206).

World-systems theory is particularly useful in understanding the rapid rise of Zhao to dominance over the various states. They shared a common culture and were linked by economic and military ties. Zhao easily moved from one to the next, taking advantage of existing administrative frameworks that were familiar to him.

World-system theory is even more important in understanding the subsequent rapid rise of peripheral groups, from the Khitan to the Mongols, to the level of semiperipheral marcher states and then to full power as conquest dynasties. Song, uniquely in Chinese history, wound up sharing the country with non-Chinese regimes.

Zhao decided to subordinate the generals to his personal control and weaken the power and independence of the military. This had the immediate effect of saving Song from the fate of Later Zhou and other predecessors, but doomed Song to military weakness (Kuhn 2009; Mote 1999: 103). Having taken over Zhou by a coup, Zhao wished to prevent other generals from doing likewise.

His policy was "strengthening the civil sector, weakening the military" (*zhong wen, qing wu*) The Chinese literally means "strengthen the literate, weaken the military." This is a typical Chinese comment on the military mind. *Wen* "cultured, literary" had the secondary meaning of "nonmilitary" throughout Chinese history. The military was brought to heel. It could not fight off the northern and central Asian groups that slowly took over more and more of China. In war with the Xixia Tangut state, "the armor built by the enemy is all cold-forged, strong, smooth, and impenetrable to even the strongest cross-bow. By contrast, the suits of armor shipped from the capital are all too yielding…the Qiang [Tangut]…approach war-preparations with dedication and spirit, while we are undisciplined and lax" (Tian Kuang [1005–1063], quoted in Smith 2017: 82).

We are fortunate in having particularly good works on climate during Song times. Not only have Joseph McDermott (2013; McDermott and Yoshinobu 2015) and Ling Zhang (2016) done major research on China, but we now have Bruce Campbell's detailed work *The Great Transition* (2016), covering Europe, with comparative comments on the rest of Eurasia.

Demography, however, cursed the regime. Of 183 children born to emperors, 2 died in war, 82 died young, and 99 survived—almost 50% infant mortality (L. Zhang 2016: 266). In old China, being born in the richest and best medically-served family was not much protection. In fact, in later centuries at least, farm families in the more remote, less epidemic-prone parts of China had better survival rates than the Imperial family, placed in the urban center (Bengtsson et al. 2004).

China prospered exceedingly for over 100 years. Agriculture and tea-growing became commercialized. Copper and silver money expanded its circulation. More fatefully, paper money was locally tried, leading to inflation. The great clans and families of earlier times shrank in importance; the newly-invented formal Chinese lineage system spread rapidly. This was part of a shift away from domination of the realm by familial elites, who went back into the unknown past and were not organized as chartered corporations. Related to this was domination by bureaucrats picked by civil service examinations—the *shidafu*, "government-service masters."

6.6 A Climatic Interruption

At the same time, the climate turned harsh. The beneficent Medieval Warm Period began to swing erratically. Cold cycles struck in the late 1000s and after 1100. These briefly but drastically interrupted the warm era. They gave China some stunningly bad years. The floods were apparently not alleviated by drier weather. The 1100–1102 period brought serious hardship. John Brooke (2014: 369–370) reports that China was relatively cold again during the major conquests by Qin in 1127 and the Mongols in 1234 and 1279. On the other hand, the fact that the Jurchen and Mongol homelands were benefiting from global warming has major relevance. Cold, wet periods in 1030–1080 and 1260–1280 tracked volcanic eruptions (Kong et al. 2017), as did those during the early 1100s (Campbell 2016: 55 lists the volcanoes in all these episodes, where known). The late 1100s, by contrast, were a climatic golden age, warm and moist.

Ling Zhang (2016) has studied the climatic and human circumstances leading up to and following from the catastrophic flood of 1048, when the Yellow River changed its course from flowing out through northern Shandong (as it does today) to a much more northerly mouth in Hebei near modern Tianjin. This was the last stage in a series of ever more catastrophic floods from 722 onward, especially in the period just before 1048 (see her table, L. Zhang 2016: 111). A further flood, following breaching the dykes in a monumentally stupid plan to stop the Jurchens, occurred in 1128, and led to the river changing to a course south of the Shandong Peninsula, a course it generally followed (with fluctuations) until 1855, when it shifted back to something close to its pre-1048 position (L. Zhang 2016: 281). Cannibalism was reported (Zhang 2016: 193–198), but cannibalism is a standard exaggeration in Chinese reports of disasters. Rainier weather, and floods, returned through most of the 1100s after the 1120s.

The government's later efforts to tame the river on the model of Yu the Great were not up to Yu's alleged level of success. The great Song scientist Shen Gua (1031–1095) advised the government to restore wetlands to give the river room to spread. This was the best advice possible under the circumstances, and it was tried, but the ponds had their own problems, including the production of mosquitoes that carried diseases (L. Zhang 2016: 164–172). Inevitably, the river got caught up in the faction fighting that dominated the empire in the late 1000s. The ultimate irony came when a flood in 1099 swept away a monument to Great Yu on top of a high point above the raging river (L. Zhang 2016: 156). Hebei remained a devastated, impoverished region—a doormat for the invading Liao and later Jin, as Song rulers eventually realized (L. Zhang 2016: 143–144). The Song state apparently made it a sacrificial zone, allowing it to lose its former independence and prosperity (L. Zhang 2016: 131–138).

Not for nothing was the Yellow River long known as "China's Sorrow." Yet, the problems it caused (and still causes) were the result of human deforestation, dense and impoverished rural population, and poorly planned hydrological works. Ironically, one of the major causes of deforestation was cutting timbers and collecting grass and brush to use in levees and fascines to shore up the river management works (L. Zhang 2016: 184, 268–279). The Chinese knew that deforestation contributed to floods, but bureaucracy took its own often-mindless way.

Here and throughout history, the effects of climate change were so thoroughly shaped, directed, redirected, and exaggerated by human management and mismanagement that we cannot say that "climate change" by itself did anything. Climate change stimulated people to do a vast number of different things, some of which led to flooding, some of which controlled it for a while. It was the human decisions and actions that caused the actual occurrences on the ground and in the water. The Li family in Sichuan, back in the Qin Dynasty, had shown that Chinese rivers can be controlled with premodern technology.

Meanwhile, the shift of China's population from north to south went on apace. This is shown in two maps in Richard von Glahn's *Economic History of China* (2015; see also Kuhn 2009). The first (p. 210) shows Tang population distribution: highly concentrated in the north, though already dense in the Yangzi delta. The next (p. 211) shows the situation in 1102: population much sparser in the north, much denser and more widespread in the south. Classically described by Herold Wiens in *China's March toward the Tropics* (1954), the rise of the south became the dominant demographic shift in Chinese history, and progressively lessened the effect of cold, dry times on the realm (Fan 2010).

By 1200 there may have been 124,000,000 Chinese (Campbell 2016: 59). About 5% of the population was urban (Mote 1999: 165). Industry flourished to the point of deforesting north China to feed the iron industry (Hartwell 1962, 1982) and, according to legend, the writing and printing industry—pine trees were burned for soot to make ink.

Trade grew, continuing the process begun in Han and Tang of making southeast China a world center of merchandising. Fujian emerged slowly, throughout Song and later Yuan, as a truly mercantile society, importing food while exporting made goods from a wide hinterland (So 2000). People of all the Old World empires and religions thronged ports in Fujian and Guangdong.

6.7 Social Changes

Key to understanding cycles is understanding the peaks, when, in Holling's terms, the r phase gives way to the K phase. Success leads to cultural brilliance, but usually to political problems, as we observed in Han and Tang. It is thus desirable at this point to provide an overview of the triumph of Song, because it shows how a peak could lead by its very successes to a later collapse.

A case study at the height of the cycle is provided by Fan Zongyan (989–1052). Besides writing notable poetry and philosophy, he began a tradition of zealous political and legal reform. Fan's father had died early, and Fan set out with few resources to make his way. He was brilliantly successful, reaching the highest bureaucratic positions (Smith 2017). He also led a successful expedition against the Xixia Tangut state, proving that good organization could overcome military power. He became famous for his outspoken and fearless criticism of the emperor and imperial policies. This led to his frequent demotion to minor posts, but he won his way back, and attracted such brilliant followers as Ouyang Xiu (1007–1072) and Mei Yaochen (1002–1060), not only reformers but two of China's greatest writers. They in turn attracted Su Shi (1037–1101), generally considered the greatest Song writer, and his brother Su Che (1039–1112).

This reform bloc was soon overshadowed by Wang Anshi (1021–1086), a genuine radical, who instituted far-reaching changes in taxation and governance. Some backfired seriously, leading to major attacks by Ouyang's group. A high rate of extraction by the government of the people's wealth occurred. Wang was idealist and hopeful, and clearly believed his reforms would help ordinary people, but they succeeded largely in increasing government revenue at the expense of the population.

Outspoken parties alternated power, depending on the will of the emperor at any given time. Never was speech more free or statesmen more outspoken in the history of agrarian empires. The reforms are still much debated in the historical literature. (See reviews in Chaffee and Twitchett 2015, Twitchett and Smith 2009, and Kuhn 2009; this story has naturally attracted a huge amount of attention. As usual, I find Mote's *Imperial China* the most judicious summary; see Mote 1999: 126–144.) This was a major one among the problems that led to the decay of Song and the eventual fall of the north to invaders. A sidelight is that Wang and his major attackers, including Sima Guang and Su Shi, were among the greatest poets and writers in the history not only of China but of the world, and they managed to exchange more than a few friendly poems through it all.

One point of contention questioned whether education should be for what we would now call a particular "major" (Wang) or in classical Chinese culture (Sima). The latter view prevailed, and history seems to indicate it was the better one. In the civil service exams, Wang fought for specialized exams in specific subjects. Sima argued for keeping the old uniform exams in shared culture. Many Chinese statespersons and historians held that having a common cultural heritage, including memorizing a core of books, united the Chinese elite as nothing else could do. Everyone had a common moral and intellectual background, whatever they may have done with it. This may have kept factional fights from being even worse than they were. By encouraging scholar solidarity but discouraging expertise, it profoundly affected the rest of Chinese history. It makes many Chinese of today wonder seriously about the present highly-specialized college education system, producing narrow experts who cannot communicate across disciplinary boundaries. These controversies and conflicts weakened the dynasty and led to increasing vacillation by emperors and bureaucrats. The catastrophic war with the Jurchens was soon to destroy Song's golden age.

The bureaucratic reforms were only part of a profound cultural change. Another symptom was the rise of women's status to probably the highest in all Chinese history until very recent times. Women writers and artists won respect equal to men. Song even invented a "good" foil to Yang Guifei: Mei Fei ("flowering-apricot consort"), not so beautiful, but modest, virtuous, brilliantly talented—and thus ignored by the swinish Xuanzong (Liu et al. 2017). Alas, she was fictional, but she deserves to be adopted by modern feminist movements.

Medieval China's greatest woman poet, Li Qingzhao (1084–1155), was already famous in her 20s. She and her husband Zhao Mingcheng (1081–1129) devoted themselves to art, writing, and collecting books and antiquities. They were true *zhiyin* ("soulmates," lit. "understand-music," with reference to a famous pair of music-loving friends in earlier dynasties). In the fall of northern Song, first the collections were lost and destroyed, then her husband was killed. An impoverished and lonely widow, she wrote agonizing songs of grief, often bitterly echoing her earlier poems of joy and happiness (Li Ch'ing-chao 1979; Mote 1999: 328–330; Liu et al. 2017).

Women's status began to decline after this time, however. Neo-Confucianism began a move toward less tolerance of womanly success. One suspects the rise of autocracy was more significant than Zhu Xi's thought in the decline of women's status. A bitter symbol was the appearance of foot-binding, though it remained rare in Song. Under the grim, woman-suppressing Ming dynasts, it became more and more widespread.

Neo-Confucianism, known in its time as Daoxue, "study of the way," was developed by the Cheng brothers and later by Zhu Xi, in the twelfth century. Zhu Xi's lifespan, from 1130 to 1200, makes him a close contemporary of Al-Ghazālī (1058–1111) the Islamic puritan reformer, Thomas Aquinas (1225–1274) the Christian theologian, Moses Maimonides (ca. 1135–1204) the Jewish philosopher, Namadeva (ca. 1270–1350) the Hindu teacher, and other philosophers who were foundational to the modern forms of the major religious traditions. This was a second Axial Age, as important culturally as the original one around 600–300 BCE.

It is no coincidence that these spiritual movements sweeping East Asia came at roughly the same time as equivalent movements in the west. Unlike the first Axial Age, whose causes remain mysterious, this second age fits well with the expansion of land and maritime contacts, the rise of commerce, the consequent widespread transmission of texts, and the fusion of Eurasian societies into one world-system. In fact, the western three were in the same universe of discourse: Aquinas and Maimonides were using and advocating the revival of Aristotelian philosophy, Ghazālī was attacking it. Zhu Xi seems to fit well with the pattern. He probably never heard of Aristotle, but he too was trying to balance tradition against new philosophic enterprises. One does wonder how much the world-systems were in touch. Certainly, religious and scientific ideas were flowing back and forth along the Silk Road. Philosophy accompanied them, but the hardening boundary between Islamic and Confucian civilizations came to separate western from eastern cultural and philosophical realms.

Zhu Xi's philosophy was in some ways forward-looking, but set in stone the rigid hierarchy of old China, with the ruler above, the oldest male the authority figure of a family, and with women firmly kept down. On the other hand, Neo-Confucianism taught social responsibility, telling people in no uncertain terms to work together for the common good. It also held firmly that human nature is inherently prosocial, with bad actions the result of bad teaching. (See Chu Hsi and Lü Tsu-Ch'ieh, *Reflections of Things at Hand,* 1967, for a selection of texts, translated; Mote 1999: 340–346 for a good brief discussion.) By the end of Song, supporters of Neo-Confucianism established an orthodoxy that was to last (with moderate changes), but Mote (1999: 317) points out that it was nothing like Christian orthodoxy in the west: "It meant little more than establishing a particular set of texts...for...scholars in the civil service examinations...it did not demand the banning of non-orthodox texts, beliefs, or practices in general, or punishing the unorthodox."

This was also a period of major scientific development, both in Song China and in the west. Song writers produced innovative work in agriculture, metallurgy, mathematics, and medicine. At the same time, Greek learning, supplemented enormously by Arabic and Persian learning, was introduced to China over the Silk Road, and to western Europe by such translators as the incredibly productive Gerald of Cremona (1114–1187), who translated the major medical and scientific works of the Greeks and Arabs into Latin. The Tibbonid family of translators in Spain did the same for Hebrew, translating, among others, Maimonides, who wrote in Arabic. The Tibbonid translations were soon rendered into European languages. A pan-Old World world-system was emerging. By the 1300s, we find essentially the same detailed descriptions of and treatment for stroke in use from France to Cairo to Beijing (Anderson 2014). Galenic humoral medicine was known from Japan to England. The spread of medical lore had reached the corners of the known world.

Civil service examinations had begun under Han, grown under Tang (partly because the Empress Wu needed them to build a power base). They truly came into their own in Song, with its drive for imperial authority. Since, at first, the examinations were under the emperor's close control, they became another means for imperial bureaucratic domination at the expense of other elites. The judging came increasingly under the influence of Neo-Confucianism.

Song was surprisingly rich compared to Ming. "A comparison of government revenue for 2 years, 1064 and 1578, reveals that, although revenue from agricultural sources was virtually identical, revenue from nonagricultural sectors under the Sung was an astounding nine times greater than under the Ming" (Hartman 2015: 23). The years are cherry-picked—1064 was the height of Northern Song and 1578 was well into the decline phase of Ming—but the difference is substantial. In later dynasties taxes amounted to "6–8% of national income." Contemporary European states took 4–6%. Song may have taken 13–24% (Hartman 2015: 23–24), though the latter figure is hard to credit, since the people operated at too low a margin above subsistence to afford it.

In the Song Dynasty, the Chinese developed the lineage system. Fan Zhongyan developed or perfected a system in which a family would identify their earliest provable male ancestor and incorporate all descendants from him (focally, those in the male line) into a corporation with an actual written legal charter (Mote 1999: 349–50). This new institution, the Chinese lineage, owned land and buildings—at least a lineage hall for meetings, records, and ancestor worship. Not only the lists of members and the rules of organization, but the tablets that held a bit of the ancestors' souls, were kept here. This institution caught on, dominating life, especially in southeast China, until the present. Great lineages became large enough to fill whole towns.

Much of the income from the lineage estate was used to help less affluent members, especially to giving quality education to the young who could not otherwise afford it. Even women could get an education. A benefit of the lineage system was recognition that educating women was highly desirable for itself, but also because educated women attracted better husbands. They produced more successful children, since women controlled at least the earlier stages of children's development, and were the ones who had to make sure that the children valued education and had a good start.

The importance of this for upward mobility, especially of talented youth from less affluent branches, can easily be imagined. The entire lineage benefited when a poor but brilliant child grew up to succeed, so everyone had a stake. Skimping on educating the young would have been suicidal. Lineages were extremely pervasive, especially in the richer area of central and southeast China. They shaped Chinese society after Fan's time. (They are called "clans" in the older English sources, but they are not much like the Scottish clans, which were less organized and less relentlessly patrilineal.)

The lineage system deserves so much treatment here because of its effect on Chinese politics and society. Lineages became more and more important, and more and more differentiated in success and social ranking. They were superb instruments for consolidating wealth and power. They advanced their members, but the powerful lineages could keep down change and mobility in society. They played a role in the conservatism of Chinese society after 1300.

The liberalizing changes weakened Song; the conservative reaction was to weaken all China, increasingly, after 1200.

6.8 Northern Invaders and Neighbors

Song (960–1279) lasted for over 300 years. However, it lost ground progressively to polities traditionally called "nomad" or "pastoral" regimes, but were semiperipheral marcher states, with mixed economies, including large and rich agricultural areas. The long period during which Song shared China with non-Han conquest states has been memorably described as "China among equals" (Rossabi 1983), though Song remained by far the most populous and rich state as long as it lasted.

The Medieval Warm Period strengthened these northwestern regimes, enabling them to expand their population of warriors and above all of horses (Anderson 2014). No longer were the nomads so afflicted by the dread *dzud*—snow and ice storms that cover the grass in late winter and early spring, starving the horses. The

East Asian world-system suddenly achieved greater connectivity, through the military revival of the steppes and mountains in the warm, moist years.

In Korea, the new Koryŏ regime emphasized Buddhism, but was also dynamic in military action and in trade. Benefiting from the warmer, moister weather, it produced a Korean golden age contemporary with Song. Korea traded very actively with China and Japan, and took advantage of their situation between those two powers. Korea had strong ties with the Siberian groups, though relations with the Khitan were increasingly unfriendly. Koryŏ civilization was brilliant, and heavily based on Chinese norms.

A pleasant insight into Korea at the time was presented by Xu Jing, a Confucian scholar and magistrate who led a mission there in the early 1100s. In 1124 he brought out a book, *An Illustrated Account of the Xuanhe Mission to Koryŏ* (Vermeersch 2016). The illustrations are long lost. Xu reports a highly sophisticated society, long following Chinese customs, and well versed in Chinese norms of behavior and ceremony. He was surprised; he had expected the "Eastern barbarians" (*dong yi*) to be more uncouth.

The first semiperipheral people to benefit from the Medieval Warm Period were the Khitan, who spoke a language related to Mongolian. As noted above, they were already harrowing China's and Korea's frontiers in the Tang dynasty. They took advantage of the fall of Tang to consolidate control of the far north, establishing the Liao Dynasty (named in 947). Their own name for their realm—*Khitai,* "land of the Khitan"—became Cathay, the medieval European term for China.

The founder of Khitan glory was Abaoji (872–926, r. 907–926), a tall, powerful man, brilliant strategist, and by all accounts a true inspirer of 'asabiyah. In this case we are not speaking of victors' history; we know of him from the records of his enemies, the Song, who were impressed (Mote 1999; Twitchett and Tietze 1994). His lineage Sinicized their surname as Yelü, a name important long beyond the fall of Liao. The Yelü produced statesmen and leaders who were critical advisors to the Mongols in China, much later. Khitan leadership also extended to women, for it was a true steppe-type society in which women had more public power than was usual in China. Of Abaoji's empress Yingtian, Frederick Mote says: "Few women in history match her independence and determination" (Mote 1999: 50). She was not alone. Another in her mold was Chengtian, a woman warrior and general who commanded field armies in her sixties (Twitchett and Tietze 1994: 91). One surmises that the Song forces were terrified. Chinese soldiers of imperial times feared the mysterious *yin* energies of militant women.

The Khitan moved progressively south, as well as east into Korea, with devastating effect (Mote 1999: 60–62). Koryŏ resisted stoutly, and managed to survive, but buffer groups between it and the Khitan were wiped out. War with Song in 1004–5, with Chengtian leading many of the troops, led to a treaty that left Liao in complete control of far north China (Twitchett and Tietze 1994: 108–110). Liao gradually increased its holdings thereafter. The rulers converted to Buddhism, and added Chinese statecraft, progressively, to what remained a steppe-type society.

At about the same time—from the mid-tenth century—another steppe group came to power: the Tanguts. They founded the Xixia state in what is now Qinghai and neighboring western China (see Dunnell 1994; Mote 1999). Usually described as Tibetan, they were actually Qiang, which means—if they were indeed ancestral to today's Qiang—that they spoke a fascinating and little-known language related to Tibetan but quite different from it. Few, if any, people who were once so important are now so little researched. The Xixia state expanded rapidly from the Ordos Desert to the deserts and oases westward. It had a relatively peaceful and surprisingly trouble-free career from around 982 to its destruction by Chinggis Khan in 1227. Instability in the final years did not contribute to its fall; Chinggis was an unstoppable force. Even so, it took him a long time to break the Xixia, and he died in the process, in the same year that Xixia fell. Farther south, Tibet was a large and powerful empire during Song times, but had rather little effect on China.

Song's factional politics steadily weakened its ability to present a united front. One odd legacy was paper money, invented in 1024 to deal with the fiscal problems (Von Glahn 2015: 233). This was the world's first fully monetized paper currency.

The Liao Dynasty had the usual evolution toward royal family increase, widening of gaps between elite and mass, and luxury consumption. Rebellion flared, serious by 1115. Weakening of Liao allowed the Jurchen to rise. They were a Tungus group, ancestral at least in part to the later Manchus. The Khitan fled to the west. A general from the royal line, Yelü Dashi, rallied, consolidated control, and formed a well-run state, Karakhitai. This state held much of central Asia for a long time It was notable—like other early central Asian states—for its extreme religious freedom (Biran 2005). Like other northwestern powers, Karakhitai was a multiethnic, multilinguistic state, though dominated by one ethnic group's royal dynasties. Liao had lasted about one Ibn Khaldun cycle. Climate seems unrelated to its fall.

The Jurchen, like the Khitan and other northeast Asians, were shamanists, and in fact introduced the Tungus word "shaman" to the world; its first appearance is in their court records. They were not nomads; they were farmers, hunters, and pig-raisers in what is now Northeast China. They had recently lived in pit-houses (Mote 1999: 213) like some Siberian groups and Plateau Native Americans, but they were rapidly learning Chinese methods of building, as well as of farming and manufacturing.

Centering on the Gold River, they declared the Jin (Gold) Dynasty in 1114, under the rapidly rising warrior Aguda (or Akuta). Aguda was a brilliant institution-builder and a charismatic dynasty-founder, much like Abaoji, whom he may have been copying. Again, we have this on the authority of less-than-friendly Chinese sources. (See Franke 1994; Mote 1999; Tao 1976; Twitchett and Tietze 1994.) He adopted Khitan warrior styles and institutions, and made his small tribe into a regional power. Under his immediate successors, the Jin "forced an alliance" on unequal terms (Mote 1999: 196) on the Tangut state of Xixia in 1124, but did not destroy it. They conquered Liao in 1125. Despite rebellions and other problems, Liao was still strong. It fell because of the sheer military force of its enemies, more than because of its own decadence (see Twitchett and Tietze 1994: 149–153).

Jin took the northern part of Song in 1127. This was an amazing achievement for a tiny, obscure group (see Mote 1999: 421–241). They continued to defeat the hapless Song on every occasion. They soon conquered all north China, forcing

Song south of the Huai and Yangzi rivers. The capital of the now Southern Song moved to Lin'an, now Hangzhou. This was a quite striking victory; unlike the Liao or even the later Mongols, the Jurchen were not a "semiperipheral marcher state" but a small peripheral tribe from the remote forests. Their victory over three huge, rich, militarily powerful states in the space of 3 years was a true case of "The Mouse That Roared." The real achievement was taking down Liao; after that, they had the Liao military machine at their command.

The Jurchen conquest was facilitated by Song's peaceful traditions. After a period of stronger rulers, the Emperor Huizong ruled from 1100 to 1126, and died in Jin captivity in 1135 (see Ebrey 2014 for a superb detailed biography). He indulged in painting (at which he was excellent) and poetry, took to meditation, and in general escaped his duties into aesthetic pursuits. His hapless successor Qinzong (r. 1126–1127; died in Jin captivity in 1161) succeeded to the throne too late to organize defense.

Conflict aversion led to the loss of Yue Fei, a superb general who was demoted and then executed in 1141 by the peace faction at court during the Jurchen invasion. The peace faction was led by Qin Guei (1090–1155) and his wife. Revisionist historians now sometimes see Yue Fei as a hot-head, and the peace faction, including the Qins, as reasonable, but the traditional Chinese appraisal of the situation is still persuasive (Mote 1999: 299–305.) To this day, South Chinese call *yutiao* (fried dough sticks, similar to churros) "oil-fried devils," punning off Qin's given name Gui, which sounds like the word for "devil" or "ghost." These pastries are cooked with a hope that the pair is similarly frying in hell.

Other "evil ministers" held power through much of Southern Song. To see how badly they were regarded, it is sufficient to read the story "Zheng Huchen seeks Revenge in Mumian Temple" in Feng Menglong's great collection of stories (Feng 2000: 383–416). It portrays the near-to-last grand councilor Jia Sidao (1213–1275) as a totally immoral scoundrel, rising from salt smuggling to imperial bully. No doubt this is exaggerated; it is the myth of the "bad last" again, applied to a penultimate councilor rather than a last emperor.

In the case of Northern Song we have a clear Ibn Khaldun decline, exacerbated by climatic shocks. Song in the late eleventh and early twelfth centuries was weakened by wildly fluctuating and always unsuccessful economic policies, run by feckless leaders, and characterized by weak policies. The fall of Northern Song completed the process of making the lower Yangzi region the economic heart and social center of China. Future regimes only strengthened this tendency (Mote 1999; Smith and von Glahn 2003), especially when the Little Ice Age made north China a poorer, weaker region. China's history came to be written more from that area than from the dry northwest that had previously been the cockpit.

The Jin played out a typical Ibn Khaldun cycle in a short time, moving from intrepid conquerors to Sinicized and vulnerable victims of the Mongols. They could never conquer Southern Song. After a rich period accompanied by warm weather, flooding and another shift of the Yellow River's course in 1194 devastated their economy for years (Franke 1994: 246). Mote finds an ironic evidence of decay in the fact that the once-mighty hunters were warning their emperor by about 1200 that "'it was dangerous for an emperor to ride in the wilderness and it was detrimental to the economy of the state'" (Mote 1999: 241). At this time Chinggis Khan's Mongols, nominal vassals of Jin, turned against Jin as soon as they thought they could defeat it. Dissent and court politics gave them the opportunity (Franke 1994: 250–265).

Even so, Jin held out, fighting the Mongols with tenacity, losing only slowly to their inexorable advance. The Jin held a steadily shrinking core of north China from Chinggis' first attacks in 1211 until final conquest in 1234. They meanwhile had to stand off Song, which allied with the Mongols at first. The great historian Herbert Franke wrote Jin's epitaph: "The end of a dynasty has always been a favorite topic for Chinese historians and historical philosophers. They usually try to explain the fall of a state by deviations from the moral principles embodied in the ethical code of Confucianism.... But even orthodox historians would have to admit that the cardinal virtue of loyalty was alive during the last stages of [Jin]" (Franke 1994: 265). Like the Xixia, Jin fell not because of their weakness but because of the sheer unstoppable power of Chinggis Khan.

Until faced by the Mongol challenge, which caused confusion and dissention, Jin had managed a successful, tenacious, and competent rule. They assimilated Chinese governance without much difficulty. A measure of their competence is their careful censusing; for instance, we know the population of Jin in 207 was 55,532,151 people, with 6.33 persons per household (Franke 1994: 278). Granted that the figure is impossibly exact, evidence shows that it is close to truth. The main capital, Kaifeng, had an incredible 1,746,210 households (this was the whole metropolitan district, including surrounding country and suburbs; Franke 1994: 279).

6.9 The Fall of Song

Thus, after the fall of Jin, the Mongols turned their attentions to Song and slowly conquered it, finally triumphing in 1279. Its pacifism undid it. Mote (1999: 321) contrasts the "vigorous polo-playing" aristocrats of old with the new Song scholars, "sedan-chair-riding, soft-voiced." Marco Polo (ca. 1254–1324) succinctly wrote the epitaph of Song: "...the country never would have been lost, had the people but been soldiers. But that is just what they were not; so lost it was" (Polo 1927: 207).

The Song Dynasty was in a unique situation, facing several powerful states in its core region. This was the only time in Imperial China's history when it faced equally powerful neighbors, holding a great deal of the realm, yet speaking different languages and coming from different cultural backgrounds.

Song held on as long as other major dynasties. Part of the reason was that the northern regimes had difficulties governing, withstanding Song military pressure, and transitioning from nomadic or thinly-settled agrarian regimes to north China's densely populated and intensively farmed expanses. But Song could barely hold on. "For most of the Northern Song, the state financed a standing army of 1 million soldiers, from a general population of 60 million people. Military expenses for pay, supply, and armaments regularly consumed 80% of the entire state budget" (Hartman 2015: 29). Military overextension was inevitable even in a peace-oriented state—a typical situation of agrarian empires. Song was trapped in the fatal Ibn Khaldunian feedback loop: more and more military need, more and more taxation, more and more impoverishment, weaker and weaker governance, leading to more and more challenge from enemies, and thus more and more military need.

The fate of Song after the fall of the north is quickly told. The throne passed to Gaozong (r. 1127–1162), who managed to hold the line at the Yangzi and stop the Jin. His successor Xiaozong (r. 1162–1189) oversaw a stable, prosperous period, but it gave way to a succession of weak emperors. Duzong (b. 1248, r. 1264–1274) was the last to rule with any degree of effectiveness. He was succeeded in 1274 by the 3-year-old Gongzong. In 1276, the Mongols took the capital, Hangzhou, and captured the boy emperor. He spent the rest of his life in captivity, becoming a Buddhist monk. Loyalists fled south, installed another young boy as the new emperor, "the third such hapless child to fill that role" (Mote 1999: 465). The loyalists were chased off the land into the sea near Hong Kong, where the Mongol ships caught up with them. "One venerated old official of the court took the boy on his back and leaped into the sea, committing suicide to avoid capture" (Mote 1999: 465).

References

Anderson, E. N. (2014). *Food and environment in early and medieval China.* Philadelphia: University of Pennsylvania Press.
Bengtsson, T., Campbell, C., & Lee, J. A. (Eds.). (2004). *Life under pressure: Mortality and living standards in Europe and Asia, 1700–1900.* Cambridge, MA: MIT Press.
Biran, M. (2005). *The empire of the Qara Khitai in Eurasian history: Between China and the Islamic World.* Cambridge: Cambridge University Press.
Brooke, J. L. (2014). *Climate change and the course of Global History.* Cambridge: Cambridge University Press.
Campbell, B. M. S. (2016). *The great transition: Climate, disease and society in the late-medieval world.* Cambridge: Cambridge University Press.
Chaffee, J. W., & Twitchett, D. (Eds.). (2015). *The Cambridge history of China. Vol. 5, Part Two: Sung China, 960–1279.* Cambridge: Cambridge University Press.
Chu, H., & Tsu-Ch'ieh, L. (Eds.). (1967). *Reflections of things at hand: The neo-confucian anthology. Trans. Wing-tsit Chan.* New York: Columbia University Press.
Collcutt, M. (1990). Zen and the Gozan. In *The Cambridge history of Japan. Vol. 3, Medieval Japan* (pp. 583–652). Cambridge: Cambridge University Press.
Dunnell, R. (1994). The Hsi Hsia. In *The Cambridge history of China, Vol. 6, Alien Regimes and Border States, 907–1368* (pp. 154–214). Cambridge: Cambridge University Press.
Ebrey, P. (2014). *Emperor Huizong.* Cambridge, MA: Harvard University Press.
Fan, K.-w. (2010). Climatic change and dynastic cycles in Chinese history: A review essay. *Climatic Change, 101,* 565–573.
Feng, M. (2000). *Stories old and new: A Ming Dynasty collection.* Tr. Shuhui Yang and Yunqiu Yang (Chinese original ca. 1600.). Seattle: University of Washington Press.
Franke, H. (1994). The Chin Dynasty. In *The Cambridge history of China, Vol. 6, Alien regimes and border states, 907-1368* (pp. 215–320). Cambridge: Cambridge University Press.
Frédéric, L. 1972. *Daily life in Japan at the time of the Samurai, 1185-1603.* E. M. Lowe (Trans.). Tokyo: Charles E. Tuttle Co.

References

Hartman, C. (2015). Sung Government and Politics. In J. W. Chaffee & D. Twitchett (Eds.), *The Cambridge history of China. Vol. 5, Part Two: Sung China, 960-1279* (pp. 19–138). Cambridge: Cambridge University Press.

Hartwell, R. (1962). A revolution in the Chinese iron and coal industries during the Northern Sung, 960-1126 A.D. *Journal of Asian Studies, 21*, 153–162.

Hartwell, R. (1982). Demographic, political, and social transformations of China, 750-1550. *Harvard Journal of Asian Studies, 42*, 365–442.

Heine, S. (2018). *From Chinese Chan to Japanese Zen: A remarkable case of transmission and transformation*. New York: Oxford University Press.

Heng, D. (2009). *Sino-Malay trade and diplomacy from the tenth through the fourteenth century*. Athens, OH: Ohio University. Research in International Studies, Southeast Asia Series, No. 121.

Hurst, G. C., III. (1999). Insei. In D. Shively & W. H. McCullough (Eds.), *The Cambridge history of Japan, vol. 2, Heian Japan* (pp. 576–643). Cambridge: Cambridge University Press.

Iannone, G. (2016). Release and reorganization in the tropics: A comparative perspective from Southeast Asia. In R. K. Faulseit (Ed.), *Beyond collapse: Archaeological perspectives on resilience, revitalization, and transformation in complex societies* (pp. 179–212). Carbondale, IL: Southern Illinois University Press.

Kiernan, B. (2017). *Viet Nam: A history from earliest times to the present*. New York: Oxford University Press.

Kong, D., Wei, G., Chen, M.-T., Peng, S., & Liu, Z. (2017). Northern South China Sea SST changes over the last two millennia and possible linkage with solar irradiance. *Quaternary International, 459*, 29–34.

Kuhn, D. (2009). *The age of Confucian rule: The song transformation of China*. Cambridge, MA: Harvard University Press.

Li, C. (1979). In K. Rexroth and L. Chung (Trans./Eds.), *Complete poems*. New York: New Directions.

Lieberman, V. (2003). *Strange parallels: Southeast Asia in global context, c. 800-1830. Vol. 1: Integration on the mainland*. Cambridge: Cambridge University Press.

Liu, H., Yu, X., Gao, C., Zhang, Z., Wang, C., Xing, W., & Wang, G. (2017). A 4000-yr multi-proxy record of holocene hydrology and vegetation from a peatland in the Sanjiang Plain, Northeast China. *Quaternary International, 436*, 16–27.

Mass, J. P. (1990). The Kamakura Bakufu. In *The Cambridge history of Japan. Vol. 3, Medieval Japan* (pp. 46–88). Cambridge: Cambridge University Press.

McCullough, H. C. (1990). *The Tale of the Heike*. Stanford, CA: Stanford University Press.

McCullough, W. H. (1999a). The capital and its society. In D. Shively & W. H. McCullough (Eds.), *The Cambridge history of Japan, vol. 2, Heian Japan* (pp. 97–182). Cambridge: Cambridge University Press.

McCullough, W. H. (1999b). The Heian Court, 794-1070. In D. Shively & W. H. McCullough (Eds.), *The Cambridge history of Japan, vol. 2, Heian Japan* (pp. 20–96). Cambridge: Cambridge University Press.

McDermott, J. P. (2013). *The making of a new rural order in South China. I. Village, land, and lineage in Huizhou, 900-1600*. Cambridge: Cambridge University Press.

McDermott, J. P., & Yoshinobu, S. (2015). Economic change in China, 960-1279. In J. W. Chaffee & D. Twitchett (Eds.), *The Cambridge history of China. Vol. 5, Part Two: Sung China, 960-1279* (pp. 321–436). Cambridge: Cambridge University Press.

Mote, F. W. (1999). *Imperial China 900-1800*. Cambridge, MA: Harvard University Press.

Polo, M. (1927). *The book of Ser Marco Polo the Venetian. Tr. Henry Yule (French original ca. 1300)*. New York: Macmillan.

Pratt, K., & Rutt, R. (1999). *Korea: A historical and cultural dictionary*. Richmond, Surrey: Curzon.

Rizō, T. (1999). The rise of the warriors. In D. Shively & W. H. McCullough (Eds.), *The Cambridge history of Japan, vol. 2, Heian Japan* (pp. 644–709). Cambridge: Cambridge University Press.

Rossabi, M. (Ed.). (1983). *China among Equals: The middle kingdom and its neighbors, 10th–14th centuries.* Berkeley: University of California Press.

Seth, M. J. (2011). *A history of Korea, from antiquity to the present.* Lanham, MD: Rowman and Littlefield.

Smith, H. A. (2017). *Forgotten disease: Illnesses transformed in Chinese medicine.* Stanford, CA: Stanford University Press.

Smith, P. J., & von Glahn, R. (Eds.). (2003). *The Song-Yuan Ming transition in Chinese history.* Cambridge, MA: Harvard University Press.

So, B. K. L. (2000). *Prosperity, region, and institutions in maritime China: The South Fukien pattern* (pp. 946–1368). Cambridge, MA: Harvard University Press for Harvard University Asia Center.

Tao, J.-s. (1976). *The Jurchen in twelfth-century China.* Seattle: University of Washington Press.

Twitchett, D., & Smith, P. J. (2009). *The Cambridge history of China. Vol. 5. The Sung dynasty and it precursors, 907-1279.* Cambridge: Cambridge University Press.

Twitchett, D., & Tietze, K.-P. (1994). The Liao. In *The Cambridge history of China, Vol. 6, Alien regimes and border states, 907-1368* (pp. 43–153). Cambridge: Cambridge University Press.

Vermeersch, S. (2016). *A Chinese traveler in medieval Korea: Xu Jing's illustrated account of the Xuanhe embassy to Koryŏ.* Honolulu: University of Hawai'i Press and Korean Classical Library.

Verschuer, C. v. (2006). *Across the Perilous Sea: Japanese trade with China and Korea from the seventh to the sixteenth centuries.* K. L. Hunter (Trans.). Ithaca, NY: East Asia Program, Cornell University.

Verschuer, C. v. (2016). *Rice, agriculture, and the food supply in modern Japan.* New York: Routledge.

Von Glahn, R. (2015). *The economic history of China.* Cambridge: Cambridge University Press.

Wiens, H. (1954). *China' march toward the tropics.* Hamden, CT: Shoe String Press.

Zhang, L. (2016). *The river, the plain, and the state: An environmental drama in Northern Song China, 1048-1128.* New York: Cambridge University Press.

The Mongol Conquests of China and Korea and Invasion of Japan

7

7.1 Climate: From Warm to Frigid

Rarely in history has one man led a remote peripheral state up to semiperipheral status and then on to conquer core states. The Germans who sacked Rome, the Aztecs who conquered central Mexico, and the original Incas may have had such histories. But probably never in history did one man accomplish so much after losing his father to enemies and fleeing with his mother to the forest to survive on roots and herbs.

The Mongol Empire and other steppe and northern empires clearly benefited from the Medieval Warm Period. Their victims could thus be said to have lost out partly because of climate. During Genghis Khan's time, climate turned warmer and wetter in Mongolia, warmer and drier farther west in the real center of Asia (Anderson 2014; Marcarelli 2014; Pei and Zhang 2014; Zhang et al. 2016). This allowed the raiders to wax powerful and their horses—the truly essential ingredient in raiding and war—to wax fat and numerous. North China was warmer and wetter too (Fagan 2008), but was overwhelmed as much by the superior horses as by the hard-bitten fighters riding them. They conquered Jin and Xixia, as well as all central Asia. Grandsons of Genghis divided the empire. Hŭlegŭ Khan conquered the Near East, taking Baghdad in 1248, but was stopped by the Mamluks on the Levant, and never conquered the southern Near East. The Golden Horde took Russia and conquered eastern Europe. With almost all Eurasia under their control, Khubilai could turn his attentions east. The Mongols first reduced Korea to tributary status, then turned their attentions on China. Safe from the dzud, the Mongols flourished and rode out to conquer the world, including the hapless Song.

However, the standard Chinese historical explanation for this has always been Song fecklessness, and this explanation has much truth. This is especially true since the weather turned sharply colder and drier at the time the Mongols began their conquest, and once again when they finished it. The warming peaked in the 1240s and 1250s. Cold weather had interrupted the warm period, around 1220, and was at

© Springer Nature Switzerland AG 2019
E. N. Anderson, *The East Asian World-System*, World-Systems Evolution and Global Futures, https://doi.org/10.1007/978-3-030-16870-4_7

least intermittently serious after 1260 (Brook 2016; Campbell 2016: 198–199). Song finally fell in 1279. The Mongols then made two disastrous expeditions from Korea to conquer Japan, but violent storms destroyed their fleets; the second in particular was the famous *kamikaze*, "Divine Wind," that gave its name to airplane fighters in WWII. The storms were not all, however. The Japanese were powerful and competent soldiers. The Mongols never tried again.

Sharply drier weather began in 1285. Volcanic activity blocked the sun. Mongolian and Siberian larch growth rings and Tibetan juniper rings narrowed sharply, showing cold dry weather from 1283 into the early 1300s (Campbell 2016: 201). Larches grew well in the 1330s (as they had in warm periods in the 250s, 550s, and 980s; Campbell 2016: 202, 250). Tree rings shrank to practically nothing in the 1340s, with minima in Qinghai in 1344, Mongolia and Tian Shan in 1348 (Campbell 2016: 285; southern hemisphere tree ring minima are in 1348). The time was not only difficult; it was variable and unpredictable. Larch growth rings show loose cycling during the whole period from Han to 1500, at which time they leveled off at a low rate during the rest of the harsh Little Ice Age (Campbell 2016: 204). Colder weather forced the rain bands south, giving the parched warm-temperate parts of Central Asia more moisture, but also more winter suffering. Spikes of cold in the 1300s dropped north Atlantic sea surface temperatures by up to 2 °C, leading to expanding sea ice (Campbell 2016: 207), and similar conditions would have occurred in northeast Eurasia. Glaciers expanded for the next few centuries.

The Mongol empire fell apart in the mid-1300s, losing China in 1368. This followed the instability of the 1330s through 1350s, "an almost uniquely disturbed and climatically unstable period when long-established atmospheric circulation patterns were on the cusp of lasting change" (Campbell 2016: 277; see chart, p. 278).

Brian Fagan (2008), misled by the old and long-disproved theories of Ellsworth Huntington (1945), saw droughts in Central Asia as driving desperate Mongol hordes out to pillage. In fact, the Mongols and Turkic groups did confront drought in the center of Asia during the Medieval Warm Period, but recruited their core troops and horses from the wetter east and north. The western enemies of the Mongols were weakened by the Medieval Warm Period. Much of west and central Asia became drier, with presumably serious effects for both rainfed and irrigated agriculture, as well as grazing and horse-raising (Campbell 2016; Pei and Zhang 2014; Zhang et al. 2016, 2018).

The countless theorists throughout history who have held that drought, famine, and want send the warriors out to conquer have never stopped to think or explain how starving, weakened troops on starving, dying horses could ever conquer anything. The same is now true of those who hold that shortages of water, food, and the like will cause the wars of the twenty-first century. Even militarily powerful nations facing a water shutoff have resorted to diplomacy rather than war (Barnaby 2009). How much more true is this for nations like the Mongols, that could not fight, or even mount plausible diplomacy, without plentiful grass and water.

All this does not detract much from the organizational genius of Chinggis Khan (ca. 1162–1227), known as "Genghis" in the west. His actual name was Temujin; Chinggis Khan means "ocean ruler" or "universal ruler," a title conveyed on him when he reached power. Despite countless studies of his life, we still do not fully understand how he did this. One reason is that we must rely on the *Secret History of the Mongols* (De Rachewiltz 2004 is the best English edition), the only major historical source on his life. It assimilates him to Mongol epic folklore, starting with descent of the Mongols from a gray wolf and a fallow doe. Oral and folkloric tradition can be strikingly accurate, but it is not equal to sober documentation, especially when one is dealing with the apotheosis of a hero of mythic proportions.

His sons and grandsons carried the empire to rule over all Eurasia. As a pan-Eurasian empire united under one ruler (Genghis' son), it lasted a mere 12 years, but the grandsons and their heirs kept up a semblance of political unity and a genuine economic and social unity for another century. It fell apart only slowly, with Genghis' heirs still ruling much of Central Asia and east Europe into the 1500s. The rise and glory of the Mongol Empire has been told so often that it needs no introduction (Allsen 1994, 2001; Di Cosmo et al. 2009; Ratchnevsky 1991; Weatherford 2004, 2010).

The Medieval Warm Period was also a golden age for the Silk Road. Active since very ancient times, owing its name to silk trade already established in Han times, this road had suffered decline in the post-Tang chaos. It had not fully recovered in Song. It reached its peak importance during the warming centuries from 500 to 750 and again from 1200 to 1350. The flow of religions described above for the Tang Dynasty was renewed, adding to the religious awakening we have noted for the twelfth and thirteenth centuries. Eurasia may have been politically disunited, but it was religiously open to flows and changes. The Mongols continued the earlier tradition of total religious freedom. Marco Polo and other European travelers report the free-wheeling debates before the Khans, in which Christians tended to be at disadvantage because they criticized not only other religions but even other Christian sects. Mongol tolerance had no respect for such factionalism.

Unfortunately for Central Asia and the Silk Road, the MWP came to a sudden and dramatic end. The Wolf solar minimum (ca. 1282–1342) put a stop to the warm weather. Insolation did not recover for centuries. The period from 1300 to 1850 comprised the Little Ice Age, probably the coldest time since the Younger Dryas at the end of the Pleistocene. Glaciers formed on the mountains of Asia. Rivers became torrents. Central Asia became wetter (see the striking graph in Campbell 2016: 49), but this hardly made up for the coming of brutal winters. Settlements that had proliferated and flourished all over far northern China in Liao and Jin were abandoned, indicating a social collapse triggered by climate but running well beyond what climate would predict. The critical mass for maintaining a dense, prosperous settlement network was gone.

With the Little Ice Age, the Silk Road declined. The standard—and partially correct—explanation for its decline is the rise of cheaper and faster shipping, thanks to Chinese and Portuguese expansion. But, also, the timing, and the complete eclipse of the Silk Road and the great Central Asian states, is clearly due in part to the Little Ice Age. Surely, Central Asia could have continued to flourish as a local center of activity, even if the Indian Ocean routes siphoned off much of its business. One immediate effect of the LIA was to end Tamerlane's threat to China; while on his way to attack it, he died of cold in one of the first of the dreadful LIA winters.

The early fourteenth century was wet, leading to problems with rivers. The Mongols tamed the Yellow River for the last time in Imperial Chinese history—inevitably comparing themselves to Great Yu (discussing Song, but with obvious self-referent; see Zhang 2016: 133, 142). By 1330, the river was beginning to shrink down to normal after wet times, and the Mongol regime took advantage of this to dyke it successfully and leave it some room to spread—room sometimes cleared by earlier floods that washed farms away.

However, by 1344, the Yellow River was running out of control again, shifting its lower course from a southern route to the current northward one with an outlet north of Shandong. The areas devastated by invasions, floods and droughts in Song (Zhang 2016) were devastated again. This area was the core of Mongol power, and could not afford it. Grain shipments to the north were interrupted. In 1351, control of the Yellow River was restored, along its new course (Dardess 1994: 576). The extreme cold of that time period presumably helped, by reducing rainfall and keeping much of the mountain rainfall bound up as ice and snow.

7.2 The Mongol Rise

The Mongols began as a small tribe near Lake Baikal. "Mongol," pronounced something like "Mung-nguet" in medieval Chinese, first appears in 1084 in Khitan records, becoming common in Jin documents (Mote 1999: 404),

They expanded rapidly southward. The Mongol empire was the biggest "upward sweep" of them all, but it did not conquer China under Genghis, who died in the conquest of Xixia in 1227. Jin fell in 1234. The south under the Song fell in 1279. Under Kublai Khan (1215–1294), it united China, and even though the Mongol empire split among Chinggis' grandsons, Yuan was an upward sweep for China itself. Kublai's China was far bigger than Song's (Rossabi 1988, 1994. Kublai would have pronounced his name something like Qubilai Qan, now Khubilai Khan in Mongolian).

The Mongols had to produce not one but two superb managers of 'asabiyah, Chinggis and Kublai, before they could conquer China. We will meet this situation again in the Qing conquest. Something similar seems to have occurred in the Khitan and Jurchen takeovers, and one might even compare the rise of Han under Liu Bang and its rebirth under Wen Di. Two successive 'asabiyah-rich conquerors may be a frequent feature of upward sweeps.

The Yuan Dynasty did not officially begin till 1272, but it is more reasonable to begin the cycle of empire in 1215, when Chinggis began the drive into China (following Mote 1994). The problem with this is that the Mongol empire was one vast realm until it split up under Chinggis' grandsons, leaving China a separate realm. That happened before 1272 and was recognized in that year by Kublai's declaration of a new dynasty. Imperial dynamics, however, were well developed by then.

The Mongols claimed that Heaven had chosen them—Genghis in particular—to rule the world, a "mandate" concept that has as much in common with Turkic views as with China's classic theory (Allsen 1994: 348). They crushed the long-surviving Xixia empire. When Chinggis died in the process, the Mongols concealed his death from the Xixia (Allsen 1994: 365). They brought in Yelu Chucai, heir of the Liao royal family, as advisor. Among other things, he put to scorn a sour suggestion by one hardened old Mongol to drive the Chinese out of north China and turn it into pasture land (Allsen 1994: 375–376; the suggestion is recorded in the *Yuanshi*, the official Chinese history of the period. It has been taken too seriously. The gentleman was being preposterous, probably for effect).

The Jin Empire was conquered under Chinggis' grandson Ögödei, but he relaxed after that by hunting and drinking, dying in 1241 at the age of 56—one of many Mongol rulers whose death was associated with alcohol. His queen Töregene actually ruled before and after his death, one of many extremely tough and independent Mongol women to hold power (Allsen 1994: 382; Weatherford 2010). She managed to put their son Güyüg on the throne. He also was said to be a hard drinker, and "profligate in the extreme" (Allsen 1994: 388). He died in 1248. Another powerful queen, Sorghaghtani Beki (d. 1252), maneuvered her son Möngke into power, and he took over the whole of the Mongol empire in 1251. In the same year, he assigned his younger brother Kublai the task of subduing the rest of China. "In 1259, Möngke died, of either dysentery or wounds inflicted by a [Song] catapult.... For the [Song] dynasty, Möngke's death meant a twenty-year reprieve, and for the Mongolian empire, a new and divisive succession crisis, from which it would never fully recover" (Allsen 1994: 410–411). The empire soon split into four parts, under Chinggis' grandsons.

Khubilai was now the independent monarch of the east. In 1256 he began work on his summer capital, Shangdu, the "Xanadu" of Samuel Taylor Coleridge's poem. It was destroyed by rebels in 1359. In 1266 he moved his capital to Beijing, called Dadu in Chinese and, in Mongol, Khanbaliq (Marco Polo's Cambaluc). The Mongols were so unused to city life that they pitched *ger* (yurts) in the park and planted steppe grass around them (Endicott-West 1994: 609). In the 1260s he pacified Korea, making it a tributary state by 1273.

In 1274 he invaded Japan, to be repelled by hard fighting and the famous "divine wind," *kamikaze*, an opportune typhoon (Rossabi 1994: 44); another invasion in 1281 met the same fate (Rossabi 1994: 484), removing all doubt in Japanese minds that the divine winds were protecting them. This was the true incorporation of Japan in the political-military world-system of East Asia; before that, it had been safe, and not prone to meddle in mainland affairs. After 1274, it was still protected by remoteness, but did a great deal of meddling, especially with Korea, but also with the China coasts and with the Ryukyus and Taiwan.

In Yuan, Chinese trade with Japan continued, but cultural influences were muted. The Japanese were as uninterested in learning from the Mongols as they had been passionately interested in learning everything from Song. Still, they imported much. A wreck in 1323, returning to Japan after trading with China, yielded to archaeologists in 1984. Incense woods, spices, medicines, coins, and 20,000 pieces of Chinese porcelain appeared (Batten 2003: 196). This was merely a normal boatload, unsignaled in historical accounts. It was evidently typical of countless such voyages. Others, however, are undocumented.

In 1276 the respite for China was over. Mongol armies struck south along the major interior rivers, and then toward the coast (Rossabi 1994: 432–433), ending the last Song resistance in 1279. Kublai's rule lasted until his death in 1294. He effectively put down rebellions and pacified the realm, but failed in advances into Vietnam and Java as well as Japan. Once the Mongols had conquered all Song, they returned to Vietnam, taking most of it between 1282 and 1288. They claimed victory, but ferocious resistance and knowledge of the landscape enabled the Vietnamese to mount a stunning defense. The Vietnamese lured the Mongols, who were using Chinese troops but still unused to waterscape fighting, into the maze of channels and swamps of the Red River Delta. Here they could break up the Mongol force, said to be a million strong, and cut the dispersed troops to pieces. (The heroic Vietnamese resistance to the enormous force of the Mongols—as well as to Han and Ming—was cited by American experts advising the United States not to go into Vietnam in the 1960s. Their advice was unheeded, and the Vietnamese repeated history.)

Kublai recruited foreigners of every stripe to serve in his court. The most famous in the west is Marco Polo (1254–1324), from Venice, who journeyed to China with his father and uncle in 1271–1295, spending much of the time in Kublai's entourage. Despite claims to the contrary, there is no question that he went to China and served the Khan (Haw 2006). His accounts of China are far too circumstantial and accurate; from giant pears to species of cranes, to come from anything but direct experience. His notorious failure to mention the Great Wall may have something to do with the fact that it was built over 100 years after his death. Minor discrepancies are easily explained by the fact that Marco narrated his story to one Rustichello, who had a reputation for exaggerating. (The claim that Marco governed a Chinese city is wrong, and suspect of being a Rustichello exaggeration.) Marco managed to get a straight story out despite Rustichello's occasional interventions.

More important, however, were the thousands of *semu*—non-Han Central Asians—serving the court. Most were Turkic or Iranic. Many were Tibetan. They contributed everything from medicine to food lore. The Tibetan monk Phagspa designed a phonemically sophisticated script for the Mongols, to write their language, imperfectly rendered by Turkic and Chinese writing. (After Yuan, the Mongols abandoned it in favor of an improved Turkic-derived script.) The Phagspa script became the inspiration for the Korean *hangul* alphabet.

Other important *semu* contributions came later. The court nutritionist in the 1330s, the Turkic doctor Hu Sihui, compiled an excellent book of nutrition and dietetics that draws on both Chinese and Central Asian sources, from the Daoist Ge Hong to the great Islamic medical scholar Avicenna (Buell et al. 2010). New foods were introduced (Laufer 1919).

Such knowledge transfer was a positive effect of the Mongol linkage of east and west. Under the Mongols, knowledge was transferred in both directions. The west received printing, gunpowder, and other inventions that changed it profoundly. In

China, western knowledge flourished, with Persian astronomy influencing such Chinese scientists as Guo Shoujing (1231; 1316; see Mote 1999: 511).

At about the same time, the Mongols compiled an enormous medical encyclopedia, the *Huihui Yaofang* ("Muslim Medical Formulas"), based on Persian and Arabic medicine. It is in classical Chinese, with Arabic and Persian glosses written accurately next to many drug names and other terms. It is fiendishly difficult to translate, however, because of the distinctive usage of medical terms that arose in Song. Paul Buell has prepared a translation (see my website www.krazykioti.com for materials).

Despite such efforts, the Chinese went on with their own traditions. The long, thorough, and sophisticated section on trauma in the *Huihui Yaofang* had no influence on the excellent and important work on that subject by the Ming doctor Xiu Ji (1548). The latter work shares nothing with the earlier, except use of several Near Eastern or widespread drugs like frankincense and myrrh that were well known long before either book appeared. Printing, gunpowder and the compass transformed the west, but western knowledge brought in by the Mongols had very little impact on China.

7.3 Bubonic Plague Goes West, Not East

In the meantime, bubonic plague ravaged the western world from 1346 to 1348, in probably the most devastating epidemic the world has ever known (the Spanish flu of 1918–20 is the main competitor). The plague came from Central Asia, riding out with the Mongols and others, carried in caravans and wagons. Bubonic plague is a rodent-vectored disease, normally endemic in gerbils and in ground squirrels (a group that includes susliks and marmots). "Gabriel de Mussis is the source of the oft-repeated story that at Kaffa plague was transferred from the besiegers to the besieged when the Mongols catapulted corpses into the city" (Campbell 2016: 300); scholars dismiss this story today, since neither the catapulting of corpses nor the ability of fleas to transmit plague after such an action are particularly credible.

It seems to have mutated into a virulent epidemic form in the 1340s, and spread through rodent and human populations that had increased to high density in the Medieval Warm Period and then reduced to suffering by the Little Ice Age. The horrific speed of its spread in the 1340s gave rise to a theory that bubonic plague was not the problem, since it must spread from rodents to humans. DNA analysis of burials has disproved this theory; we have learned, meanwhile, that shipping and land transport was far more advanced in the 1300s than was thought, so the mystery is reduced in scope. Human-to-human direct transmission was in fact common, and other insects besides fleas could serve as vectors. Also, most recent authors have not realized just how pervasive rats were in the old days. Having lived on Chinese waterfronts decades ago, dealing with rats running over my sleeping body and everything else in the household, I am more aware of how easily rat-borne diseases spread. (Bubonic plague was fortunately absent there, but leptospirosis, also rat-vectored, was general.)

The plague has, from time immemorial, had an endemic focus in high Central Asia, and the Qinghai-Tibet area has been identified as a likely home from which it radiated in various forms. Bruce Campbell in *The Great Transition* (2016: 227–331) and Kyle Harper in *The Fate of Rome* (2017: 206–245, referring to the earlier plagues from 541 to the 900s in the west) provide extensive monographs on plague, covering recent materials very well. Plague, in a mutant form that spread extremely easily and rapidly, with the terrible climate of the 1340s driving it. Quarantine was invented by 1377 in Ragusa. They imposed a 30-day rule; true "quarantino"—40-day rule—was then imposed by Venice a few years later. This was the beginning of the end for plague; it took more than three centuries, till the final 1666-67 epidemic, for Europe to complete the pattern of quarantines, solid rat-proof buildings, better rat control on ships and elsewhere, and better public health in general that finally ended epidemic plague in most of the world (Cipolla 1979; note that this was long before modern medicine).

This raises the question of why it never ravaged China with huge epidemics. Contra speculations by Morris (2010) and other historians, it never did. (See Buell 2012; Campbell 2016: 241.) The chief reason is that its local origin allowed it to come into something of an ecological adjustment with the rodents and humans of China, so that it remained endemic, rarely flaring up (Benedict 1996; the same situation has developed in today's California, where plague is endemic in ground squirrels but almost never jumps—literally, with fleas—to humans.) Epidemics usually occur in "virgin soil," populations in which almost no one has inherent immunity. Disease moves fast, killing many, affecting most of the rest, such that they acquire immunity. Thus the disease puts itself out of business, and dies out till a new cohort of children come of age and can sustain an epidemic. Where disease is endemic, there is always some death, selecting for inherent resistance among the survivors. Genetic resistance evolves quite rapidly. Plague reveals this pattern. Such a pattern is more clearly demonstrated by the fate of measles in Iceland, the Faroes, and Europe. Iceland and the Faroes do not have populations large enough to sustain the disease as endemic, so in pre-inoculation days it appeared as an epidemic every generation, causing many fatalities. In mainland Europe, however, it was endemic, infected almost all children, and was not usually serious. Plague in Europe between 1346 and 1667 appeared on roughly generational cycles. In China, it was always present, as a low-level epidemic (Benedict 1996).

China also had some advanced public health measures at the time, and although these were based on traditional medical theories that did not recognize microorganisms, they did show some understanding of contagion, and may have had some protective value against diseases. China had frequent epidemics, but they were usually local, and they are always hard to identify, since Chinese descriptions of illness do not map out on modern clinical terms (Smith 2017). Local epidemics did indeed affect Yuan China, but we do not know what they were, other than knowing that the unmistakable symptoms of plague are rarely, if ever, mentioned. Many of them must have been flu strains, from the vast flu factory of southeast China, where duck-pig-human transmission cycles and rapid viral mutation rates produce new strains of influenza every year. Those strains now go worldwide and bedevil "flu shot" designers (Cohen 2018); in premodern times they must have run through China annually. They always appear slightly different, confusing describers of symptoms. I suspect that this is why Chinese medicine always focused on strengthening the body rather than on classifying, identifying, and developing cures for specific diseases. Flu was ever-present, shape-shifting, and incurable.

Few things more clearly united the world-systems of medieval Eurasia as dramatically as bubonic plague. Its epidemic behavior in western Eurasia and endemic behavior in the east remained a striking contrast until the late nineteenth century, when, ironically, a final epidemic swept East Asia but only locally affected the rest of the world.

7.4 Mongol Decline and Fall

Population crashed before and at the beginning of the dynasty, declining perhaps 30%. It rose healthily and rapidly after that, but then declined again by 23% in the wars that created the Ming Dynasty (Mote 1994: 618–620; Smith 2003, citing several sources). Population thinned out greatly in the north but expanded in the south, making the Yangzi drainage the center of Chinese demography (see map, Mote 1994: 619).

Kublai was followed by succession conflict. There followed several effective and competent but short-lived rulers. Court factions and intrigues characterized the period. China's population declined due to war from perhaps 140,000,000 in 1200 to about 83,000,000 by 1300, 70,000,000 by 1400 (Campbell 2016: 59).

The period from 1320 to 1332 was chaotic. Regional warlords and bandits controlled local armies. "Many accounts report that bandits were able to bribe local officials to ignore their banditry, or to release them when caught.... The struggle to protect lives and property could draw on little beyond desperate initiatives and makeshift ingenuity" (Mote 1999: 520, 522). By this time the Little Ice Age was in full swing, creating untold misery.

A violent coup in 1323 brought in a young ruler, Yesün Temür. He died only 4 years later. His death unleashed a long and complicated succession crisis. There were now many lines of descent from Chinggis, and even from Kublai. Steppe vs. China-based individuals fought for control. The war was not resolved by the takeover of Tugh Temür (r. 1328–1332). The military expenses of the period were ruinous, and local rebellions broke out, as they so often did during troubled times in Chinese history. Tugh Temür was followed by a 6-year-old child who died after 53 days (Hsiao 1994: 557). Fortunately for Yuan, this pathetic boy's elder halfbrother Töghön Temür (Chinese title Shun Di; rumored to be only adopted by his titled father; 1320–1370, r. 1333–1368) was chosen to replace him, and proved physically and politically tough. He was considered a competent and strong ruler, but came too late to save the dynasty.

Yuan was ultimately brought down by a combination of local warlordism, bandits, sectarian movements (Dardess 1994; Mote 1988: 18), and, once again, succession and court politics. Violent rebellions broke out again in 1351. They were suppressed, but left much power in the hands of regional military leaders. By this time, the usual accumulation of dynastic family members, corrupt officials, self-serving bureaucrats, and other hangers-on had appeared. Too many people could reasonably put their immediate self-interest ahead of regime stability.

The main sectarian movement was the millennial and mystical Red Turban (more correctly "red headband") rebellion, a wild and multi-sited one characterized by magic with Daoist, Buddhist, and Manichaean elements. Correlation with climate is poor, though Little Ice Age conditions certainly weakened Yuan control, and caused enough misery to undercut popular support.

The Ming rulers assessed the fall of Yuan as due to weak rule, regionalism, and corruption, and set about instituting authoritarianism modeled in part after earlier Mongol practice. John Dardess (1994) and many others add the terrible climate. Others have held that the Mongols did not successfully assimilate Chinese ways of governance and order.

Yuan (1279–1368) as empire of all China lasted only one IK cycle of roughly a century, partly because of resurgent ethnic-Chinese defiance of "barbarian" conquerors. The Mongol period, however, really started with Genghis' early triumphs just after 1200. It thus lasted over 150 years, a respectable run. The Chinese had usually been quite accommodating to the Mongols as people (Endicott-West 1994: 610–611), but they developed something close to ethnic bias at the bitter end.

Disproving the myth that China always absorbed its conquerors, the Mongols rode back to the steppes when Yuan fell. The great historian Frederick Mote emphasizes this. He also comments: "That was 104 years after Kublai had moved his capital to the site [of Beijing], 432 years since the Khitan emperor Abaoji's son, the Liao emperor Taizong…in 936 took over the Sixteen Prefectures…The long era of the conquest dynasties had at last ended" (Mote 1999: 563; see also Mote 1994). The Mongols continued to harry the frontiers for generations. In Ming they even sacked Beijing, captured the emperor, and held him for ransom. Their power was broken only by the increasing savagery of the winters, which probably climaxed in severity in the seventeenth and eighteenth centuries. During this time, the Manchus conquered the Mongols, ruling them as a rather ambiguous entity, not quite part of Qing "China" but certainly not independent (see Perdue 2005).

Verdict: There is every reason to believe that all the reasons mentioned by the authorities combined to destroy Yuan (see Dardess 1994; Mote 1994). From alcohol to Central Asian-style succession battles, from imperfect conversion to Chinese governance to indifferent leadership, from sudden climate cooling to warlordism, all the direct and indirect causes were effective. Nativism and even real ethnic prejudice seem to have surfaced in the Red Turban rebellion. Elite overproduction took its usual course, with the added problem that Mongols could rarely deal with succession without family conflict.

References

Allsen, T. (1994). The rise of the Mongolian Empire and Mongol rule in North China. In H. Franke & D. C. Twitchett (Eds.), *The Cambridge history of China, Vol. 6, Alien regimes and border states, 907-1368* (pp. 321–413). Cambridge: Cambridge University Press.

Allsen, T. (2001). *Culture and conquest in Mongol Eurasia* (Cambridge studies in Islamic civilization). Cambridge: Cambridge University Press.

Anderson, E. N. (2014). *Food and environment in early and medieval China*. Philadelphia: University of Pennsylvania Press.

Barnaby, W. (2009). Do nations go to war over water? *Nature, 458*, 282–283.

Batten, B. (2003). *The end of Japan: Premodern frontiers, boundaries, and interactions*. Honolulu, HI: University of Hawai'i Press.

Benedict, C. (1996). *Bubonic plague in nineteenth-century China*. Stanford, CA: Stanford University Press.

Brook, T. (2016). Nine sloughs: Profiling the climate history of the Yuan and Ming dynasties, 1260-1644. *Journal of Chinese History, 1*, 27–58.

Buell, P. D. (2012). Qubilai and the rats. *Sudhoffs Archiv, 96*, 127–144.

Buell, P. D., Anderson, E. N., & Perry, C. (2010). *A soup for the Qan* (2nd ed.). Leiden: Brill.

Campbell, B. M. S. (2016). *The great transition: Climate, disease and society in the late-medieval world*. Cambridge: Cambridge University Press.

Cohen, J. (2018). Universal flu vaccine is 'an Alchemist's dream'. *Science, 362*, 1094.

Cipolla, C. (1979). *Faith, reason and the plague in seventeenth-century Tuscany* (M. Kittel, Trans). Ithaca: Cornell University Press.

Dardess, J. (1994). Shun-Ti and the end of Yuan rule in China. In H. Franke & D. C. Twitchett (Eds.), *The Cambridge history of China, Vol. 6, Alien regimes and border states, 907-1368* (pp. 561–586). Cambridge: Cambridge University Press.

De Rachewiltz, I. (2004). *The secret history of the Mongols: A Mongolian epic chronicle of the thirteenth century*. Leiden: Brill. 2 vols, 1347 pp.

Di Cosmo, N., Frank, A. J., & Golden, P. B. (Eds.). (2009). *The Cambridge history of inner Asia: The Chinggisid age*. Cambridge: Cambridge University Press.

Endicott-West, E. (1994). The Yuan government and society. In H. Franke & D. C. Twitchett (Eds.), *The Cambridge history of China, Vol. 6, Alien regimes and border states, 907-1368* (pp. 587–615). Cambridge: Cambridge University Press.

Fagan, B. (2008). *The great warming: Climate change and the rise and fall of civilizations*. New York: Bloomsbury Press.

Harper, K. (2017). *The fate of Rome: Climate, disease, and the end of an empire*. Princeton, NJ: Princeton University Press.

Haw, S. (2006). *Marco Polo's China: A Venetian in the realm of Khubilai Khan*. London: Routledge.

Hsiao, C.-C. (1994). Mid-Yuan politics. In H. Franke & D. C. Twitchett (Eds.), *The Cambridge history of China, Vol. 6, Alien regimes and border states, 907-1368* (pp. 490–560). Cambridge: Cambridge University Press.

Laufer, B. (1919). *Sino-Iranica*. Chicago: Field Museum of Natural History.

Marcarelli, R. (2014). Genghis Khan rose to power during drought, expanded empire during rainy season. *HNCN News Online*, March 11.

Morris, I. (2010). *Why the west rules…for now*. New York: Farrar, Straus and Giroux.

Mote, F. W. (1988). The rise of the ming dynasty, 1330-1367. In F. W. Mote & D. Twitchett (Eds.), *The Cambridge history of China* (The ming dynasty, 1368–1644, part I) (Vol. 7, pp. 11–57). Cambridge: Cambridge University Press.

Mote, F. W. (1994). Chinese Society under Mongol Rule, 1915-1368. In H. Franke & D. C. Twitchett (Eds.), *The Cambridge history of China, Vol. 6, Alien regimes and border states, 907-1368* (pp. 616–664). Cambridge: Cambridge University Press.

Mote, F. W. (1999). *Imperial China 900-1800*. Cambridge, MA: Harvard University Press.

Pei, Q., & Zhang, D. D. (2014). Long-term relationship between climate change and nomadic migration in Central Asia. *Ecology and Society, 19*(2), 68–84.

Perdue, P. (2005). *China marches west: The qing conquest of Central Eurasia*. Cambridge, MA: Harvard University Press.

Ratchnevsky, P. (1991). *Genghis Khan: His life and legacy*. Oxford: Blackwell.

Rossabi, M. (1988). *Khubilai Khan: His life and times*. Berkeley, CA: University of California Press.

Rossabi, M. (1994). The reign of Khubilai Khan. In H. Franke & D. C. Twitchett (Eds.), *The Cambridge history of China, Vol. 6, Alien regimes and border states, 907-1368* (pp. 414–489). Cambridge: Cambridge University Press.

Smith, H. A. (2017). *Forgotten disease: Illnesses transformed in Chinese medicine*. Stanford, CA: Stanford University Press.

Smith, P. J. (2003). Impressions of the Song-Yuan-Ming transition: The evidence from Biji Memoirs. In P. J. Smith & R. von Glahn (Eds.), *The Song-Yuan-Ming transition in Chinese history* (pp. 71–134). Cambridge, MA: Harvard University Asia Center.

Weatherford, J. (2004). *Genghis Khan and the making of the modern world*. New York: Three Rivers Press.

Weatherford, J. (2010). *The secret history of the Mongol Queens: How the daughters of Genghis Khan rescued his empire*. New York: Crown.

Zhang, L. (2016). *The river, the plain, and the state: An environmental drama in Northern Song China, 1048-1128*. New York: Cambridge University Press.

Zhang, Y., Meyers, P. A., Liu, X., Wang, G., Li, X., Yang, Y., & Wen, B. (2016). Holocene climate changes in the Central Asia mountain region inferred from a peat sequence from the Altai Mountains, Xinjiang, Northwestern China. *Quaternary Science Reviews, 152*, 19–30.

Zhang, Y., Yang, P., Tong, C., Liu, X., Zhang, Z., Wang, G., & Meyers, P. A. (2018). Palynological record of Holocene vegetation and climate change in a high-resolution peat profile from the Xinjiang Altai Mountains, Northwestern China. *Quaternary Science Reviews, 201*, 111–123.

Long-Lived Dynasties: Ming and Its Contemporaries

8.1 Climate and Environment

The warm days of the 1200s began to decline in the late part of that century, then declined precipitously in the fourteenth. Not only did the world become cold, but it also became more variable; sudden extremely cold peaks every century or so added to the misery of the period. There were wild fluctuations in weather in Vietnam in the sixteenth century (Kiernan 2017: 213), and in China in the seventeenth. With cold, in East Asia, comes drought, which in the warmer parts of the region were more destructive than the cooling. Vietnam suffered the worst drought cycles in its history (Kiernan 2017: 177). India too suffered (Campbell 2016), hurting China's and southeast Asia's trade. Japan and Korea dealt with arctic winters.

Central Asia became wetter, as the westerly wind belt was forced south by arctic air, but that did not help the Central Asian regimes. The horribly bitter winters more than cancelled out any benefits. Frederick Burnaby, writing in the mid-nineteenth century when warming had just begun, provided a harrowing description in *A Ride to Khiva* (1877) of the unbearable cold, far below freezing. He neglected to put on heavier gloves over his already thick mittens, and almost lost his hands. Central Asia, and especially Mongolia, were knocked out of serious contention for power. Their golden age was over.

The Ming Dynasty (1368–1644) coincided with the worst of the Little Ice Age. Timothy Brook, who set himself the task of thoroughly researching the resulting suffering, finds nine particularly bad periods, which he calls the Nine Sloughs (Brook 2010: 50–78, 2016; Brooke 2014: 439, 444–446). The nine were 1268–72, 1324–30, 1343–46, 1403–06, 1450–55, 1554–5, 1586–88, 1615–17, 1637–43. These were cold, or cold and dry. Probably all were relatively dry except the wet periods 1343–46 and 1403–06. (See Brook 2016: table 5; see also Campbell 2016: 344–345.) In addition, southern China and southeast Asia suffered drought in 1439–45, and the volcanic explosion of Kuwae around 1458 had a cooling effect

(Campbell 2016: 346). Korea and Japan would have suffered the same problems, with Korea tightly linked to China in climate and weather.

To this may be added the Spörer solar minimum of 1416–1534, which kept temperatures down and the north very dry. Essentially all the Yuan and Ming Dynasties were cold except for scattered years in the fourteenth, fifteenth and sixteenth centuries. There were famines at regular intervals, rising to a peak (with spiking rice prices) after 1600. Many of the famines, especially the worst ones, coincide with famines in England (Brook 2016: table 4; Campbell 2016), because both are influenced by worldwide oceanic events. Brook finds that cold was more damaging to agriculture in China. Cold years were generally dry, but the cold kept evapotranspiration stress down. When cold years were also wet, the floods of cold water did not dry up, producing disproportionately serious flooding.

This period of cold to very cold years ruined the Mongols in their homelands of Central Asia, and in north China. Cold weather in central Asia later made Ming safe from most invasions. Drought in Vietnam hurt that country's ability to stand up to Ming pressure. Ming suffered, but its enemies suffered more, leaving Ming too big to fight. This is part of the reason why the dynasty, with its troubled rulers, endured so long.

The climate was particularly cold in the seventeenth and eighteenth centuries, with some years much worse than others. Full Little Ice Age cold set in around 1350, intensifying in the following two centuries. Despite generally very dry winters, north winds could raise enough moisture to produce rain and snow, especially in central and south China. Even in the north, flooding could result. In the southeast, winds that started dry in the far north can pick up enough moisture on the way south to produce dismal drizzling rains that stress people and animals severely. In Huizhou, Anhui, where almost miraculously good preservation of Ming records allows detailed year-by-year and even season-by-season analysis, the late fifteenth century was cold and dry, the sixteenth even colder, but wetter (McDermott 2013). In this mountainous region, that meant a great deal of snow.

Epidemics followed cold, drought and famine. These were probably the normal epidemics of old, especially smallpox, influenza, typhoid and its relatives, dysentery, and measles. Bubonic plague, often alleged, is improbable (Brook 2010 64–68; cf. Benedict 1996; Buell 2012).

Floods were commoner in the earlier half of that century (McDermott 2013: 285–289), and especially in the middle 1400s, when extreme drought in the northwest alternated with floods in the center and east, including outbreaks of the Yellow River (Twitchett and Grimm 1988: 310–312)—a classic Little Ice Age pattern. This led to renewed efforts to control the Yellow River and replace hazardous and uncertain river and sea traffic with canals. The Grand Canal was restored, improved, made far more impressive, and driven (via complicated locks) over high hills (Mote 1999: 646–653). It became China's great artery, and a wonder to the whole world when China was opened to wider contacts.

Ming achieved fruition (literally) of agricultural improvements begun in Song, including the coming of high-yield rices, improved beans, and better sugarcane varieties. Probably far more important, but only beginning to affect East Asia by

1600, were the New World food crops. The Portuguese and Spanish introduced these in the late 1500s. Better cultivation techniques and water management also improved livelihoods (Anderson 2014; Li 2003). Superb studies by Sucheta Mazumdar (1998, 1999) reveal how new crops changed China, and how China managed to grow sugar without the usual plantations (complete with slavery and abuse) by strongly favoring individual small farmers. This reached the point of mounting mills on boats so they could go from farm to farm, thus avoiding the situation in the New World, where only the plantation owners could afford the necessary machinery. Ming had an amazingly long run of peace and relative prosperity. Ming did not reach a Malthusian limit. The country could feed itself easily, despite local shortages, until about 1630.

If grain often failed, people could turn to forestry on the mountain slopes. The trials of managing forests take up much of Joseph McDermott's meticulous work *The Making of a New Rural Order in South China* (2013). The villagers of Huizhou produced China fir (*Cunninghamia lanceolata*) and bamboo. China practiced scientific forestry long before Germany. Huizhou lineages managed seeding, transplanting, clearcutting to plant even-aged stands, and other management devices, as well as constantly fighting tree theft.

8.2 The Rise of Ming

The only successful mass movement led by a lower-class person in imperial Chinese history was the anti-Mongol rebellion that produced the Ming Dynasty (1368–1644). Zhu Yuanzhang, the leader in question, was the child of poor farmers in Anhui, central China (see Dreyer 1988; Mote 1988a, b, 1999: 549–582). He rose with the mystical and religious movements of the time, taking part in—but later repudiating—the Red Turbans, a mystical millennial sect budding from the White Lotus movement, which had Manichaean roots.

Zhu Yuanzhang rose through a career of wandering, fortune-telling, Daoist magic, shady operations, and unpredictability. The Red Turbans' wild behavior included extremism, incitement to riot, apparently ethnicized anti-Mongol hate, and violence (Mote 1988b, 1999: 558).

Zhu Yuanzhang was a careful and brilliant war leader, planning successfully how to take on formidable enemies. After winning the empire, he showed great care for rural life, and exemplary responsibility by such things as fasting and sleeping in the open—in the hot sun—during a drought; this was generally believed to be effective in getting Heaven to take mercy on the sun-scorched realm. He also fasted rigorously to insure safety and well-being in the realm (Langlois 1988: 122–125). He restored and nurtured an imperial academy and benefited education (Mote 1999: 572). Through the 1370s he seems to have been a reasonably caring, hard-working ruler. He was sharp enough to see through a con job when flatterers presented two melons on one stalk—an auspicious omen, but faked in this case—and write an insightful memo about it (Schneewind 2006). He moved the capital from Nanjing to Beijing in

1380, and focused his attention on the northwest. This was to have fateful consequences.

He was to become "scornful, even bitterly resentful, of those who had known only the refined, comfortable life of the elite" (Mote 1988b: 49). He succeeded only when he got sufficient momentum to attract a vast mass of disaffected elites, few of whom cared for the rotting Mongol leadership. However, Zhu Yuanzhang's behavior as ruler was increasingly erratic and violent (Mote 1999: 576–582), until he declined into frank paranoid schizophrenia. The wise historian normally resists armchair psycho-diagnosis, but there is simply no other way to make any sense out of Zhu's later career. Even the cautious historian John Langlois, not given to loose verbiage, called him "paranoid" (Langlois 1988: 179).

After 1380, he turned murderous, killing perhaps 40,000 people in one purge alone. It was occasioned by an unproven coup allegation. He repented and issued an amnesty a few years later, and then seems to have been erratically virtuous or violent in the 1380s. He ordered a scholar who mildly criticized him to be beaten to death with bare hands on the palace steps. Thereafter he declined into increasingly irrational violence, slowly ridding the world of all his old friends and associates (Brook 2014; Langlois 1988: 140–181).

He left 26 sons and 16 daughters. Through the Ming rules he created, the imperial family members were not to work, but to be supported by the state. They eventually reached over 100,000 people, creating an enormous drain on the economy (Mote 1999: 565–566) as well as a target for rebel propaganda. Peter Turchin's "elite overproduction" was never more literal.

Many of the later emperors of Ming acted in a manner that suggests familial schizophrenia. The Wan-li emperor (r. 1573–1620) went on strike and refused to sign edicts for decades, thus paralyzing the government. The Tianqi emperor (r. 1620–1627), who was "perhaps mentally deficient," spent all his time woodworking instead of attending to governance (Atwell 1988: 595). Several other Ming rulers had quirks of this kind.

Ming also displays a long history of palace crises. Thanks to Zhu's demented behavior, Ming collapsed by 1400, being run by a young grandson of Zhu who was a meek and scholarly youngster, unable to manage the empire. Zhu Di, perhaps Zhu Yuanzhang's most military son, intervened, taking over through a civil war in 1399–1402. He may have been the son of a Mongol consort (Langlois 1988: 164) or a Korean court woman (Mote 1999: 594). He had been a leader in conquering and chasing the Mongols in the northwest, and like many another Chinese dynast he used the northwest as a power base from which to sweep down and conquer the rest of the country. He ruled as the Yongle Emperor from 1402 to 1424. He too seems to have tended toward paranoia, killing on a scale like his father's.

It was under him that Ming sent out the seven famous voyages that explored the entire Indian Ocean and traded with Africa. (Neither they nor any other early Chinese voyagers reached North America, *contra* an urban legend without supporting evidence.) The Muslim eunuch Zheng He commanded these. The ships were enormous. "The largest of the ships...were as large as 440 feet in length and carried nine masts...with 2500 tons cargo capacity and 3100 tons displacement; in comparison," Columbus' largest ship "was 125 feet long and had a capacity of only 280 tons" (Mote 1999: 614). Zheng He could almost have loaded Columbus' ships as lifeboats. There were 62 of these giant "treasure ships" in his first fleet, with 255 smaller ones. The total manpower was in the vicinity of 20,000–32,000.

The fleets brought back everything from giraffes to gems. They did not enrich China, but they secured Ming dominance over much of coastal Southeast Asia, and led to profitable trade. More and more Chinese settled in the "Nanyang" ("Southern Ocean," i.e. Southeast Asia), establishing colonies at Melaka, Palembang, and elsewhere. This failed to incorporate southeast Asia in the East Asian world-system or to inject Ming into the Indian Ocean one. However, it increased the visibility of the China trade in the eyes of the western and south Asian realms (Tan 2009). The Portuguese looked and plotted.

This could have been the moment when all Eurasia and even East Africa were incorporated in one world-system. The time was premature. After his time, the voyages were abandoned; they were expensive and unprofitable. China locked itself solidly into a west-facing, land-based strategy, and turned away from developing the burgeoning ocean trade. It continued to develop despite lack of government concern, but without equivalents of Portugal's Prince Henry the Navigator and his many successors in the west, it had little chance of leading East Asia into progress or of expanding the world-system westward. However, southeast Asia was all integrated into the East Asian world-system. The Indian Ocean became increasingly incorporated into the wider world, as trade from India, southeast Asia, and Indonesia led to progressive conquests by the Portuguese and then the Dutch (on the Indian Ocean in world-system history, see Beaujard 2009, 2012, 2017).

By this time the richest parts of Europe—the Low Countries and northern Italy— had pulled ahead of China, probably even ahead of the Yangzi Delta. By 1400 the GDP has been estimated at approximately $1200 for the Low Countries, $1600 for northern Italy, and $960 for China, in 1990 dollars. By 1500 the figures were $1600, $1400, and $1127 respectively. The figure for China was comparable with that for Europe in general (Campbell 2016: 378). Japan was probably making less. The Yangzi Delta was well ahead of most of China. Europe had benefited from the demographic release caused by the Black Death. West European rulers were fortunate and clever enough to use this wealth to expand commerce, exploration, and communication.

The Yongle Emperor's successors cut back and saved money. Many of the emperors after this time—and later in the dynasty—were an incompetent lot, "feckless young men whose brief lives tended to be dominated by their consorts, their mothers and grandmothers, and their eunuch servants" (Mote 1988a: 343). Emperor Yingzong rashly attempted military leadership in the field. He was captured by the Mongols under Esen in 1449. Eventually they gave him back, having no use for him. His brother Prince Cheng (the Jingtai Emperor) took over until Yingzong could be recovered in 1450 (Mote 1999: 626–628), presumably with substantial payment. Prince Cheng did not take kindly to being replaced, and Yingzong had to stage a coup to regain the throne.

This coincided with an especially cold, dry period, which weakened both the Mongols and the Chinese. Fortunately for the latter, it weakened the former more, another case of climate affecting world-system dynamics. Yingzong died at 37 and was succeeded by Xianzong, "a stolid, somewhat doltish man...easily misled by his consort, by his eunuchs, by charlatans claiming to have spiritual powers, and by innumerable sycophants" (Mote 1999: 630).

The Great Wall was finally built to keep the Mongols out, or to keep the Chinese in (according to another theory). The Wall was started in 1475, and not completed until well into the sixteenth century (Mote 1988b: 401, 696). This, of course, is the reason Marco Polo did not comment on it.

It seems incredible that the Ming Dynasty survived all this. It survived partly because Yingzong's successor, Emperor Xiaozong, was better. He has the distinction of having insisted on monogamy and made it stick; he adored his wife and "remained completely devoted to her" during his reign from 1487 to 1505, though she was said to be "foolish and demanding" (Mote 1988a: 353, 1999: 631). This had its costs; it annoyed elite families that wanted to marry their daughters to the emperor. Perhaps it shows more mental quirkiness than monogamous morality.

A long, slow dynastic decline set in. There followed some emperors who succeeded as teenagers, and acted as one would expect of not-very-gifted teenagers running an empire. The autocratic system established by Zhu Yuanzhang was deadly when this occurred. In earlier dynasties, courtiers and imperial relatives could have better controlled the youths. Frederick Mote points out that the emperors spent their early years in an "unnatural environment" and never "knew what might be called 'normal' family relationships" (Mote 1999: 727), let alone normal human society. Moreover, the "bureaucratic system...had become bogged down in procedures and forms" (p. 728). Factioneering reached usual end-of-dynasty levels; factions were labeled *dang*, a term equivalent to "party" in James Madison's negative sense.

There were some nasty small wars in between, but not enough to slow the growth of population and prosperity. The steady and dramatic rise of population (probably around 192 million in 1630, perhaps many more; there are no accurate figures) was accommodated by agricultural intensification, migration to the south and southwest, and the Malthusian horsemen of famine and plague. Intensification was dramatic, but innovation was not; Ming merely developed, consolidated, and built on earlier advances (Anderson 2016).

The one real innovation was adoption of New World crops. These slotted so easily into traditional agronomic niches that they were not even given new names; they were simply called "foreign *x*," *x* being whatever crop they most resembled. (The actual word was *fan*, a term for southern non-Chinese, including the Portuguese and Spanish who brought these crops.) Tomatoes were foreign eggplants, maize was foreign millet, sweet potatoes were foreign (or golden) yams, pineapples were foreign jackfruit, papayas were foreign quinces.

After the mid-1500s, "imperial incompetence triumphed over the best efforts of many dedicated public servants to bring about lasting reversal of the degenerating patterns" (Mote 1999: 723). The self-sequestered Wanli Emperor, who spent most of his 48-year reign "on strike" and refusing to do even the most routine actions, reduced the dynasty to the edge of collapse (Brook 2010: 100–105; Huang 1982, 1988; Mote 1999:723–742). His reign was also characterized by some of the worst weather in Ming (Brook 2010: 241–244; Brook 2016).

He was followed by the mentally challenged Tianqi Emperor, all of whose children died very young—an all too typical demographic story in old China. A brother thus succeeded: the boy Chungzhen Emperor (r. 1627–1644), who was apparently decent enough but completely unable to stem what had become a full-scale collapse. Corrupt eunuchs and secretaries dominated court and ruined governance.

The society of late Ming is described in exquisite detail in 120 brilliant and detailed stories by Feng Menglong (1574–1646; Feng 2000, 2005, 2009). Some are old tales reworked, but all are set in late Ming environments. They reveal a society with tremendous trust in the ultimate justice of government, but sharp awareness of the corruption, abuse of law, and banditry that characterized the times. Feng was a wise and skeptical observer, convinced of the innate goodness of most people but aware of the badness of many. He portrays a society where people could count on universal awareness of, if imperfect following of, broadly Confucian virtues of loyalty, mutual support, and charity. Above all, it was a world of family solidarity, but solidarity all too often at the expense of women and children.

Feng anticipated early Qing writers in portraying clever, independent women struggling within a horribly patriarchal order; sometimes they won, often they lost. Similarly, ordinary people struggled against a wider system that was sometimes protective, often oppressive. People could be incredibly generous and charitable under harsh circumstances; this would seem romanticizing were it not for Joanna Handlin Smith's meticulous study of the strikingly high level of charity in Ming (Handlin Smith 2009). Feng's combination of sharp awareness and brilliant literary skill makes these stories particularly valuable as source material. He shines a brilliant series of lights on the process of decline in a great dynasty.

Ming lasted a long time by sheer institutional inertia. The Ming Dynasty's ruling family declined, but "Great Ming" went on, administrators did their job, and only slowly did system errors build up to an intolerable level (see Huang 1982, 1988; Mote 1999). The sixteenth-century Portuguese explorers found a stunning loyalty to "Ta Me." (They transcribed the Hokkien pronunciation of Da Ming, "Great Ming," with a strongly nasalized *e*. See accounts in Boxer 1969, 2004). Since the Portuguese were generally critical of Asian regimes, this is telling testimony. They settled at Macau, leasing it from Ming in 1557. The settlement is there still, reverting to full Chinese control in 1999 but maintaining some cultural independence. The Portuguese-Cantonese creole developed there, never widely spoken, is now almost extinct.

The bureaucracy could operate without a functional emperor. Not only was the system self-maintained through loyalty and expedience; also, an ambitious and amoral contender always found it easier to go for power within the system than to go against it. With Mongol insurgencies, Muslim activity, millenarian movements (pursuant to the one that started Ming in the first place), and ethnic battles in the south, Ming was not free from strife, but before 1630 there was no enemy big enough to compete with the Ming governmental machine.

The countryside was so peaceful that little in the way of policing was found. "European travelers remarked on the mannerliness, good humor, and social graces of the common people" (Mote 1999: 759–760), as well as on the peace, good order, and law-abidingness of the citizenry in general. Mote quotes Matteo Ricci as praising " 'a gaiety of spirit among the people, who are well-mannered and nicely spoken' " (Mote 1999: 763). Portuguese found even the jails less awful than Portuguese jails. Society and economy resembled those of Europe at the time. Speculations on Ming "sprouts of capitalism" have occurred, but Ming was not really a proto-capitalist society (Mote 1999: 769); it lacked several of the necessary institutions, notably those for security of private investment. The tributary mode of production, perfected by this time, gave ultimate ownership or control to the bureaucratic state. Landlords needed to place members of the family in government service to maintain security of tenure. A mix of rentier landlords, entrepreneurial businesspeople, and all-powerful government bureaucrats ran the system, and the wise family played all three of these games, as memorably recounted in China's great novel *The Story of the Stone* (written in Qing but set in Ming; Cao 1973–1986).

The high level of literacy and culture was protective. Late Ming was a Golden Age for prose fiction, drama, and painting, but also for science, such as botany and technology. It was one of those odd cultural golden ages that occurred during an otherwise dismal time, rather than during a prosperous, upbeat time like the Golden Ages of mid-Tang and Northern Song. Ming fits Turchin and Nefedov's theory that in declining times elites will seek out literacy to get into government service (Turchin and Nefedov 2009, e.g. p. 64). That is exactly what happened in Ming and later in Qing. The government was no longer expanding or hiring more scholars, but elite families multiplied, education increased. Recall that the imperial family eventually included over 100,000 persons. (Ming emperors would occasionally deal with this by reducing imperial remote cousins to commoner status by fiat.) Eunuchs increased "from 100 eunuch retainers under Zhu Yuanzhang to 10,000 by the end of the fifteenth century and 100,000 by the end of the dynasty" (Smith 2003: 101).

8.3 The Final Stroke

When Ming finally did fall, it fell during a period that was cold and dry even by Little Ice Age standards. Population—already fairly high by 1400—peaked around 1630 around 192 million or more, but then fell by perhaps 40 million, or even more, by 1644, before the wars that finally ended Ming (von Glahn 2015: 311; see also Mote 1999: 743–745). Droughts caused horrific epidemics and famines, with rumored cannibalism, in the 1630s and early 1640s (Atwell 1988: 621–632).

The conquering Manchus who created the Qing Dynasty held a very small semiperipheral marcher state from a bitterly cold part of northeast China— "Manchuria." There remains some mystery as to how they could so successfully organize during what may have been the coldest period in recorded history. As Fan

(2010) points out, the Manchus should have been more weakened by cold than the Ming. Part of the reason is that the earlier Ming dynasts had nurtured them, thinking them a valuable counter to the Mongols and the rising power of Korea. Still, they could not have prevailed unless Ming's government had collapsed.

Ming after 1600 suffered the interaction of both a climatic trough and a dynastic deterioration. Collapse might have happened from the former; it certainly would have come from the latter. Between 1630 and the triumph of Qing in 1644, and then throughout the rebellions ending in 1690, perhaps a hundred million people died of violence or famine. Census figures are shaky; toward the end of a dynasty, census figures tended to mask long-range declines by reporting that the number of taxpayers was the same as ever. In late Ming, some districts reported exactly the same population for decades. By contrast, the beginning of a dynasty was generally a time for accurate counts. A long decline could seem, artifactually, to be compressed into a very few years.

Mote (1999: 781–786, 800–803) does not see the fall as inevitable or due to governmental moral and institutional decline. He may be correct in saying that "intelligent and forceful management" would have saved the country, and "the Ming collapse...[was] not brought about by any general disintegration of government and society.... Those fatal circumstances were brought about carelessly, by an administration that simply could no longer manage its resources, utilize its strengths, and maintain its focus" (Mote 1999: 784, 802). However, that was probably structurally impossible. It would have required a superhuman level of competence.

Unlike other dynasties, Ming did not have regular crises at rough Ibn Khaldun intervals (60–70 years); the only major crises were the coup and civil war of 1399–1402 and the Mongol capture of the emperor in 1449. The latter does fit with Ibn Khaldun cycling, coming 81 years after the start of the dynasty. After that, minor rebellions occurred, but not even the ineptness or shirking of the emperors could provoke a major revolt or crisis. The 276-year survival of Ming, in spite of its emperors and the Little Ice Age, remains hard to explain.

Bandits multiplied and coalesced into greater and greater armies. These were not true millenarian rebellions, but millenarian rhetoric was abundant among their ranks (Parsons 1970). James Tong (1991), writing on Ming bandits, noted that people with a lot to lose are less likely to rebel than people with nothing to lose, but eventually most people simply decided that their fortunes were better—or at least more hopeful—with almost any rebel than with the Ming. The bandit Li Zicheng conquered most of China, only to fall to the Manchus with whom he had foolishly allied. Beijing fell in 1644. The Chungzhen Emperor hanged himself on a tree in the Forbidden City gardens.

8.4   Loyalty After the Fall

Incredible loyalty to Ming continued after its fall (Struve 1988). Desperate rearguard action kept much of the south in Ming hands. The last emperor, or pretender, took refuge in Burma, where he was hunted down by Wu Sangui and his army, captured, and killed in 1662 (Brook 2010: 255–256).

A more successful and formidable resistance was led by the brilliant naval commander Zheng Chenggong, but he declined, fell ill, and died, also in 1662 (see Mote 1999: 833–835). His loyalty caused a last Ming ruler to allow him to use the Imperial surname Zhu, at least in some contexts, so he was nicknamed Coxinga—Hokkien *kok seng a*, "National Surname Kid." (*A* is a slangy Hokkien diminutive, exactly equivalent to "kid" in English. The Mandarin equivalent of *kok seng a* is *guoxingzi*.) This is the name by which he became known in the west after forcing the Dutch out of Taiwan and taking it over.

Even when he died and Taiwan was regained by China, the amazing strength of Ming was not run out. Ming loyalism reached almost fanatical levels (see e.g. Mote 1999: 850–851). A huge, extremely widespread and powerful resistance movement, often covertly led at first by descendants of the Zhu family, kept on throughout the entire Qing dynasty. Its motto, "overthrow the Qing and restore the Ming," was still being chanted by the rebels that drove out Qing in 1911. Hand signals with the same meaning (pointing down and then up in a specific way) were still current as underground signs on the Hong Kong waterfront in the 1960s (personal observation). Few things in history are more surprising than this undying loyalty to a hapless, star-crossed regime.

8.5   Disasters in and After Ming

A grand summary of China's sorrows and their antecedents has recently been provided by Harry Lee, of the Lee-Zhang team noted above. He analyzed "5368 natural disasters" including "1478 famines, 5700 epidemics, 456 nomadic invasions, and 1315 internal wars in the agricultural region of China in AD 1470–1911" (Lee 2018: 1079). The internal wars tended to follow closely after epidemics in the wheat region, after famines in the rice region. The rice region is less prone to famine, so more aware of disruption and trouble when famine occurs. The epidemics and famines tend to follow natural climatic changes, but not very closely, with imperfect coupling. He notes that "the effect of climate deterioration is translated into human crises via climate-driven economic downturn and its associated pathways" (Lee 2018: 1073), causing a gap that may be large. He continues: "This may make some observers suspicious about the causal relationship" (Lee 2018: 1073). Lee has moved from an earlier, more determinist position to this one, closer to my own but still attaching more importance than I do to the climate. Wars come when people feel they have more to fear from the system and its breakdown than from violence, or, put another way, when they feel they can get more from going for themselves against the system than from following the system's rules. Clearly, the catastrophes of famine and disease make such feelings more likely, but so does general bad governance, especially corruption and banditry. All are linked; climate, epidemics, and local violence can all feed back on each other, cause famine, and lead to social breakdown.

8.6 The East Asian World-System in Ming Times

David Kang et al. (2016) have counted up the wars in Ming and pre-1841 Qing, finding a total of 822, the vast majority of which were western and northern border skirmishes or internal conflict rather than real wars. The major wars in the whole period were largely Qing's conquests of the west (Perdue 2005). Ming-Mongol war was also important. Attempted conquest or interference in Korea and Vietnam occurred, but involved a total of very few incidents.

These authors suggest that the states dominated by Chinese cultural norms—China, Korea, Vietnam and Japan—had come to an accommodation that preserved peace. A cynic would surely respond that it was more an issue of cost than of Confucian world order. China had all too much experience with the difficulty of conquering and holding Vietnam and Korea, and Ming experiences there were even worse than those of previous dynasties. Japan-China warfare was unthinkable to both parties. Japan-Korea war was more a matter of equals, and was savage and uncompromising during Ming times. Japan's thriving and burgeoning pirate economy (Shapinsky 2014), however, spilled over into China, where the *wako* (Japanese pirates) harassed the China coasts mercilessly, followed by many more Chinese adventurers.

Vietnam constantly rebelled or stonewalled to maintain as much independence as possible. Unlike the major Thai-speaking regions of south China, the Vietnamese world stoutly resisted cultural assimilation. Vietnam had begun to suffer from climate change during the late Mongol period, but its fall from prosperity is better recorded under Ming. By 1320 the Little Ice Age had forced the Intertropical Convergence Zone back to the south, and dry weather followed. From 1340 into the 1360s, the worst drought in Vietnam's history led to widespread disaster (Kiernan 2017: 177). Another burst of Little Ice Age difficulty occurred in the early 1500s, when extremely wet and extremely dry years followed in quick succession (Kiernan 2017: 213). Victor Lieberman (2003, 2009) stresses the serious effects of these climatic problems on the Vietnamese state.

This weakened the Tran dynasty, which fell to a coup by the chief minister, Ho Quy Ly, in 1395. He officially ended the Tran in 1400. China took this as an excuse to invade and conquer Vietnam in 1407, "on the pretext of restoring the Tran dynasty" (Kiernan 2017: 194; see also Lieberman 2003: 377). Ming rule lasted a mere 21 years. Resistance succeeded yet again, as the Vietnamese under Le Loi drove the Ming forces from the country. Le explained his success where Ho Quy Ly had failed: "Ho had a million soldiers, but they were torn by a million different opinions; my men are only a few hundred thousand, but they all fight with one mind"

(Kiernan 2017: 196). One can be sure that Ho Chi Minh and Vo Nguyen Giap knew that line.

Firearms, relatively new as a significant force, figured largely in this war. Champa, with few firearms, inevitably suffered from Vietnam's new power. Continual war led to annexation of northern and central Champa by Vietnam in 1470. Increasingly brutal repression of the conquered Cham people led to full-scale genocide in the conquered provinces in 1505–1509, under the "devil king" Le Uy Muc who had murdered "his grandmother and two ministers...ushering in an era of instability" (Kiernan 2017: 212). Actual genocide—governmental mass murder of conquered subjects whose only crime was their ethnicity—is exceedingly rare in East Asian annals.

The Le dynasty by this time was weak and vacillating, thanks in large measure to the "devil king," and the fatal omega stage of the cycle had been reached: more and more individuals saw their best option as serving themselves and their faction rather than the imperial state. Mac Dang Dung (1483–1541) rose to power as a faithful servant of the Le, but then took advantage of a fraternal conflict in the Le family. Mac seized power in 1527, proclaiming a new dynasty. The Ming rulers became aware of this and threatened another invasion; Mac prudently submitted, and Vietnam dropped back to tributary status.

The Le rallied (no doubt holding up Mac's caving in as a bad precedent) and fought a long and largely inconclusive war with the Mac dynasts. The Le succeeded in 1600, but by then the real power devolved into the Trinh family in the north and the Nguyen in the south. These "contending northern and southern rulers...still recognized the powerless Le emperors as the legitimate symbolic authority.... The concept of a united Viet kingdom persisted"; Kiernan explains the chaos and disunion of the period as following in part from "the new and extreme climate fluctuations...with alternating droughts and floods.... Droughts dried out, cracked, and weakened earthen dikes and critical embankments, and then in flood years, fast-flowing water broke through and swept them away" (Kiernan 2017: 221).

8.7 Korea Changes Dynasty

A major knock-on effect of Ming victory was a change of dynasty in Korea. Korea had become a part of the Mongol empire. With the fall of Yuan, Ming invaded. Korea fought hard for the Mongols, then for independence. The Koryŏ Dynasty (938–1392) was replaced via military coup by Yi Sŏng-gye (1335–1408), who initiated the Yi dynasty and renamed the country Chosŏn (or Joseon; 1392–1910). Yi, as general of the Koryŏ army, had been ordered to attack Ming, which he saw was foolhardy, so he simply turned around and took over Korea instead (Seth 2011: 117). The transition was remarkably peaceful, with little of the bloodshed that attended Chinese dynastic changes. Basically, the government simply went on, except that the military ruler had taken over the kingship instead of leaving a puppet king on the throne, as in latter-day Koryŏ. Climate played no role in this; it was strictly military politics, in the wake of Koryŏ's humiliation by the Mongols and then by Ming. "Yi's revolt against the Koryŏ court was achieved by his own military skill but still more through the determination of a group of scholars to assert the authority of Confucianism over the Buddhism and shamanism which had dominated the Koryŏ court" (Pratt and Rutt 1999: 63). Since this change of dynasty was a direct result of the fall of Yuan and rise of Ming, it cannot be explained by climate, except in the indirect sense that the coming of the Little Ice Age did contribute to the collapse of Yuan.

Under Yi's descendant King Sejong (1418–1450), the Chiphyonjon Academy created the Han'gul script, commissioned in 1446. It was and remains a thoroughly scientifically designed script, the most phonetically and phonemically sophisticated script in ordinary use in the world (see Lee and Ramsey 2011). It traces back to the script created for the Mongols by Phagspa. He and his Korean successors were brilliant linguists far ahead of their time. Sejong also promoted Chinese medicine. A work of 264 volumes was completed, and had to be summarized for actual use (Han 1970: 292). Even music progressed and was updated; a standard work for court music came out under the later King Songjong (Han 1970: 294).

The Yi was a highly Confucian regime that imitated China faithfully; in fact, it was more Confucian than China ever was. In early Yi times, Neo-Confucian philosophy replaced Buddhism as the foundation of Korean society. It split into several branches, ranging from extremely conservative (like so much of Chinese Neo-Confucianism) to highly practical and quite liberal in the modern sense; this latter trend later gave rise to the Sirhak ("practical learning") movement. In the meantime, Christianity came to Korea at the very end of the Ming Dynasty.

Koreans continued to travel to China; one Ch'oe Pu, stranded there in the 1480s, left a good account (Ch'oe 1965). Some measure of how Sinicized Korea became is the abundance of Chinese poetic images in Korean poetry. Zhuangzi had used seagulls as symbols of free and independent mind, and a walking-stick made from a goosefoot (pigweed) stalk as a symbol of the ultimate in rustic simplicity. These caught on and became clichés in Chinese poetry. From there they passed into Korean verse, showing that even the most casual or arcane Chinese references would inevitably be picked up by Korean literati. Clichés can be excellent indicators of world-system linkages.

A man that can stand for much of Korean culture at this time was Chŏng Ch'ŏl (1536–1593), statesman, poet, military commander, and moral thinker. A strict Confucian dedicated to loyalty above all, he wrote impassioned poetry in devotion to the dynasty. He was also capable of savage and bitter satire and irony, attacking the faction-ridden court with merciless intensity. Other poems speak of rustic retreat and escape into the world of song and art (see e.g. poems in O'Rourke 2002: 65–75, complete with seagulls, and a goosefoot staff on p. 71). The existence of men like Chong goes far to explain the long career of the Yi dynasty.

Korea found itself allied with Ming against Japan; Ming was happy to let Korea be a peaceful tributary state, drawing on Ming for alliance and support when necessary, trading otherwise. Korea became more and more heavily influenced by Confucianism.

In 1592, Toyotomi Hideyoshi, who had just succeeded in conquering and uniting Japan, invaded Korea with 150,000 troops (Elisonas 1991; Han 1970: 270; Naohiro 1991; Seth 2011: 146–149). The Koreans organized defense, most famously the ironclad ships of Admiral Yi Sunshi (1545–1598). These were a technological marvel—far ahead of their time—and Admiral Yi deployed them brilliantly, destroying the Japanese fleet. By this time the Ming Dynasty had become interested, seeing a tributary attacked, and Ming Chinese armies entered Korea in 1593.

Peace negotiations failed to satisfy Hideyoshi. War broke out again in 1597. Ming and Korea drove Japan out in 1598. Ming troops shored up the Korean state, and the intrepid Yi Sun-sin cut Japan's navy to pieces. He was killed at the end by a Japanese bullet. His memorials, including full accounts of his battles, have been preserved, and make some of the most gripping reading in military history (Yi 1981).

By this time, much of Korea was a smoking ruin. Hideyoshi burned, looted, and destroyed wherever he could. Among other things, he abducted vast numbers of Korean artists and craftsmen, to the great and lasting benefit of Japanese industries. The Yi Dynasty remained weak and faction-ridden. Korean and Japanese mutual hatreds were virulent from then on (they are far from reconciled today). Both Ming and the Toyotomi regime of Japan were greatly weakened, leaving the way open for rebellion and collapse in the former and Tokugawa Ieyasu's takeover in the latter.

Then only a few decades later, the Manchus invaded Korea and China. Korea fought to save itself and to be loyal to Ming, who had so recently saved the Korean state, but the Manchus were unstoppable, and eventually Korea became a tributary of Qing. The Yi Dynasty survived, battered but proved in trials by fire. It remained nominally tributary to Qing until the Japanese moved toward occupying it in the late 1800s. Korea became a Japanese colony from 1910 to 1945.

8.8 Japan Collapses into War

In Japan, the Kamakura bakufu declined slowly and painfully (Susumu 1990). This appears to have had little to do with climate (Susumu 1990 does not even mention it) and everything to do with the back-and-forth rivalry between the imperial family in Kyoto and the bakufu in Kamakura. Finally, the highly effective emperor Godaigo managed to recruit enough force to defeat the latter. One of his military leaders, Ashikaga Takauji, turned against him and eventually won a long back-and-forth war, crushing both Kamakura and Godaigo. The Muromachi bakufu, ruled by the Ashikaga family (a branch of the Minamoto) from Kyoto, began in 1336, but was not really consolidated for several years (Hall 1990).

The old power base in Kamakura remained refractory. Drought and famine (especially in 1457–60) accompanied the Little Ice Age (Hall 1990: 228). The Ōnin War (1467–1476) devastated the country, and left disorder. The shogun of the period, Yoshimasa, seems particularly ineffectual (Hall 1990: 228–229).

The Muromachi regime slowly fell apart, weakened by military challenges. Climate was a constant; the Little Ice Age went on, not impacting southern Japan very seriously. Weak leadership and the remaining power of the emperor had more to do with Muromachi fall than distance from the masses or corruption. Military overextension included confrontation with Korea and China, and troubles with great families who had independent power bases. Elite overproduction occurred as the Ashikaga and rival families proliferated and branched.

Its fall led to a "Warring States period"—the Japanese borrowed the old Chinese term. At this time, Portuguese came to Japan, beginning in 1543, bringing Jesuit missionaries (Elisonas 1991). The unifier of Japan, Toyotomi Hideyoshi, and the Tokugawa shogunate founder Ieyasu eventually concluded (with considerable evidence) that the Portuguese were planning to take over, or at least attempt to take over, some of the country. They knew enough of what the Portuguese had done in India, Indonesia, China, Malaysia, and elsewhere. Christianity was banned and the missionaries expelled by Hideyoshi in 1587, but the ban was not enforced until after 1569, when the Franciscans appeared and entered into rivalry with the Jesuits. Their competition, and military and other activities of unruly Christian lords, alienated the power structure. Further troubles led to complete ban and suppression by Ieyasu from 1614 onward (Elisonas 1991). Western contacts were confined to Dutch traders on a small island, Deshima, within the city of Nagasaki.

Significantly, Japan did not adopt most of western science or medicine, but did enthusiastically adopt harquebuses. The Japanese quickly learned to make them, and used them to advantage in the unification wars. It is characteristic of human history everywhere that implements of destruction attract far more attention and imitation than instruments of peace. The Japanese were trading widely by this time, and Chinese merchants in Japan were managing their own trade. Japan followed China even as far as southeast Asia (Batten 2003: 190). Japanese trade in Song, Yuan, and Ming became a major part of East Asian commerce. Japan remained to a significant extent a semiperipheral country. Japan exported raw materials—including, importantly, gold and silver—and simple processed goods, and imported skilled manufactures (Keiji 1990; Shōji 1990). Diplomacy and learning about Buddhist developments continued to be the most visible links, but trade may have been more significant to the economy and people.

The Warring States period wound down with the conquests of Oda Nobunaga (1534–1582), Toyotomi Hideyoshi (1537–1598), and Tokugawa Ieyasu (1542–1616). Nobunaga used to full advantage the guns he had learned to make from the Portuguese. The nobility of Japan was decimated by low-born soldiers with guns, just as it was in Europe. This was not his only foray into modernization; he tried hard to create a more rational order, and pursued "the introduction of Western medicine, the construction of castle towns filled with a new entrepreneurial spirit, the establishment of free markets," reduction of tolls, and other ideas anticipating Europe's Enlightenment (Naohiro 1991: 44). He also succeeded in the very important step of monetizing the economy, producing gold and silver coins (Naohiro 1991: 61–62); recall that Japan had depended on importing Chinese coins.

Nobunaga was assassinated by a follower, leaving the final conquest to Hideyoshi, who united Japan in 1588. He extended control from the Ainu of Hokkaido to the Ryukyus in the far south, and brought the Portuguese missionaries and sailors under control. Hideyoshi seems to have been one of those genuinely psychopathic killers that populate history. He had a strange cultured side: he made himself hated by the local lords by taking their beautiful ornamental stones and other treasures for his own palaces. Ornamental stones are enormously important, even spiritual, in East Asian culture, and this seemed an unnecessary insult.

More sensibly, he disarmed the local farmers, continuing Nobunaga's policy. The farmers were told that their swords and guns were made into clamps and nails for Buddhist temples, thus winning better rebirth for the deprived farmers (Naohiro 1991: 49–50). Firearms caused horrific damage in the civil wars, and the Tokugawas effectively controlled them, bringing the sword back as the samurai weapon (Perrin 1988). At least according to some stories, this involved convincing the samurai that firearms were a coward's weapon and a real man would fight with the sword...meanwhile making sure the Tokugawa law and order forces had plenty of firearms.

He also suppressed piracy, chronic scourge of the local seas. His next step was to continue a cadastral survey, following Nobunaga's beginning. This was no mere survey; in the process he broke the daimyos' power by reassigning their land bases, and reduced the farmers to peasants, bound to the land, disarmed, and subject to samurai and lords. This ended the *ikki*—armed bands of farmers agitating for their rights and for justice. After Hideyoshi, the farmers had little hope. Tax rates ranged from 30 to 80% of crop. One saying had it that "farmers should be taxed at a rate that keeps them suspended between life and death" (Osamu 1991: 107). China often charged only 3% (though, admittedly, this was a nominal rate, and extortion pushed the real rates higher). Reduced in status and recourse, highly taxed by the bakufu and the daimyo, the farmers suffered.

As in other Confucian societies, merchants ranked even lower theoretically than farmers, but had considerable and increasing power thanks to wealth. Free market policies led to spectacular flourishing of trade and production in cities, especially Osaka, which became the economic capital.

Japan emerged in ruinous state. Tokugawa Ieyasu was a brilliant organizer and leader, a true Khaldunian founder. He took over the unification effort when Hideyoshi died, substantially united the country by 1600, and was declared shogun in 1603. He established peace and order. Under the Tokugawas, Confucianism and Shinto tended to grow at the expense of Buddhism, but refugees from the fall of Ming introduced Chinese Buddhist practice, revivifying Zen (Heine 2018: 148) and presumably other Chinese religious sects.

Relations with China continued, with "war, trade and piracy" (Elisonas 1991: 235) summing them up. Japanese pirates (Chinese *wokou*, Japanese *wakō*) had raked Ming continually, leading to a policy of withdrawing people from the coasts; this was to be repeated in Qing times. Pirates sought silk and cotton, and thread and cloth made from them, as well as metal goods, art objects, money, and slaves (Elisonas 1991: 260). Mercantile missions interested in the same commodities visited more honestly, but sometimes turned disruptive.

One might explain the vicissitudes of Japan from 1300 to 1603 as influenced by the Little Ice Age. Droughts and famines are reported and are what one would expect. However, most of the action took place in the warm, wet south, where the

Little Ice Age had minimal effect. Internal dynamics, especially the inherent instability of having two ruling regimes and two centers of power, created the instability. The troubled and erratic Ming Dynasty did not help the situation. In Japan, the immediate story was one of accidents of imperial and shogunate competence and maturity (there were the inevitable boy emperors), and jockeying for power by great lineages. Climate shows no significant role.

8.9 The Ryukyus

The Ryukyu kingdom deserves a special note. The incorporation of the Ryukyus into Japan is a good example of world-system dynamics. The islands were first an independent realm. They became a small nation locked into trade networks with bigger neighbors, then tributary to those neighbors. The Ming converted the Liuqiu (Ryukyu) kingdom into a tributary of China in 1372. (Akamine 2017: 100–126 provides a fascinating account of the tribute missions.) Ryukyu trade, already impressive, became extremely important. The Ryukyus were not a threat, and did not have restrictions on seafaring (as Ming did) or on trade with foreigners (as Japan later did). They traded actively all over southeast and East Asia, carrying every sort of goods. They became classic intermediaries (Shōji 1990: 442–444).

In 1609, Japan, or more accurately the Satsuma daimyoship, took over the Liuchiu kingdom based on Okinawa, turning it into the Ryukyu Island sector of Japan. The Ryukyu kingdom now had to pay tribute to both China and Japan (via Satsuma). It lost much trade with Ming shutdowns and retreats from the sea, and then lost the strong protection of China with the end of Ming. The islands became integrated into Japan, a fact which had to be tactfully concealed from the Chinese, though of course they realized the truth (on Ryukyu history, see Akamine 2017; Glacken 1955). Their distinctive banana-fiber cloth (from an inedible species of banana) became important in tribute.

Finally, they were victims of takeover and assimilation, between 1871 and 1884. They probably rose in wealth and stability through all this, but became progressively less politically and socially independent. They suffered a sad fate eventually; an integral part of Japan but treated as a sacrifice zone, they became a battle ground in 1945. A quarter of the population died, and the buildings of Okinawa were leveled. The gate to the old castle survived, alone among major buildings; it stood stark and alone on a barren hilltop. The castle has since been rebuilt as a memorial, and the gate symbolically integrated into the University of Okinawa.

Ryukyuan art and culture are distinctive, with a rather Korea-like quirkiness, shown for instance in the tendency of the households' protective lion-dog (*shīsā*) sculptures being set out on the open slopes of the roofs, as if they were real animals patrolling the housetops, rather than stiffly confined to the ridgepole as in China and Japan. Small lion-dogs are not confined to roofs; they are everywhere in Okinawa, usually in pairs. Okinawan stonework is unexcelled in the world, and stone walls sometimes serve as demon-blocking good luck structures rather than mere practical walling.

In short, the Ryukyus show the typical fate of a semiperipheral polity: maintaining some important cultural independence, but integrated more and more into the world-system, and thus acculturating more and more to the core states.

Today, the Ryukyuan language, a relative of Japanese that separated from the Japanese language about the time the Japanese Empire arose, is threatened with extinction. The Ryukyus continue to have the longest life expectancy of any place in the world, a fact they attribute to their diet, including its distinctive purple sweet potatoes as well as the abundant vegetables and sea foods.

8.10 Sidebar: Mazu the Empress of Heaven

Another insight into the East Asian world-system at its peak is provided by the spread of the cult of Mazu.

The Ryukyus borrowed from the south China coast the cult of Mazu Tianhou, Mazu the Empress of Heaven, the leading protective deity of south Chinese seafarers. She is said to have been a real girl who lived in the Song Dynasty. In the Ryukyus, she is said to have been named Lin Mo, to have had Daoist training, and to have come from Fujian (Akamine 2017: 103–105). She is honored as Bōsa in Ryukyuan, Maso in Japanese. Other stories are told about her in China; she is said to have been a fisherman's daughter in Fujian or Guangdong who saved her father and brothers from storms, and eventually lost her life trying to save others. She has taken on some of the iconography of Kuanyin, and is sometimes equated with her in the Ryukyus.

Mazu is in fact very widely revered in south coastal China and the Chinese diaspora. Temples to her suffered in the 1960s in China, but beautiful ones survive in Taiwan, Hong Kong, Macau, and the Nanyang (the Chinese-diaspora world of southeast Asia). Particularly venerable and beautiful temples to her exist in Hoi An and Cholon, the old Chinese merchant communities in Vietnam. Macau is named for the old temple there ("cau" from a local word for "temple"). Her festival day, the 23rd of the 3rd lunar month, is celebrated by millions of people in Hong Kong and Taiwan. Temple societies save money to hold lavish ceremonies on this day.

She is thus emblematic of the Chinese trading world, embedded in the East Asian world-system. As an internationally recognized patron of south Chinese fishermen, seafarers, merchants, and boat people, as well as local Japanese and Ryukyuan communities, she is a perfect image for the steady incorporation of wider areas in that world-system—perhaps for the East Asian world-system as a whole.

References

Akamine, M. (2017). *The Ryukyu Kingdom: Cornerstone of East Asia*. Honolulu, HI: University of Hawai'i Press. Trans. Lina Trerrell; Ed. Robert Huey.

Anderson, E. N. (2014). *Food and environment in early and medieval China*. Philadelphia, PA: University of Pennsylvania Press.

Anderson, E. N. (2016). Agriculture, population, and environment in late imperial China. In T.-j. Liu & J. Beattie (Eds.), *Environment, modernization and development in East Asia: Perspectives from environmental history* (pp. 31–58). Houndsmill, Basingstoke, UK: Palgrave MacMillan.

Atwell, W. (1988). The T'ai-ch'ang, T'ien-ch'i, and Ch'ung-chen Reigns, 1620-1644. In F. Mote & D. Twitchett (Eds.), *Cambridge history of China, vol. 7, The Ming dynasty, 1368-1644, Part 1* (pp. 585–640). Cambridge: Cambridge University Press.

Batten, B. (2003). *The end of Japan: Premodern frontiers, boundaries, and interactions*. Honolulu, HI: University of Hawai'i Press.

Beaujard, P. (2009). *Les mondes de l'océan Indien. Tome I: De la formation de l'État au premier système-monde Afro-Eurasien (4e millénaire av. J.-C. – 6e siècle apr. J.-C.)*. Paris: Armand Colin. 613 pp.

Beaujard, P. (2012). *Les mondes de l'Océan Indien. Tome II: L'océan Indien, au coeur des globalisations de l'ancien Monde du 7e au 15e siècle*. Paris: Armand Colin.

Beaujard, P. (2017). *Histoire et voyages des plantes cultivées à Madagascar avant le XVIe siècle*. Paris: Karthala.

Benedict, C. (1996). *Bubonic plague in nineteenth-century China*. Stanford, CA: Stanford University Press.

Boxer, C. R. (1969). *The Portuguese seaborne empire 1415-1825*. London: Hutchinson.

Boxer, C. R. (2004). *South China in the sixteenth century*. Bangkok: Orchid Press. Orig. 1953.

Brook, T. (2010). *The troubled empire: China in the Yuan and Ming*. Cambridge, MA: Harvard University Press.

Brook, T. (2014). Commerce: The Ming in the world. In C. Clunas & J. Harrison-Hall (Eds.), *Ming: 50 years that changed China* (pp. 252–291). London: British Museum.

Brook, T. (2016). Nine sloughs: Profiling the climate history of the Yuan and Ming dynasties, 1260-1644. *Journal of Chinese History, 1*, 27–58.

Brooke, J. L. (2014). *Climate change and the course of global history*. Cambridge: Cambridge University Press.

Buell, P. D. (2012). Qubilai and the rats. *Sudhoffs Archiv, 96*, 127–144.

Burnaby, F. (1877). *A ride to Khiva: Travels and adventures in Central Asia*. London: Cassell Petter and Galvin.

Campbell, B. M. S. (2016). *The great transition: Climate, disease and society in the late-medieval world*. Cambridge: Cambridge University Press.

Cao, X. (1973–1986). *The story of the stone* (Vols. 4–5 with Gao E). New York: Penguin. Translated by David Hawkes and John Minford (Chinese original, 18th century).

Ch'oe, P. (1965). *Ch'oe Pu's diary: A record of drifting across the sea*. Edited and translated by John Meskill (orig. 1488). Tucson, AZ: University of Arizona Press.

Dreyer, E. (1988). Military origins of Ming China. In F. W. Mote & D. Twitchett (Eds.), *The Cambridge history of China. Vol. 7, The Ming dynasty, 1368-1644, Part 1* (pp. 58–106). Cambridge: Cambridge University Press.

Elisonas, J. (1991). Christianity and the Daimyo. In J. W. Hall & J. L. McClain (Eds.), *The Cambridge history of Japan. Vol. 4: Early modern Japan* (pp. 301–372). Cambridge: Cambridge University Press.

Fan, K.-w. (2010). Climatic change and dynastic cycles in Chinese history: A review essay. *Climatic Change, 101*, 565–573.

Feng, M. (2000). *Stories old and new: A Ming dynasty collection*. Seattle, WA: University of Washington Press. Tr. Shuhui Yang and Yunqiu Yang (Chinese original ca. 1600).

Feng, M. (2005). *Stories to caution the world: A Ming dynasty collection*. Seattle, WA: University of Washington Press. Tr. Shuhui Yang and Yunqin Yang (Chinese orig. ca. 1600).

Feng, M. (2009). *Stories to awaken the world: A Ming dynasty collection*. Seattle, WA: University of Washington Press. Tr. Shuhui Yang and Yunqin Yang.

Glacken, C. (1955). *The Great Loochoo: A study of Okinawan village life*. Berkeley, CA: University of California Press.

Hall, J. W. (1990). The Muromachi Bakufu. In *The Cambridge history of Japan. Vol. 3, medieval Japan* (pp. 175–230). Cambridge: Cambridge University Press.
Han, W.-k. (1970). *The history of Korea*. Seoul: Eul-Yoo. Tr. Lee Kyung-shik.
Handlin Smith, J. (2009). *The art of doing good: Charity in late Ming China*. Berkeley, CA: University of California Press.
Heine, S. (2018b). *From Chinese Chan to Japanese Zen: A remarkable case of transmission and transformation*. New York: Oxford University Press.
Huang, R. (1982). *A year of no significance: The Ming dynasty in decline* (Vol. 1587). New Haven, CT: Yale University Press.
Huang, R. (1988). The Lung-ch'ing and Wan-li reigns, 1567-1620. In F. Mote & D. Twitchett (Eds.), *Cambridge history of China, vol. 7, The Ming dynasty, 1368-1644, Part 1* (pp. 511–584). Cambridge: Cambridge University Press.
Kang, D., Shaw, M., & Fu, R. T. -m. (2016). Measuring war in early modern East Asia, 1368-1841: Introducing Chinese and Korean language sources. *International Studies Quarterly, 60*, 766–777.
Keiji, N. (1990). The medieval peasant (S. Gay, Trans.). In *The Cambridge history of Japan* (Medieval Japan) (Vol. 3, pp. 303–343). Cambridge: Cambridge University Press.
Kiernan, B. (2017). *Viet Nam: A history from earliest times to the present*. New York: Oxford University Press.
Langlois, J. D., Jr. (1988). The Hung-Wu reign, 1368-1398. In F. W. Mote & D. Twitchett (Eds.), *The Cambridge history of China. Vol. 7, The Ming dynasty, 1368-1644. Part 1* (pp. 107–181). Cambridge: Cambridge University Press.
Lee, H. F. (2018). Internal wars in history: Triggered by natural disasters or by socio-ecological catastrophes? *The Holocene, 28*, 1071–1081.
Lee, K.-M., & Ramsey, S. R. (2011). *A history of the Korean language*. Cambridge: Cambridge University Press.
Li, B. (2003). Was there a 'fourteenth-century turning point?'. In P. J. Smith & R. von Glahn (Eds.), *The Song-Yuan-Ming transition in Chinese history* (pp. 135–175). Cambridge, MA: Harvard University Asia Center.
Lieberman, V. (2003). *Strange parallels: Southeast Asia in global context, c. 800-1830. Vol. 1: Integration on the mainland*. Cambridge: Cambridge University Press.
Lieberman, V. (2009). *Strange parallels: Southeast Asia in global context, c. 800-1830. Vol 2: Mainland mirrors: Europe, Japan, China, South Asia, and the Islands*. Cambridge: Cambridge University Press.
Mazumdar, S. (1998). *Sugar and society in China: Peasants, technology, and the world market*. Cambridge, MA: Harvard University Press.
Mazumdar, S. (1999). The impact of new world food crops on the diet and economy of China and India, 1600-1900. In R. Grew (Ed.), *Food in global history*. Boulder, CO: Westview.
McDermott, J. P. (2013). *The making of a new rural order in South China. I. Village, land, and lineage in Huizhou, 900-1600*. Cambridge: Cambridge University Press.
Mote, F. W. (1988a). The Ch'eng-hua and Hung-chih Reigns, 1465-1505. In F. W. Mote & D. Twitchett (Eds.), *The Cambridge history of China. Vol. 7, The Ming dynasty, 1368-1644, Part 1* (pp. 343–402). Cambridge: Cambridge University Press.
Mote, F. W. (1988b). The rise of the Ming dynasty, 1330-1367. In F. W. Mote & D. Twitchett (Eds.), *The Cambridge history of China. Vol. 7, The Ming dynasty, 1368-1644, Part 1* (pp. 11–57). Cambridge: Cambridge University Press.
Mote, F. W. (1999). *Imperial China 900-1800*. Cambridge, MA: Harvard University Press.
Naohiro, A. (1991). The sixteenth-century unification. In J. W. Hall & J. L. McClain (Eds.), *The Cambridge history of Japan. Vol. 4: Early modern Japan* (pp. 40–95). Cambridge: Cambridge University Press. Tr. Bernard Susser.
O'Rourke, K. (2002). *The book of Korean Shijo*. Cambridge, MA: Harvard University Press.

Osamu, W. (1991). The social and economic consequences of unification. In J. W. Hall & J. L. McClain (Eds.), *The Cambridge history of Japan. Vol. 4: Early modern Japan* (pp. 96–129). Cambridge: Cambridge University Press. Tr. James L. McClain.

Parsons, J. B. (1970). *The peasant rebellions of the late Ming dynasty*. Tucson, AZ: University of Arizona Press (for Association of Asian Studies.

Perdue, P. (2005). *China marches west: The Qing conquest of central Eurasia*. Cambridge, MA: Harvard University Press.

Perrin, N. (1988). *Giving up the gun: Japan's reversion to the sword, 1543-1879*. New York: David R. Godine.

Pratt, K., & Rutt, R. (1999). *Korea: A historical and cultural dictionary*. Richmond, England: Curzon.

Schneewind, S. (2006). *A tale of two melons: Emperor and subject in Ming China*. Indianapolis, IN: Hackett.

Seth, M. J. (2011). *A history of Korea, from antiquity to the present*. Lanham, MD: Rowman and Littlefield.

Shapinsky, P. D. (2014). *Lords of the sea: Pirates, violence, and commerce in late medieval Japan*. Ann Arbor, MI: Center for Japanese Studies, University of Michigan.

Shōji, K. (1990). Japan and East Asia. In *The Cambridge history of Japan. Vol. 3, Medieval Japan* (pp. 396–446). Cambridge: Cambridge University Press. Tr. G. Cameron Hurst III.

Smith, P. J. (2003). Impressions of the Song-Yuan-Ming transition: The evidence from Biji Memoirs. In P. J. Smith & R. von Glahn (Eds.), *The Song-Yuan-Ming transition in Chinese history* (pp. 71–134). Cambridge, MA: Harvard University Asia Center.

Struve, L. A. (1988). The Southern Ming, 1644-1662. In F. Mote & D. Twitchett (Eds.), *Cambridge history of China, vol. 7, The Ming dynasty, 1368-1644, Part 1* (pp. 641–725). Cambridge: Cambridge University Press.

Susumu, I. (1990). The decline of the Kamakura Bakufu. In *The Cambridge history of Japan. Vol. 3, medieval Japan* (pp. 128–174). Cambridge: Cambridge University Press. Tr. By Jeffrey P. Mass and Hitomi Tonomura.

Tan, T. S. (2009). *Cheng Ho and Islam in Southeast Asia*. Singapore: Institute of Southeast Asian Studies.

Tong, J. W. (1991). *Disorder under heaven: Collective violence in the Ming dynasty*. Stanford, CA: Stanford University Press.

Turchin, P., & Nefedov, S. (2009). *Secular cycles*. Princeton: Princeton University Press.

Twitchett, D., & Grimm, T. (1988). The Cheng-Tu'ung, Ching-T'ai, and T'ein-Shun reigns, 1436-1464. In F. Mote & D. Twitchett (Eds.), *Cambridge history of China* (The ming dynasty, 1368–1644, part 1) (Vol. 7, pp. 305–342). Cambridge: Cambridge University Press.

Von Glahn, R. (2015). *The economic history of China*. Cambridge: Cambridge University Press.

Yi, S.-s. (1981). *Imjin Changch'o: Admiral Yi Sun-sin's memorials to court*. Seoul: Yonsei University Press. Tr. Ha Tae-hung, ed. Lee Chong-young.

The Early Modern Period in the East Asian World-System

9.1 Climate in the Early Modern Period

Climate reached a dismal low point in the Maunder solar minimum, 1654–1714. During the depths of this period, solar activity dropped to perhaps its lowest level in several thousand years. Cold and drought followed. Climate slowly improved from the low point of the seventeenth century, but the eighteenth century continued cold and dry, with some particularly bitter years in the mid-century. After 1800, slow and irregular warming began, retarded by the rather minor Dalton solar minimum (1790–1830).

After 1850 recovery from the Little Ice Age became steadier, and after 1900 the inexorable upward movement of heat became a world phenomenon. How much of this was natural, how much due to human release of greenhouse gases, we do not yet know. By the middle of the twentieth century, human-released greenhouse gases took over the major role, but the scope of natural fluctuation and the relative contribution of nature and culture (in the form of greenhouse gases) is still under debate. In any case, the warmer the weather, the worse the Qing Dynasty and Korea did. This is counterintuitive, and counter to the theories of climate change affecting human affairs. Japan flourished, however.

9.2 The Rise of Qing

The Manchus of the northeast founded the Qing Dynasty even before conquering China; the latter goal was accomplished in 1644. (On Qing history, see esp. Peterson 2002, 2016; Rowe 2009.) The dynasty was created by Nurhachi (1559–1626), who brought the Manchu Jurchen from a tiny tribe to a huge regional power. He formed alliances with the Mongols, conquered and incorporated other Tungus tribes, and began nibbling away at China itself. The fall of Ming presented a unique opportunity to his successors. The Manchus let the Chinese bandits unite and bring the dynasty

© Springer Nature Switzerland AG 2019
E. N. Anderson, *The East Asian World-System*, World-Systems Evolution and Global Futures, https://doi.org/10.1007/978-3-030-16870-4_9

down—allying themselves with the strongest—and then turned on their allies. Trained in Mongol-style warfare and battle-hardened and disciplined, they swept away the bandit gangs and took over China. Early reigns followed Ming policy, but the Qing rulers realized they could ally themselves with steppe peoples, drawing on their shared heritage as northern "barbarians," rather than standing against them as the Ming rulers had done. This was an important step in consolidating rule and allowing expansion (Smith 2003: 34). Nurhachi referred to his realm as the Later Jin, declaring it officially in 1616. In 1636 his son relabeled it Qing.

Within China, the Manchus developed a complex imperial ideology, involving the complexities of a small minority running a huge empire in which they were outnumbered manyfold. Ethnicity, nationalism, and imperialism intersected in complex ways (Crossley 1999). The greatest early emperor of Qing was the Kangxi Emperor (1654–1722, r. 1662–1722), who succeeded at the age of seven, becoming one of the few boy emperors who rose to greatness. His 60-year reign is almost unique in Chinese history. Many of his personal writings survive, and they reveal a brilliant, restless, dynamic man who far preferred hunting in the boreal forests to suffering through court ceremonies and petty politics (Spence 1974). Few conquerors left such personal records. As in the Mongols' rise, there were thus two superb managers of 'asabiyah, and it is hard to imagine Qing success without them.

His first major crisis came when the Chinese generals who had allied themselves with the Manchus, led by Wu Sangui (who had hunted down the last Ming pretender), decided that the Manchus had to go. Wu launched—with two local warlords—the Rebellion of the Three Feudatories in 1674–1681. It was crushed so decisively that relative peace reigned thereafter. Ming sympathy was strong in the south, and the rebellion did well there. Stereotypes of the stalwart but rough Northerners and refined but devious Southerners were revived (Mote 1999: 881), but those refined Southerners could fight hard.

The Kangxi emperor was followed by the Yongzheng emperor (r. 1723–1735). The Yongzheng and Qianlong periods were times of increasing autocracy. China had become steadily more autocratic, with power centralized in the Emperor, since Yuan. Qing, afraid of Ming loyalism and its strategic deployment by reformists and critics, launched repression. Local control became more difficult as population grew, so the Qing fell back on allying itself more and more with local power figures.

Much expansion took place under the 60-year reign of the Qianlong Emperor (1711–1799, r. 1736–1795), who maintained peace, partly through cruel oppression. His 60-year reign matched his grandfather's; he survived a few more years, but was persuaded to abdicate so that he would not reign even longer. It is thought that the Qianlong emperor was declining at the time, hence the persuasion. In any case, he reigned over China's greatest expansion. He ruled great areas of Siberia that were later lost to Russia. Under the Qianlong Emperor, the Dzungar (or Zunghar) Mongols were virtually exterminated in a genocidal campaign in 1754–1757 (Mote 1999: 876, 936–937; Perdue 2005). He extended control over Tibet (Mote 1999: 877).

The Qianlong rule saw appalling corruption by Heshen (1750–1799), who won the Emperor's favor and robbed the country before being apprehended and executed by Qianlong's successor, the Jiaqing Emperor. Factions flourished. North-South rivalry caused problems (Mote 1999: 881).

Worse was the literary inquisition, a crackdown by the Qianlong emperor on anything that might smack of disloyalty or Ming irredentism (as well as on salacious literature). Countless literati were exiled or fired, though relatively few were executed. Intellectual enquiry and critique were virtually shut down (see Mote 1999: 925–931; Rowe 2009; Wang 2016). This was the worst, but far from the only, example of extreme repression and intellectual bullying under Qing (Wang 2016).

Qing was China's last and greatest upward sweep, integrating Manchuria and neighboring Siberia, Mongolia, Central Asian Turkestan (which they renamed Xinjiang, "New Borders") and Tibet into China (Perdue 2005). Tibet was formally part of the Manchu domains but not part of the Chinese Empire, a conflicted status which today feeds both Tibetan calls for independence and Chinese calls for continued subjection. This approximately doubled the size of the Chinese Empire, expanding China to its greatest extent. This was not a peaceful business; it was classic colonialism, quite comparable to what the west was doing at the time. The spreading conquest was part of the spectacular worldwide rise of imperialism in the seventeenth and eighteenth centuries. China was responding directly to Russian expansion. The Qing rulers were also at least somewhat aware of Dutch, Portuguese, and later British and other imperialist expansions.

Expansion in the south was largely within existing borders. Ming attempts to conquer Vietnam were not renewed. Han Chinese migrated into minority areas, forcing minority people and poorer Han families up into the hills. Here they could survive by growing New World crops—maize, white potatoes, sweet potatoes, peanuts, and the like (Mote 1999: 903)—as well as more traditional buckwheat and millets. Deforestation and erosion followed. Only in the farthest south did they retain control of prime rice land. Population soared, but recall that Ming late totals are unknown, surely larger than the poor records show.

Southeast Asia was nominally tributary, in practice quite independent, using the "tribute" missions strictly for trade. Qing continued the Chinese presence in the Nanyang, but without enthusiasm. There was little imperial support for the large colonies in places like Malacca and Jakarta. Even so, an enormous amount of trade went on (Tagliacozzo and Chang 2011). Business flourished, especially after the Portuguese, Spanish, British and Dutch established major colonies. The British and Dutch imported vast numbers of Chinese workers in the nineteenth century and into the twentieth, vastly changing the demography of Malaysia and Indonesia. Chinese were already in the Philippines when the Spanish got there, but under Qing the Chinese presence increased, again with huge demographic impact. Thailand also saw a huge immigration. French Indochina was also impacted, with the Chinese settlement of Cholon (part of Saigon) and the port of Hoi An becoming major Chinese business settlements.

Minor crises failed to damage Qing control, or the peace and stability of the eighteenth century. China may have been as well off as western Europe (Pomeranz 2000) though this is controversial (Anderson 2016; Von Glahn 2015: 348–399). This involved agricultural improvements—in this case, the coming of new and highly productive and nutritious foods from the New World. Some had come during Ming, but we have little evidence. The great age of adoption of New World crops was early Qing, when maize, white and sweet potatoes, peanuts, chiles, and other crops totally remade the agricultural picture, especially in the southeast. Also, agricultural reclamation flourished: terracing, fertilizing, dyke building, recovery of at least some forest lands, and other measures.

New World food crops swept across China in the Qing Dynasty. The elite may have been resistant to new ideas, but the people knew good things when they saw them. Maize, chiles, sweet potatoes, squash, beans, peanuts, tobacco (to China's cost), and even oddities like guavas and pineapples spread. They quickly were incorporated not only into the food system but into the medical one, with maize silks being recognized (correctly) as diuretic, chiles as heating (of course), and guavas as soothing to the throat. In Fujian, "a gazetteer published in 1612 observed that sweet potatoes had been introduced to southern Fujian from Guangdong only five or six years before, but already had become the main staple food of the poor" (Von Glahn 2015: 328). Carl Thunberg, botanizing Japan in the 1770s, recorded that "Among their esculent-rooted vegetables Batatas [sweet potatoes]…are the most abundant, and the most palatable" (Thunberg 1795: 84).

Qing was also a golden age for Chinese trade. Portuguese merchants had integrated Ming into the world-system. In Qing, the whole world flocked to China to buy tea, rhubarb, ceramics (porcelain was long known as "China ware"), fabrics, metalwork, lacquerware, furniture—anything beautiful, exotic, rare, or even simply useful. China needed little. It drained the northwest of sea otter pelts and Hawai'i of sandalwood, but usually required payment in silver. This caused an increasing drain on silver for the rest of the world, eventually leading to the famous British policy of flooding the Chinese market with opium simply to have something addictive to sell the Chinese. As Qing weakened, treaty ports arose, to manage the trade on increasingly good terms for the European and American powers. Chinese ceramic shards abound from the early-day ports of East Africa to the ghost towns of Nevada. Trade integrated China with the world-system.

The same occurred in Japan, much later and to a lesser extent. Only after the Meiji Restoration did Japan fully open to international trade, following which "exotic" Japanese art of wares became the rage in Europe and the Americas. Only a couple of decades after the Restoration, Impressionist artists were collecting Japanese prints and copying their styles. More economically serious was an enormous trade in silks and in tea.

9.3 Qing's Malthusian Stress

Population grew as fast as production, and eventually faster. By this time, China—specifically the eastern river deltas—probably had the most productive land in the world. The only competition was from Chinese-style agriculture in Japan and northern Vietnam. China's fields produced about 2500 pounds of grain per acre per crop. They were cropped two or even three times a year in the south. This took incredible amounts of labor. As of 1800, a Chinese rice field absorbed eight times as much labor per acre as an English grain field, but produced nine times as much food (Von Glahn 2015: 360).

China had about 100 million people in early Ming. "A consensus estimate might place the population in 1700 at about 150 million, roughly the same as it had been a century earlier under the Ming. By 1800 it had reached 300 million or more, and then rose further to perhaps 450 million at the outbreak of the Taiping rebelling around 1850" (Rowe 2009: 91). It stabilized at around 400 million for the rest of the nineteenth century. This all meant that there was very little farmland per person. "Between 1753 and 1812, per capita acreage declined a remarkable 43 percent, to less than half an acre per person" (Rowe 2009: 150). Farmland expanded in the south and northeast, but not as fast as population.

Agriculture expanded into increasingly steep slope areas, desert margins, and other areas where farming was unsustainable. Where agriculture could not expand, people simply had to work ever harder, wringing more and more production out of less and less land. Incomes remained the same. This is Boserup's "agricultural intensification" at its most gritty. Philip Huang (1990) borrowed Clifford Geertz' term "agricultural involution" (Geertz 1972) for the process. Others have challenged the use of the term, but it describes the reality, at least after the early eighteenth century (see discussions and reviews of this literature in Anderson 2016 and Von Glahn 2015: 348–399).

Consumption by farming families in the rich rice lands stayed similar from the mid-eighteenth century to the 1930s, with food taking over 60% of income, clothing 10%, and—an interesting third—"ritual and religion" taking another 9% of income (Von Glahn 2015: 351). This last was important because the "rituals" were in fact the social cement that kept communities together and kept families cooperating with each other. Investing in public rites to create public bonds was an absolute necessity.

The question of involution was part of a wider debate on Qing economics. Bin Wong (1997) and Philip Huang (1990) had written of Chinese economic woes. Kenneth Pomeranz (2000) challenged the field by alleging that China was ahead of Europe in prosperity until around 1700. This was promptly countered by Philip Huang (2002) in a book review followed by comments by Wong and Pomeranz. (It is relevant to note that all three are part of a close friendship network that also includes von Glahn.) By contrast, *The Economist* (2017) found evidence that China, and even the rich lower Yangzi, had been behind at least some of Europe since the Song Dynasty. In any case, China could only fall backward from a high point of prosperity in the eighteenth century.

For Richard von Glahn, Qing was a *"provisioning state* [his italics]...dedicated to improving the people's livelihood through, for example, investments in famine relief and flood control" (Von Glahn 2015: 313). The state took full care of food supply, using the ever-normal granary (*changpingcang*) concept and a comprehensive nationwide system of famine relief. The ever-normal granary, an old concept, involved maintaining stocks of grain at a more or less constant level, drawing them down in famine times and building them up in good times. It was borrowed, term and all, by the United States in the Depression.

The result was an unprecedented level of peace and welfare in the long eighteenth century. It was an economic golden age, highly important for understanding cyclic dynamics within a world-system. Trade was extensive. Taxes were low; the government tried to hold them down to 3%, following early Han example. Rebellions were exceptional (Naquin 1981). During the period from around 1700 to 1820, the chronically drought-stricken northwest—the inner Yellow River drainage—saw a sustained, impressive, unbroken growth in population and no reports of serious famine. To put this in perspective, this is China's poorest region—an area where marginal rainfall tempts farmers to invest in building up their farms and having many children, only to have disaster strike at frequent intervals when the rains fail (see e.g. Lee and Zhang 2010; Lee et al. 2015; Li 2007; Mallory 1926). Only after 1970 did this area begin to be free of famines, and current erosion makes one wonder how long the current productive period will last.

Economic debates swirled around how much to free the market; Qing paralleled Europe in arguing over the benefits of markets and state control (Dunstan 1996). River control was perhaps the function that most clearly needed the state, with infrastructure in general being also a government duty.

9.4 Waning and Decline

The good times waned in 1796, when a White Lotus rebellion devastated much of China. The White Lotus groups were a mix of millennial cultism, Ming loyalism, and outright crime. The rebellion was soon crushed, but White Lotus millennials merged with the huge Qing underworld, including the triad societies of later fame.

Half a century later came the Taiping Rebellion and Daoguang Depression, caused by currency mismanagement about the same time (Rowe 2009: 137–138). Though Christian in inspiration, the Taiping Rebellion was similar in ideology and course to earlier millenarian rebellions. "Taiping," literally "Great Peace," is the Chinese equivalent of the Millennium in Christendom. Beginning in 1844, but not becoming a serious outbreak till 1851, it lasted on and off until 1864, and killed tens of millions of people in south and central China (Kuhn 1978; Rowe 2009; Spence 1997). Most of the deaths were not from combat, but from hunger and exposure, as the rebellion devastated the most densely populated parts of the country. Like so many other movements in history, the Taiping Rebellion collapsed when rivalry between its leaders spun out of control. The rebels turned on each other in 1856. Even so, the rebellion sputtered on for years.

9.4 Waning and Decline

At the same time, other brushfire rebellions sprouted in other parts of the empire, causing incredible dislocation and misery. The Nian Rebellion sputtered on in the north from 1851 to 1868 (Kuhn 1978; Von Glahn 2015: 375), assimilating remnants of the Taipings. Muslims rebelled in the northwest and southwest. Tribal minorities warred against the state.

At this point, the western powers intervened, largely to force China to accept opium as a trade good. The Opium War of 1839–42 opened China, and allowed European powers to occupy treaty ports; the major one was Hong Kong, occupied by the British in 1842, and growing slowly to become the familiar colony, now reintegrated into China (since 1997) and losing its cultural and political independence. The Arrow War or Second Opium War of 1856–60 led to yet more of the "unequal treaties" that gave the western powers special rights and access. These wars integrated China with the global world-system that was emerging at the time. A period of transition from core of the East Asian system to periphery and then semiperiphery of the global system can be dated as extending from 1842 to 1911.

Western benefits such as better medicine and better economic policies helped China slightly, but colonialism and the highly negative, conservative Chinese reaction to it doomed these benefits to insignificance (see Fairbank 1978; Rowe 2009).

The Russians then schemed to break off progressively large portions of the Siberian realms, taking over vast areas in the eighteenth and nineteenth centuries through treaties that the Chinese consider unequal. Perhaps the greatest effect Europeans had on Qing was the seizure by Russia of vast areas the Manchus had held in Siberia. Particularly notable was the Machiavellian scheming of the brilliant Russian Colonel N. N. Muraviev, who ran a virtually independent realm in east Siberia, employing political exiles such as the anarchist philosopher Mikhail Bakunin. Muraviev encouraged Russian settlement north of the Amur and east of the Ussuri, then talked the Qing government into signing these areas away in the Treaty of Beijing in 1860 (Fletcher 1978). Unlike earlier losses of unproductive Siberian wastes, in this treaty China lost thousands of square miles of extremely rich and productive territory, sites of the modern cities of Vladivostok ("Lord of the East") and Khabarovsk. This may stand as an extreme example of the "unequal treaties" signed by China and the west in the nineteenth century.

Mongolia became independent in the 1920s and became a client state of the USSR. British scheming never led to actual occupation or claim for Tibet; the British decided to leave it to the Manchus. Tibet's final attempts at independence were too late. Loss of the Siberian lands proved nothing about Chinese decline; there is no evidence that China suffered. It may have benefited, since it had to expend nothing on controlling and administering vast realms that had provided little or nothing except furs and, locally, ginseng. Soon after this, a Turkic leader, Yaqub Beg (1820–1877), carved out an independent state in Xinjiang in the 1870s, but he was soon crushed.

By the early nineteenth century, Qing China faced a difficult situation. Its population was increasing faster than agricultural production. People limited their family size, especially by infanticide or neglect of girls (Bengtsson et al. 2004), but it was not enough. Inevitably, population levels were trimmed back by famine. The birth rate stayed high. Agricultural production failed to rise because China was substantially filled up, and the autocratic, conservative government discouraged modernization or learning from the west. Corruption and elite factions were endemic. The combination of Malthusian squeeze, autocratic government, and cyclic decline was deadly.

Turchin's "elite overproduction" was conspicuous. The imperial family and the privileged Manchus increased in number. Educated elites also multiplied. By 1800 there were over a million lower degree holders, causing a "talent glut," and another 400,000 with higher degrees (Rowe 2009: 151–152). These highly trained individuals, roughly the equivalent of B.A.'s, could not all find jobs in government; many were local teachers and other far-from-affluent intellectual service workers (Mann Jones and Kuhn 1978: 111).

A brief restoration under the Tongzhi imperial period (1862–1874; Liu 1978) saved the dynasty for a few more decades. Famine in the mid-1870s, among the worst in all Chinese history, ended the brief flowering. Finally, the Empress Dowager Zixi managed the country as a succession of child emperors fell under her care. She fought against rapid modernization, and with age she became less and less enamored of the idea of progress, especially if it means Westernizing. The dynasty survived her by a few years. Overproduction of elites showed itself again, in the masses of disaffected people with good educations who argued for reforms. Advocates ranged from those wanting to go back to ancient traditions to those wanting to follow Japan into rapid, massive westernization. Amid these discordant voices, with the conservative Zixi firmly in charge, all change was frozen except the steady decline of the economy and of China's independence among nations.

Meanwhile, often-violent pressures from the west, especially the British, progressively weakened the dynasty and discredited its rulers (Rowe 2009).

Finally, as in several other dynastic declines, a boy emperor, a rebellious population, and a powerful and independent general came together in timing, bringing down the dynasty in 1911. The general Yuan Shikai lasted only until 1915 as founder of a new dynasty, before being thrown out in the name of modernity. A long period of disunion followed from 1915 to 1950. Then the Communists founded a new regime, which, at this writing, endures, but is running out of time, if Ibn Khaldun cycles are any guide.

9.5   Vietnam Declines

Vietnam suffered under its climatic nightmare, caused by the wild fluctuations of the Little Ice Age during the 1500s and on into the 1700s (Lieberman 2003). Its worst famine came in 1764. Central Vietnam's main port, Hoi An, silted up (Kiernan 2017: 252), causing it to fossilize as a very well-preserved eighteenth-century port town (Avieli 2012).

The previous chapter ended its Vietnam section with a division into two kingdoms, Dang Ngoai—"Tonkin" to the west—in the north and Dang Trong—

"Cochinchina"—in the south. A major war with Cambodia ensued in the mid-eighteenth century, with vast slaughter on both sides.

A confused war, the Tay Son Rebellion, broke out in 1771, leading to chaos until 1802. The ultimate victor was Nguyên Anh, from the deep south. He unified Vietnam for the first time in centuries, but the country had only a brief run before succumbing to French imperialism after 18 years of resistance fighting and holding out, 1862–1880. Drought in the 1860s and 1870s reached devastating levels, breaking Vietnamese spirit at a critical time. "The drought of 1877–78 and 1888–89 were both associated with El Niño 'warm episode event.' The first brought a 'once in 200-year' global drought. In 1876–79 perhaps sixteen million people died in famines in India, China, and Brazil, and as many again in 1896–1902" (Kiernan 2017: 295), thanks to an extremely strong El Niño event that dried up the monsoons.. This was probably China's worst famine except for the human-caused Great Leap Forward famine of 1958–61. Apparently in Vietnam it was little better.

The successive collapses of Le, Tay Son, and Nguyen show climate and weather mattered as part of the back story. Another part was continual meddling and frequent testing-the-water incursions by Qing, but the Vietnamese attended to these in their usual fashion.

More important in the end was the rapidly increasing French intrusion from the seventeenth century onward. At first the French wanted trade and missionary opportunities (usefully creating the Latin-alphabet Vietnamese script by Father de Rhodes in 1651). By the mid-nineteenth century they felt a need to develop a colonial empire like their foes the British. A fascinating story of Qing, French, and Vietnamese jockeying for power has recently been told by Bradley Camp Davis in *Imperial Bandits* (2017). It provides a gripping case study of life in the borderlands. The French foothold in Vietnam and Cambodia made that area an obvious place for imperialism, and Vietnam's long and glorious history of independence was over, until its stunning resurgence after WWII.

9.6 In the Northeast

Meanwhile, Korea continued under a single dynasty, the Chosŏn. It suffered interruptions from rampant factionalism and palace intrigues. These were chronicled with sorrow by Lady Hong (Hyegyŏnggung Hong Ssi; see Kim 1996 for translation; the work is the source of countless Korean historical soap operas). In the mid-eighteenth century, the crown prince, Sado, descended into schizophrenia, harrowingly described in great detail by Lady Hong, who was his wife. An interesting an important feature is that she describes his insanity as "illness," rather than as demon possession or personal evil or the other causes that were commonly used to explain madness in that day.
When the crown prince took to randomly murdering people in fits of derangement, something had to be done, and obviously he could not succeed to the throne. The near-certain fall of the dynasty was prevented only by the condign and presumably heartbreaking execution of the prince by his father, King Yŏngjo (r. 1725–1776; see

Hyegyŏnggung 1996). To avoid shedding royal blood or getting the dynasty embroiled in a full criminal trial that would have led to many more deaths, the king locked the unfortunate prince in a grain box, where he died after 8 days of desperate but gradually failing cries.

Against these dismal trends was the rise of practical learning and arts, including some awareness and sensitivity to western learning; this trend gave rise to the Sirhak ("practical learning") movement in the eighteenth century (Han 1970: 322–328) and other modernizing movements. Through it all, neither imperial madness nor active reform movements could bring down the long-lived dynasty.

In the late nineteenth century, Korea ran parallel to Qing to a fascinating degree. In 1864 the king died without male heir, and a boy from a from a collateral branch of the lineage succeeded to the throne. The government was taken over by his father, the regent, known under the title Taewon'gun, Prince of the Great Court (Han 1970: 361–362). The Taewon'gun was a brilliant, strong-willed, highly competent ruler, but also conservative—increasingly so with time. Strikingly similar to the Empress Dowager Zisi, he began by reforming and cleaning up the government, but became increasingly dedicated to resisting modernization and foreign encroachment.

The droughts that affected China also affected Korea, causing famine in 1876–77 as in China, and again in 1888–89 (Han 1970: 403–404). These and other woes led to the Tonghak Rebellion of 1894–95 (Han 1970: 404–415). Tonghak ("eastern learning") was a nationalist but somewhat modernizing school of thought that morphed into a cult-like revolutionary ideology. The Japanese were all too eager to help in defeating it, thus gaining more power in the peninsula.

Through the latter half of the nineteenth century, not only China and Japan but also America and the European powers were taking interest in Korea—far too much interest for most Koreans, who knew what was happening in China, thanks to spillover problems from the Taiping Rebellion and the Opium Wars. Many Taiping refugees had fled to Korea. The Taewon'gun tried, ultimately without success, to stop the unequal treaties and wanton exploitation. Ironically, at the same time China was suffering from these, it was imposing them on Korea. Full-scale semiperipheralization had come, complete with massive exports of food grain at a time when Koreans were starving. Modernization of transport and communication allowed full-scale mobilization for resource extraction.

The Japanese were the major problem. Korea's fall was due to increased pressure by Japan, which took more and more power in the late nineteenth century, notably with the Treaty of Shimonoseki in 1895, which drove China out of the picture. Japan finally took over Korea, as a colony, in 1910. Korea remained a brutally and cruelly exploited colony until the end of WWII, after which it suffered yet again from war between communist and anti-communist factions; this led to a de facto world war. None of this had any discernible connection with climate.

Japan's Hideyoshi period ended with the rise of the Tokugawa shoguns, brilliant master administrators who brought Japan to world leadership in areas ranging from agriculture to literacy. They reforested the country (Totman 1989), improved agriculture, regulated fisheries successfully (Ruddle and Johannes 1983), and ran a very tight ship. The level of peace and lawfulness in the country, and the freedom of the people from imposts and arbitrary governance, deeply impressed foreign observers. These included many Chinese, but also the Swedish natural historian Carl Thunberg (1795), who had the prejudices of his time and place, yet praised Japan as being well ahead of Europe in such respects.

An idea pioneered earlier but enforced by the Tokugawas was to keep the daimyos' wives and children in the capital city, and make the daimyos reside there in alternate years. This was a very stringent and effective method of control. It was made permanent in 1633.

The fifth ruler, Tokugawa Tsunayoshi, endears himself by his love of dogs. He even executed some who mistreated dogs (Bodart-Bailey 2006). He also arranged for care of less fortunate humans, including the sick and of infants. He was notably less pleasant to adults that caused difficulties, evidently following Ming and Qing examples, but his condign rule led to new heights of prosperity and creativity in Japan. Tokyo reached a population of a million, making it possibly the world's biggest city.

Tokugawa isolationism did not stop trade from bringing rich cargoes of all sorts (Batten 2003: 198–200). Trade remained important in spite of government suppression. Japan's prosperity in early Tokugawa times contrasts strikingly with the awful times of China in the early 1600s and Europe in the entire seventeenth century; the three regions are thoroughly decoupled in that difficult century (Batten 2003: 230–232; Parker 2013).

Famines occurred, some very severe (in 1732, 1782–88, 1833–37), but nothing like the Chinese continual near-famine situation. In 1749, wild boars multiplied and competed for food with humans. The worst famine, the Tenmei famine in 1782–88, was caused or exacerbated by Northern Hemisphere-wide bad weather associated with the catastrophic eruption of Grimsvötn and Lakagigar in 1783 in Iceland, which released millions of tons of dust. It killed a quarter of Iceland's people and most of the livestock, because of deadly hydrofluoric acid. The dust darkened the skies and bring poor weather as far as Japan (Walker 2015: 132–135; the year was not exceptionally awful in China). Also extremely serious was the Tempō famine of 1833–37, which does not seem to coincide with a major volcanic explosion or any major famine elsewhere in the world. At least in northern Japan, it raised rice prices even higher than the Tenmei had (Tsuya and Kurosu 2014: 354). Significantly, the terrible famines that struck most of the Northern Hemisphere after the eruption of Tambora in 1815 did not affect Japan, nor did the cold wet weather of 1846–48 (possibly associated with Hekla's eruption in Iceland in 1845) that gave Europe the Potato Famine. Farmers had to pay taxes in rice, which the bakufu and samurai then sold for money, so famines amplified power disparities: the farmers starved to pay taxes, the rulers made more money because of skyrocketing grain prices.

The eighteenth century was notable for corruption, famine, intrigue, shifting economic policies, boy shoguns, and other problems. Neither these nor the bad weather brought down the government or even slow its progress, though the problems laid the groundwork for subsequent collapse after 1850 (Tatsuya 1991). As of 1800, Japan may have had the highest agricultural productivity, literacy rate (see Dore 1965), and forest and fishery management successes of any country in the world. It was an amazing record. Farming communities managed their own lands and waters (Ruddle and Johannes 1983; Toshiyo 1991; Totman 1989). They met Elinor Ostrom's conditions for successful communal management (Ostrom 1990, 2009; see also Janssen and Scheffer 2004). Rights were established, defined, and held collectively. Enforcement was adequate, and backed up by the Tokugawa regime, which needed the rice, timber, and fish, and was quite aware of the need to manage sustainably. Perhaps better than most farmers worldwide, the Japanese of the era had a thorough knowledge of how to do this. Moreover, Japanese peasants had evolved into independent farmers, though still bound to their village communities and given fewer rights than the nobles. "The shift from 'peasant' to 'farmer'...involved a major shift in the institutional context within which the Japanese farmer lived" (Toshio 1991: 517–518).

Hokkaido was brought into the Japanese national realm at this time (1600–1900), a slow conquest with all the usual cruelties and abuses that occur when a rich, powerful people subdues a small-scale society (Batten 2003: 251–254; Walker 2001, 2015: 135–139; see also Fitzhugh and Dubreuil 1999). It is interesting to see the same trajectory of abuse, subjection, consequent alcoholism and despair, and slow mending as the Ainu became incorporated in the Japanese world—losing much of their culture in the process. The dismal parallels with settler colonialism in the Americas and Australia indicate that this is a human pattern, not a specifically cultural or economic one (Batten 2003: 157). Today's Hokkaido Japanese have a rough, open, frank quality that reminds observers of Australians and old-time frontier Americans, as noted by the Japanese cross-cultural psychologist Shinobu Kitayama (Kitayama and Cohen 2007).

Japanese bulk trade now tied the whole archipelago together, including the Ryukyus. Vast amounts of all kinds of fish and sea food flowed back and forth (Batten 2003; with true Japanese love of fish, he lists them on pp. 166–169). Timber, rice, mined metals, and other land-based resources were equally conspicuous in trade (Walker 2015). The Japanese islands were a world-system of their own, encapsulated in the wider East Asian one, but preserving much more independence than they once had.

The Tokugawa cut off Japan from the rest of the world, as Korea was increasingly cut off by the Yi dynasty. The Tokugawa tried to hold trade to the Four Mouths: Matsumae, the south tip of Hokkaido, was the mouth or gateway thither; Tsushima Island was the mouth to Korea; Nagasaki was the mouth to the western and Chinese world; Satsuma was the mouth to the Ryukyus. Major trade through other ports was forbidden, though much piracy and clandestine small-scale trade went on.

Commodore Perry famously opened Japan "with a can opener" (as historians traditionally say) in 1852, entering with his "black ships" and forcing contact and trade. The Japanese quickly decided this was not a bad thing, and took to modernization with enthusiasm. The Japanese even copied Perry by opening Korea forcibly in 1876, but the result of that was more sinister: Japan slowly moved in on the unfortunate "hermit kingdom," finally turning it into a repressed and exploited colony.

However, the Tokugawa shogunate was in deep trouble. The Tempō famine noted above was one case of bad weather clearly influencing history (Bolitho

1989). It allowed the many reformers who had been critical of the Tokugawa regime to burst out of the shadows and recommend dozens of measures, some adopted, all ultimately futile. Another factor forcing reform was the presence of more and more European ships, and the increasing visibility of the Chinese on the water also. The Japanese became aware of the Opium War and its devastating consequences for China. They beefed up defenses and began to move away from self-isolation.

The famine initiated the *bakumatsu* period: the breakdown of the Tokugawa dynasty (Tsuya and Kurosu 2014: 354). The demilitarization of the country had been highly successful for over 200 years at maintaining peace, but now it prevented Japan from defending itself (Bolitho 1989). Decentralization—devolving power from the state to the 264 daimyos—had also reached a point at which necessary state functions became hard to maintain.

Turchin's "elite overproduction" even had its day when Tokugawa Ienari (1773–1841) produced 55 children, apparently a record for Japan. Revenues were inadequate after the famines. Peasants and samurai grew worse off. Since almost everyone married early and had few other birth control options, massive infanticide resulted. Families "returned" (to the other world) up to 10% or more of their babies (Tsuya and Kurosu 2010), locally and in famines to as high as perhaps 50% (Smith 1977). Sex ratios were manipulated, often in the direction of an ideal one girl to two boys. These are the highest rates of infanticide reported in the world. The Chinese at the time (and since) practiced considerable female infanticide, but usually only in circumstances of poverty, and disposing of only one or two girls per family. (Contrary to some reports, evidence including research interviews showed that mothers, at least, are normally heartbroken and never got over the need to eliminate a child. Confirming evidence from other cultures shows that infanticide is not an easy option.)

The Tokugawa shogunate finally fell in the famous Meiji restoration in 1868, when modernizers and conservatives united to displace the shoguns and restore administrative power to the emperor (Jansen 1989a–c). Five of the largest and richest daimyoships—Satsuma being the richest of them and second richest in Japan—united against the bakufu. Loyalist supporters of the emperor against the shogunate multiplied and terrorized the bakufu supporters. The bakufu was running out of money, maintaining itself in large part by debasing currency and racking the samurai. The daimyo had increasing power at Tokugawa expense. The classic pattern of a dynastic collapse appeared: people worked and fought for their own local groups and interests, not for the state.

The inherent weakness of a situation in which an emperor, a shogun, and 264 local lords divided power finally brought down the dynasty (Jansen 1989b). "If the definition of the Meiji restoration is limited to the events of 1867 and 1868, it constituted little more than a coup that shifted rule from one sector of the ruling class to another" (Jansen 1989b: 360). What happened after that, though, was all-important. The reformers and modernizers took over, and Japan quickly westernized.

The dream was to restore the emperor to power. In fact, shoguns turned bureaucrats ran the state, and the emperor went back to his traditional ceremonial status. The striking difference between enthusiastically modernizing Japan and a China locked into conservatism by the reactionary Empress Dowager Zixi and her court has been analyzed countless times; suffice it to say that climate is obviously not involved, nor do we need to resort to "national character" or other mysticisms. Personalities and opportunities had more to do with the different fates. Climate, especially through causing the Tempō famine, was one part of the back story, but the fall of Tokugawa was a classic Ibn Khaldun case.

9.7 East Asia Fails to Modernize

East Asia's failure to modernize more rapidly in Ming and Qing seems clearly due to governance. Asia and Europe were united by Smithian growth: growth in integration, diversity, specialization, and skill. Europe had an advantage in that it was difficult to unify, as noted above. This, among other things, made it more warlike, with arms, war horses, military organization, and related matters highly developed. Victor Lieberman, like Turchin, sees warfare as often the driver of modernization (see esp. Lieberman 2009: 73–74).

One explanation is Europe's rising democracy versus autocracy in China and elsewhere. Perry Anderson's study of the rise of autocracy in Europe (Anderson 1974) disabuses us of any notion that European nations contemporary with Ming were glorious showcases of freedom.

Michael Mann (2012) argued that the roots of European progress lay in the autonomous lords and yeomen of the early medieval period. They set the scene for independent economic enterprise, free from the state and from communal land decisions. (On these matters see e.g. Anderson 2016; Arrighi 2007; Elman 2005, 2006; Lieberman 2009; Pomeranz 2000.)

A fateful Ming decision was to move the capital to Beijing and face westward toward a still-threatening Mongol realm, instead of following the Treasure Fleets into focus on seaborne expansion. If the Ming had kept its capital in Nanjing and pursued ocean trade, elites of the lower Yangzi region would have guaranteed a forward-looking, innovative, technologically sophisticated, and economically capitalist or proto-capitalist China. This realization led in the 1990s to speculation in China about the continuing role of "yellow China" (the dry northwest with its yellow soil) and "blue China" (the maritime east), and much popular pressure to go with the latter as the twenty-first century dawns (Mote 1999: 615–616).

Late Ming was hardly a time to modernize; the government could not even control banditry. Qing was far too concerned with maintaining power. The Qianlong Emperor was especially reactionary in the pursuit of security. Korea was similarly afflicted with centralized, reactionary rulers, especially in the nineteenth century. Japan was the great contrast: modernization had already begun before Commodore Perry's famous entry in 1852, and then the Meiji Restoration opened the floodgates. That and the modernization of Vietnam under the French show that East Asia was no more inevitably "backward" than rural Europe.

The enormous growth of population over time—from 150 to 400 million in China, with similar increases in Korea and Japan—put pressure on land and led to agricultural involution (Huang 1985, 1990, 2002). This fed into the massive, devastating Chinese rebellions of the 1840s and 1850s, from which Qing never recovered. Its autocratic, tradition-bound governance strategy, paranoid about open expression or advocacy for change, prevented adaptation. If Qing and Korea had rapidly modernized and adopted western techniques, like Japan, they might have survived. Dissention among elites, and reactionary politics at the top, prevented that.

Qing took 40 years to consolidate. Then it went without a major crisis until 1844, 200 years after Qing capture of Beijing. The final collapse in 1911 came 67 years later. The moral seems to be that good governance combined with rising productivity and population can give an empire a long span of rule. Perhaps this was helped by the specific mix of autocratic rule that still rested with a relatively light hand on the ordinary people. A strong, centralized government with rigid orthodox views had to accommodate to lack of real power, especially after 1800. It fell back on low taxes, lack of interference with the grassroots, and preference for moral suasion over terror, and strong local institutions. It could never get rid of the hydra-headed Ming loyalist movements. It used indirect control (via appointed *tusi* chiefs) to manage isolated minorities. The result was considerable local freedom, but a government that could crush anything problematic above the local level until 1844.

From 1844 onward, Qing follows the well-worn path of other Chinese dynasties: increasingly unresponsive and backward-looking government as response to increasing unrest; failure to produce economic growth; finally, a succession of child emperors, making a mockery of centralized governance, until a general finally stepped in to call game.

We then have:

Rise. Nurhachi and the Kangxi Emperor were classic arousers of 'asabiyah.

Expansion. Qing expanded more than any other Chinese dynasty.

Widening gap. Obvious; rarely has anyone been so far above the masses as were the later Qing emperors.

Elite overproduction. Obvious in both the imperial family and the examination successes.

Increasing corruption. Clear; may have reached a peak under the Qianlong Emperor.

Decline. Followed the classic Ibn Khaldun pattern.

Widespread unrest. Extreme by 1850.

Military overextension. Made clear by failures to deal with the west, or, at the last, Japan.

Collapse.

Fall. The common pattern of a boy emperor displaced by a coup by a leading general.

References

Anderson, E. N. (2016). Agriculture, population, and environment in late imperial China. In T. -j. Liu & J. Beattie (Eds.), *Environment, modernization and development in East Asia: Perspectives from environmental history* (pp. 31–58). Houndmills, Basingstoke, UK: Palgrave MacMillan.
Anderson, P. (1974). *Lineages of the absolutist state*. London: NLB.
Arrighi, G. (2007). *Adam Smith in Beijing: Lineages of the twenty-first century*. London: Verso.
Avieli, N. (2012). *Rice talks: Food and community in a Vietnamese town*. Bloomington, IN: Indiana University Press.
Batten, B. (2003). *The end of Japan: Premodern frontiers, boundaries, and interactions*. Honolulu, HI: University of Hawai'i Press.
Bengtsson, T., Campbell, C., & Lee, J. A. (Eds.). (2004). *Life under pressure: Mortality and living standards in Europe and Asia, 1700-1900*. Cambridge, MA: MIT Press.
Bodart-Bailey, B. (2006). *The dog shogun: The personality and policies of Tokugawa Tsunayoshi*. Honolulu, HI: University of Hawai'i Press.
Bolitho, H. (1989). The Tempō crisis. In *The Cambridge history of Japan. Vol. 5. The nineteenth century* (pp. 116–167). Cambridge: Cambridge University Press.
Crossley, P. K. (1999). *A translucent mirror: History and identity in Qing imperial ideology*. Berkeley, CA: University of California Press.
Davis, B. C. (2017). *Imperial bandits: Outlaws and rebels in the China-Vietnam borderlands*. Seattle, WA: University of Washington Press.
Dunstan, H. (1996). *Conflicting counsels to confuse the age: A documentary study of political economy in Qing China, 1644-1840*. Ann Arbor, MI: Center for Chinese Studies, University of Michigan.
Dore, R. P. (1965). *Education in Tokugawa Japan*. Berkeley: University of California Press.
Economist, The. (2017). China has been poorer than Europe longer than the party thinks. https://www.economist.com/news/china/21723459-how-will-affect-xis-chinese-dream-china-has-been-poorer-europe-longer-party
Elman, B. (2005). *On their own terms: Science in China, 1550-1900*. Cambridge, MA: Harvard University Press.
Elman, B. (2006). *A cultural history of modern science in China*. Cambridge, MA: Harvard University Press.
Fairbank, J. K. (Ed.). (1978). *The Cambridge history of China. Vol. 10, Lante Ch'ing, 1800-1911, part I*. Cambridge: Cambridge University Press.
Fitzhugh, W., & Dubreuil, C. O. (Eds.). (1999). *Ainu: Spirit of a Northern people*. Washington, DC: Smithsonian Institution in Association with University of Washington Press.
Fletcher, J. (1978). Sino-Russian relations, 1800-62. In J. K. Fairbank (Ed.), *The Cambridge history of China. Vol. 10, Late Ch'ing, 1800-1911, part I* (pp. 264–330). Cambridge: Cambridge University Press.
Geertz, C. (1972). *Agricultural involution*. Berkeley, CA: University of California Press.
Han, W.-k. (1970). *The history of Korea*. Seoul: Eul-Yoo. Tr. Lee Kyung-shik.
Huang, P. C. C. (1985). *The peasant economy and social change in North China*. Stanford, CA: Stanford University Press.
Huang, P. C. C. (1990). *The peasant family and rural development in the Yangzi Delta, 1350-1988*. Stanford, CA: Stanford University Press.
Huang, P. C. C. (2002). Development or involution in eighteenth-century Britain and China? A review of Kenneth Pomeranz's *The Great Divergence: China, Europe, and the Making of the Modern World Economy*. *Journal of Asian Studies, 61*, 501–538.
Hyegyŏnggung, H. S. (1996). *The memoirs of lady Hyegyŏng*. Berkeley, CA: University of California Press. Ttr. JaHyun Kim Haboush.
Jansen, M. B. (1989a). Japan in the early nineteenth century. In *The Cambridge history of Japan. Vol. 5. The nineteenth century* (pp. 50–115). Cambridge: Cambridge University Press.

References

Jansen, M. B. (1989b). The Meiji restoration. In *The Cambridge history of Japan. Vol. 5. The nineteenth century* (pp. 308–366). Cambridge: Cambridge University Press.

Jansen, M. (Ed.). (1989c). *The Cambridge history of Japan. Vol. 5: The nineteenth century.* Cambridge: Cambridge University Press.

Janssen, M. A., & Scheffer, M. (2004). Overexploitation of renewable resources by ancient societies and the role of sunk-cost effects. *Ecology and Society, 9*(1), article 6.

Kiernan, B. (2017). *Viet Nam: A history from earliest times to the present.* New York: Oxford University Press.

Kim, J. H. (1996). *The memoirs of lady Hyegyong.* Berkeley, CA: University of California Press.

Kitayama, S., & Cohen, D. (Eds.). (2007). *Handbook of cultural psychology.* New York: Guilford.

Kuhn, P. A. (1978). The taiping rebellion. In J. K. Fairbank (Ed.), *The Cambridge history of China* (Late Ch'ing, 1800–1911, part I) (Vol. 10, pp. 264–317). Cambridge: Cambridge University Press.

Lee, H. F., Pei, Q., Zhang, D. D., & Choi, K. P. K. (2015). Quantifying the intra-regional precipitation variability in Northwestern China over the past 1,400 years. *PLoS One.* https://doi.org/10.1371/journal.pone.0131693.

Lee, H. F., & Zhang, D. D. (2010). Natural disasters in Northwestern China, 1270-1949. *Climate Research, 41*, 245–257.

Li, L. (2007). *Fighting famine in North China: State, market, and environmental decline, 1690s-1990s.* Stanford, CA: Stanford University Press.

Lieberman, V. (2003). *Strange parallels: Southeast Asia in global context, c. 800-1830. Vol. 1: Integration on the mainland.* Cambridge: Cambridge University Press.

Lieberman, V. (2009). *Strange parallels: Southeast Asia in global context, c. 800–1830* (Mainland mirrors: Europe, Japan, China, South Asia, and the Islands) (Vol. 2). Cambridge: Cambridge University Press.

Liu, K.-C. (1978). The Ch'ing restoration. In J. K. Fairbank (Ed.), *The Cambridge history of China. Vol. 10, Late Ch'ing, 1800-1911, part I* (pp. 409–490). Cambridge: Cambridge University Press.

Mallory, W. H. (1926). *China: Land of famine.* New York: American Geographic Society.

Mann Jones, S., & Kuhn, P. A. (1978). Dynastic decline and the roots of rebellion. In J. K. Fairbank (Ed.), *The Cambridge history of China. Vol. 10, Late Ch'ing, 1800-1911, part I* (pp. 107–162). Cambridge: Cambridge University Press.

Mann, M. (2012). *The sources of social power* (A history of power from the beginning to AD 1760. New edn. (orig. 1986)) (Vol. 1). Cambridge: Cambridge University Press.

Mote, F. W. (1999). *Imperial China 900-1800.* Cambridge, MA: Harvard University Press.

Naquin, S. (1981). *Shantung rebellion: The Wang Lun rebellion of 1774.* New Haven, CT: Yale University Press.

Ostrom, E. (1990). *Governing the commons: The evolution of institutions for collective action.* New York: Cambridge University Press.

Ostrom, E. (2009). A general framework for analyzing sustainability of social-ecological systems. *Science, 325*, 419–422.

Parker, G. (2013). *Global crisis: War, climate change and catastrophe in the seventeenth century.* New Haven, CT: Yale University Press.

Perdue, P. (2005). *China marches West: The Qing conquest of Central Eurasia.* Cambridge, MA: Harvard University Press.

Peterson, W. J. (Ed.). (2002). *The Cambridge history of China, vol. 9, part 1: The Ch'ing dynasty to 1800.* Cambridge: Cambridge University Press.

Peterson, W. J. (Ed.). (2016). *The Cambridge history of China. Vol. 9, Part two: The Ch'ing dynasty to 1800.* Cambridge: Cambridge University Press.

Pomeranz, K. (2000). *The great divergence: China, Europe, and the making of the modern world economy.* Princeton, NJ: Princeton University Press.

Rowe, W. T. (2009). *China's last empire: The Great Qing.* Cambridge, MA: Harvard University Press.

Ruddle, K., & Johannes, R. E. (Eds.). (1983). *The traditional knowledge and management of coastal systems in Asia and the Pacific*. Jakarta: UNESCO, Regional Office for Science and Technology for Southeast Asia.

Smith, P. J. (2003). Impressions of the Song-Yuan-Ming transition: The evidence from Biji memoirs. In P. J. Smith & R. von Glahn (Eds.), *The Song-Yuan-Ming transition in Chinese history* (pp. 71–134). Cambridge, MA: Harvard University Asia Center.

Smith, T. C. (1977). *Nakahara: Family farming and population in a Japanese village, 1717-1830*. Stanford, CA: Stanford University Press.

Spence, J. (1974). *Emperor of China: Self-portrait of K'ang-Hsi*. New York: Random House.

Spence, J. (1997). *God's Chosen son: The Taiping heavenly Kingdom of Hong Xiuquan*. New York: W. W. Norton.

Tagliacozzo, E., & Chang, W.-C. (Eds.). (2011). *Chinese circulations: Capital, commodities, and networks in Southeast Asia*. Durham, NC: Duke University Press.

Tatsuya, T. (1991). Politics in the eighteenth century. In J. W. Hall & J. L. McClain (Eds.), *The Cambridge history of Japan. Vol. 4: Early modern Japan* (pp. 425–478). Cambridge: Cambridge University Press. Tr. Harold Bolitho.

Thunberg, C. P. (1795). *Travels in Europe, Africa, and Asia, made between the year 1770 and 1779. Vol. 4, Containing travels in the Empire of Japan, and in the Islands of Java and Ceylon, together with the voyage home*. London: F. and C. Rivington.

Toshio, F. (1991). The village and agriculture during the Edo period. In J. W. Hall & J. L. McClain (Eds.), *The Cambridge history of Japan. Vol. 4: Early modern Japan* (pp. 478–518). Cambridge: Cambridge University Press. Tr. James L. McClain.

Totman, C. (1989). *The green archipelago: Forestry in preindustrial Japan*. Berkeley, CA: University of California Press.

Tsuya, N., & Kurosu, S. (2010). Family, household, and reproduction in Northeastern Japan, 1716 to 1870. In N. O. Tsuya, W. Feng, G. Alter, & J. Z. Lee (Eds.), *Prudence and pressure: Reproduction and human agency in Europe and Asia, 1700–1900* (pp. 249–285). Cambridge, MA: MIT Press.

Tsuya, N. O., & Kurosu, S. (2014). Economic and household factors of first marriage in two Northeastern Japanese villages, 1716-1870. In C. Lundh, S. Kurosu, et al. (Eds.), *Similarity in difference: Marriage in Europe and Asia, 1700-1900* (pp. 349–392). Cambridge, MA: MIT Press.

Von Glahn, R. (2015). *The economic history of China*. Cambridge: Cambridge University Press.

Walker, B. L. (2001). *The conquest of Ainu lands: Ecology and culture in Japanese expansion, 1590-1800*. Berkeley, CA: University of California Press.

Walker, B. L. (2015). *A concise history of Japan*. Cambridge: Cambridge University Press.

Wang, F.-s. (2016). Political pressures on the cultural sphere in the Ch'ing period. In J. K. Fairbank (Ed.), *The Cambridge history of China. Vol. 10, Late Ch'ing, 1800-1911, part I* (pp. 606–648). Cambridge: Cambridge University Press.

Wong, R. B. (1997). *China transformed: Historical change and the limits of European experience*. Ithaca, NY: Cornell University Press.

Lessons: Factors Driving the Rise and Fall of Dynasties

10.1 Climate and Chinese Dynasties

> History and anthropology very often can make little use of Occam's razor. (Khazanov 1984: 278, on the infuriating complexity of historical causation)

Bret Hinsch (1988), and others since, have predicted that better times—warmer and wetter, with stronger monsoons—will predate the rise of dynasties, while worse times—colder and drier—will predate their fall. On the other hand, colder times may lead to consolidation. Difficulties may lead to reliance on central organization.

The Xia Dynasty began at a time of dramatic cooling, 2400–2000 BCE, but apparently also a time of flooding—a rare and portentous combination in Chinese history. Some historians theorize the cold may have led to consolidation in what was the economic center of northern China. Since not only Xia but also other large towns and extensive polities appeared at this time, the evidence supports that conclusion. Worldwide, the same time period saw rise of empires and consolidation of power in Mesopotamia and Egypt, and appearance of large towns and extended borders in the Near East and elsewhere, confirming a link between hard times and centralization.

The Shang, and Zhou Dynasties are too poorly known for meaningful correlations. The rise of Qin and especially of Han accompanies the Roman Republic/Empire Optimum. The interregnum of 9–23 CE followed some bad years that may partially explain it, though history suggests that imperial household dynamics were far more important than the quite minor climatic fluctuation. The fall of Han tracks the beginning of the end of the Roman Empire Optimum.

The spectacular rise and Sui and Tang, and the beginning of Tang's glory days, coincide with a sharp deterioration in climate. Not only did China rise, but the conquest by the founders of Sui and Tang came from the hardest-hit area, the north edge of China where it fringes into Central Asia. Tang benefited from good weather after 650. The fall of Tang accompanies drought and heat associated with the uneven beginning of the Medieval Warm Period. One would have expected Tang to flourish more from the better monsoons of the 800s, but the dynasty was weakened by political factors. The same time period saw the rapid rise of civilization and empire in Korea and Japan.

The rise of Song is associated with a more strong and reliable monsoon at the start of the Medieval Warm Period. The Liao, Jin, and Yuan (Mongol) Dynasties rose during the Medieval Warm Period, which made it far easier for them to increase their herds and manpower, increase their food supply, and conquer outward. Korea unified and reached a golden age. But the reduction of Song and its eventual fall took place in relatively cool times, which should have weakened the northern regimes in relation to Song. Sharp and sudden cooling breaks in the Medieval Warm Period, especially around and after 1100, certainly hurt Song (McDermott and Yoshinobu 2015; Yin et al. 2016; Zhang 2016).

Yuan took power when the Medieval Warm Period was still in its favor, but it declined as that age gave way. The succeeding Ming Dynasty had a difficult job: running the empire during a period of unprecedented cold and dry conditions. It succeeded, not losing power for centuries. However, even worse cold and drought supervened in the early 1600s, probably hastened its fall (Brooke 2014; Parker 2013). At the same time, Japan suffered chaos in the sixteenth century, unification and rapid growth in the seventeenth.

Then comes the strangest thing of all. Ming was conquered not by a powerful regime, not by internal unrest, but by the tiny Manchu state—a state that was based in China's frigid and snowy northeast, and that suffered miseries from the exacerbation of the Little Ice Age in the early 1600s. Outside of traditional Chinese ascriptions of success to the personalities of the early Qing emperors, there is no way to explain this. Ibn Khaldun would surely have found Nurhachi and the Kangxi Emperor model dynastic founders, able to arouse *'asabiyah*. Similarly, the decline and fall of Qing took place during a period of steadily ameliorating climate. This warming trend both produced more floods (the monsoon strengthened) and more droughts (heat exacerbated dry weather), stressing a densely populated, politically stagnating empire.

In short, Xia, Sui, Tang, and Qing rose during cold times. Qin, Han, Song, Liao, Jin, and Yuan rose during warm ones. Han, Sui, Yuan, and Ming fell during cooling to cold times. Tang, Northern Song, Liao, Jin, and Qing fell during warm times. It would be hard to be more evenly balanced. Clearly, climate is not the only factor. Cold seems to lead to consolidation by rising central powers at the expense of margins, but to weakening of central powers that are already downward-bound. Warmth leads to overall dynamism, but especially to strengthening margins at the expense of the center. However, such things as the rise of Qing make even these vague correlations seem hollow.

One could, of course, come up with contrived post-hoc explanations for the seemingly perverse rises of Sui, Tang, and Qing, and the perverse weakness of Song, but there is no way to save climatic change as a critical driving variable in those cases.

Many key papers emerge from a group of climate historians in Hong Kong, led by David Zhang and Harry Lee. For example, Jianyong Li, writing with them and several others (Li et al. 2017), finds rough warm/cool cycles of around 22, 50, and 100-year durations, associated with solar output. They find precipitation—specifically dryness—correlates twice as well with dynastic changes as temperature does. We have already noted associations of drought or drying trends with the decline of Han, Tang, Yuan, and Ming. Wetter periods seem more associated with glory days than with dynastic rises; the best times of Han, Tang, Song, and Ming were during rather moist periods. Still, correlations are far from perfect—not good enough to imply simple direct causation. Good times are probably enablers, but not determiners.

Harry Lee and his colleagues looked at dry and drought-prone northwest China (Lee and Zhang 2010; Lee et al. 2015, 2016). They found that wild fluctuations in rainfall characterized the Little Ice Age, with many droughts, but that the famously peaceful period of the middle Qing Dynasty from 1700 to 1820 saw a lack of famines and a rise in population, because of successful land management and the coming of New World crops (Lee et al. 2016). In fact, bad weather, especially in the south, was more disruptive to food production than the steady rise of population, which was accommodated well (Pei et al. 2016a).

In far northwestern China, westerlies and north winds dominated, totally decoupling that region from the rest of China and making its climate countercyclical (Zhang et al. 2016, 2018), but the area was not part of the Chinese cultural realm at the time.

Lee et al. (2017) showed that floods tend to precede armed conflicts in the rice-growing south, while both floods and droughts precede wars in the wheat-growing core of the north. Again the correlation is not strong. The pastoral and colder margins—Tibet, central Asia, northeast China—do not show clear correlations with either one. Rising population to the point of high density is associated with rising risk of war in all regions.

Lee (2018) analyzed a database of 1315 internal violent episodes and wars in China from 1470 to 1911. He found that epidemics tended to precede and presumably affect wars in north China. Famine was more prone to precede conflict in the south, though of course it affected the north too. Famine-caused conflict only made the famine worse. Much fighting there was desperate attempts to capture food. Disease is more a constant in the south than in the north, famine may be more constant in the north. In general, food prices are a good predictor of internal war, so invasion (basically Mongol and Manchu) is also related. The paper is a model of cautious use of a vast database to generate multicausal explanations for violent events.

Migration is similarly correlated with climate events and population increase (Pei et al. 2016a, b). It makes sense that too much water is worse than too little in the rainy south, while both flood and drought cause trouble in the agricultural north. The outer regions, China's classic semiperiphery and periphery, are more susceptible to cold dry times. We can conclude with Fan (2010), Kidder et al. (2016), and Wei et al. (2015) that climate can help or harm, but does not make or break.

Yin et al. (2016) looked at imperial China from Qin through Qing. They find that social rise was associated with warming (which normally meant wetting too) 57% of the time, and decline with cooling and drying 66.6%. (The very few warm-dry and cool-wet periods did not so correlate.) This is not compelling; the first is statistically not obviously better than chance. We shall have to look for other explanations here. They gathered 1586 data points from the standard histories of China, and parceled out even such things as particularly dynamic reign periods when China expanded its power, e.g. under Han Wu Di (140–87 BCE), who conquered neighboring areas during a relatively warm period. They miss the fact that the warm period should have, and in fact did, benefit his enemies as much as it benefited him, forcing him to fight hard and spend the empire's wealth. They also ignore the sharp cold snap at the start of the Sui-Tang period, smoothing it out into the long warm period of Tang, though their illustrated temperature curves clearly show the cold snap. They also find that historical records of good and bad times are particularly good for Han, bad for the Tang-Song interregnum and the Song Dynasty—fitting the history of war and conquest in the latter case. They find that China was peaceful 68.4% of the time, turbulent otherwise.

Qiang Chen (2015), on the other hand, thinks drought and cold did it. Cold was associated with more wars—a claim that does not explain the violent Medieval Warm Period or the long, peaceful Ming Dynasty. He finds that the main correlates were age of dynasty (older ones were weaker; but that goes for nomad regimes too) and drought. This does not check with the warlike but pleasantly warm period from 220 into the 500s, though it does coincide with the rise of Sui and Tang. It does not work for the Mongols.

Wei et al. (2015) find that climate events are related to dynastic cycles. They provide a careful, methodologically interesting assessment of troubles, with a very full bibliography. They use Holling's resilience cycle. They find a fair correlation of moist warm periods with good times, and vice versa, but they note the obvious Ming exception. They find a major crash in post-1420 Ming, though with recovery. They, like Kidder et al., focus on the Xin Dynasty interregnum in the Han Dynasty, attaching more importance to it than do most historians.

From the same group comes a recent paper by Yun Su, Wei, and others (Su et al. 2018). They see three subsystems interacting within the Chinese human-ecological system: "environment and resource subsystem, . . . the support subsystem" including production and infrastructure, and "the socio-economic subsystem." They looked at 4186 famines from Former Han to Qing. Out of 2117 years, 39% had famines—826 years. Cold units (decades) had 1.34 times as many severe and 1.16 mild famines, compared to warm decades. Warm units had 1.13 times as many good harvests and 1.28 times as many very good ones. Cold units had 2.21 times as many economically depressed decades. Proportions of mild, moderate and severe famines differed expectably: 11.7% mild, 41.2 moderate, 47.1 severe in bad harvest years, 37.8, 32.4, and 29.7 in good harvest years, 52.8, 32.9, and 8.3 in bumper years (note there were famines even in bumper years). All possible pathways existed: cold time to good harvests but followed by famine even so, for instance, actually occurred once or twice. Policy, war, and other political events could and did offset every climatic occurrence, producing great times in cold years and very bad times in warm years. The coupling of climate to economics was more direct before 1100; after that, policy made hard times more bearable when the government was strong, because relief infrastructures fairly steadily improved. Improvement was also due in part to the shift of the economic center to the south, and to building canals. Bad times were, of course, inevitable in dynastic collapses, no matter what the weather did. Periods of sharp improvement (famine succeed by good times) were 70–90 CE, 580–590, 1140–50, 1280–1300, 1400–1410, 1680–1760 (note that the last is much longer than the others—a sustained improvement in early Qing). Good governance was almost as important as climate in making a decade good.

Another approach is to look at local regions, which often had quite different climate histories from the rest of China. In northeastern China, it was the Medieval Warm Period that was problematic, causing many floods, often alternating with horrific droughts in wild swings (Lee et al. 2015). We have already considered Ling Zhang's brilliant work about the consequences: progressive breakdown in management of the Yellow River and other water sources and wetlands (Zhang 2016). Her work fits with Peter Turchin's findings on cycles. During the disintegration cycle of Northern Song, politics got more and more polarized and acerbic, and one result was failure to come up with coherent, consistent policies for the Yellow River. As Turchin says, "During the disintegrative phases…it is very difficult to generate the cooperative action needed to win a major war" (Turchin 2016: 106). That was true of Song's war with the conquest dynasties, and it was also true of Song's war with the Yellow River.

Large-scale forces act indirectly, through people. Direct causes of social events are personal decisions, and the resulting actions. This follows Anthony Giddens' idea of "structuration" (Giddens 1984, based heavily on Max Weber's ideas of individuals in society; see also Latour 2005, for similar theories). People, not climate or culture or society, do things, though people act in response to climate, culture and society.

During the well-recorded period from 770 BCE to 1912 CE, there were 3756 wars and serious conflicts in China—1.4 per year (Kang et al. 2016, citing the database "Chronology of Wars in China through Successive Dynasties," 2003; somewhat different data are used by the Harry Lee-David Zhang group; see Lee et al. 2017). Most were internal; rebellions, civil wars, conflicts between states in breakdown periods. Most of the rest were border wars and skirmishes along the Inner Asian borders. There were also many pirate conflicts.

By a standard definition of war—a conflict with over 1000 deaths within a year— only a rather small percentage of these 3756 episodes were wars (Kang et al. 2016). Still, there was no lack of genuine, serious war, and these wars could be appallingly bloody, with literally millions of deaths. Typically, in wars, more people died from famine caused by disruption than by violence. Violent deaths were often due to social disruption and breakdown more than to actual fighting.

China over time displays a rather astonishing ability to move from total war to considerable peace. Strong imperial control led to peace; when it weakened, chaos (*luan*) resulted. The contrast with Europe and west and central Asia—disunited and torn by endless brushfire wars and "barbarian" raids—is interesting.

Warfare shows some correlation with cold periods (Zhang et al. 2007, 2014). This is especially notable for interdynastic wars. Shortly after the start of cold periods came the rise of Sui and Tang and the falls of Northern Song, Southern Song, Yuan, and Ming, as well as the Taiping rebellions. From Qin onward, China saw six great dynasties, averaging 280 years each, and many shorter dynasties and kingdoms. The six ranged in length from the Mongol period at 153 years (following Mote in counting it from 1215 to 1368) to Han's slightly-interrupted 414.

Of the great dynasties (Qin through Qing), five were started by leading generals of previous regimes; two were started by semiperipheral marcher state conquest; one was a popular rebellion. The rebellions of the Yellow Scarves, Red Eyebrows, Huang Chao, and the Taipings also seriously affected national politics.

As we have seen, the early dynasties (Shang and Zhou, confined to the central area) were cases of conquest by semiperipheral marcher states. The major north-only regimes were also semiperipheral marcher state conquests (Wei, Liao, Jin, early Mongols). The minor north-only regimes, the southern regimes in the early interregnum, and the regimes of the Tang-Song interregnum were usually started by warlords or generals; typical was the revolving-door coup situation seen in the southern regimes of the Northern and Southern Period.

Ethnic wars did not release the genocidal fury typical in the west. Genocides of small minorities did occur, but rarely show up in the histories. Almost always, the Chinese preferred to keep them as taxpayers. Exceptions were largely late. Notable among these were the Dzungar genocide, and the massacres of Muslims following the Muslim rebellions of the 1840s–1850s.

Several conclusions emerge from this very quick overview of Chinese history. Definitive solution to the problems must rest on far better data and analysis, but some conclusions are clear.

First, climate is clearly a major factor in only one period, involving four dynasties: the fall of Song and the rise and fall of Central Asian conquest dynasties. The claims for Tang, Ming, and other dynasties are thin at best. For Tang, only very slight climatic fluctuations—as often as not in the "wrong" direction—occurred. In the case of Ming, the Little Ice Age did not bring down the dynasty, though a particularly cold snap was implicated in the last decades, when Ming was collapsing.

Second, Malthusian factors do not explain the timing of the various transitions. Population increased throughout much or most of every dynasty's run. Food production often did not keep pace. During the Golden Ages of the dynasties, agricultural modernization, tax reforms, and streamlined governance allowed an increasing population to be fed. This is evident in Han, Tang, Song, Ming, and early Qing (later Qing was under western influence and is hard to compare). A part of population growth was elite overproduction, especially the inexorable increase of the polygamous imperial families.

Third, imperial dynamics—bad governance, child emperors, mutually-destructive factions, out-of-control corruption, and other contingencies—are the most obvious proximate causes of dynastic collapses that allow new dynasties to emerge. The tragic spectacle of little boys ruling China was the immediate correlate of dynastic fall in Han (both Former and Later), Tang, Ming (though the boys at least

10.1 Climate and Chinese Dynasties

managed to grow up and try their best), Qing, and especially Song, with its quick succession of three young boys at the end.

Crises early in a dynasty were rarely fatal. They brought down dynasties in the cases of Qin, Sui, and several minor reigns such as Western Jin. Zhou survived early regentship, Han survived Empress Lu, Tang survived fratricidal rivalry and then the "Wu and Wei catastrophe," Song survived factions in the 1000s, Ming survived the Yongle coup, Qing survived the Rebellion of the Three Feudatories. Mid-dynasty crises were harder to manage, but could be survived: Wang Mang's brief takeover in Han, An Lushan's rebellion in Tang, the Liao and Jin victories in Song. On the other hand, a large majority of China's dynasties and kingdoms were local and short-lived. They died out after a very few generations, usually without controlling much of the realm. The dynasties with the most initial 'asabiyah could survive two or three or even four Ibn Khaldun cycles, but eventually all were brought low.

Negative-sum games among court factions expanded out of control in late Tang and in late Northern Song, among other times, with devastating effects on governance. The opposite extreme—totalitarian control—was preferable in terms of dynastic stability, but fatal to progress; it was notably strong in most of Ming and the Qianlong period of Qing. The free, open debate that leads to better policy and to progress, as seen in the United States before 2017 and widely in Europe since the Enlightenment, was rare in China, occurring in Former Han, mid-Tang, and eleventh-century Song.

Fourth, most Chinese polities were relatively limited in area and usually short-lived: the Warring States, the Three Kingdoms, the Northern and Southern Dynasties of the interregnum between Han and Sui, the many short-lived states between Tang and Song, and many marginal states (such as Nanzhao) that once flourished within what is now China. Many of these, like the more successful Qin and Sui dynasties, lasted less than one Ibn Khaldun cycle; they barely survived their founders.

Typically, a founder took over in a coup, and then he or (more often) his immediate successor was displaced in another coup. Leading generals were normally the perpetrators. Some of the limited dynasties lasted one or two cycles and flourished considerably. Wei and Jin during the Han-Sui interregnum were particularly successful, lasting a couple of typical Ibn Khaldun cycles, and would repay more study. They were as successful and important in their way as the much better studied Liao and (later) Jin dynasties. Climate changes did not correlate with their rise or fall.

Fifth, and probably most important: All the dynasties that lasted any significant time exhibit the Ibn Khaldun dynamics of increasing landlordism, stockpiling of power and wealth by elites, widening gaps between rich and poor, increasing elite factionalism, and (usually) military overreach. Ibn Khaldun showed how these combine with demographics in a single, predictable complex. Declining dynasties showed a gradual change from widespread support of the government, and relatively honest performance by bureaucrats and courts, to increasing selfishness. People move from working for the common good to working for themselves within the system to working for themselves at the expense of the system.

Why did a few dynasties conquer all China and last for centuries, when most stayed small and had brief reigns? Further research, especially on the short-lived dynasties, will have to tell.

Dynasties compare with modern totalitarian regimes in their patterns of killing. Genocide is now known to be predictable: it occurs when a totalitarian government with an ideology of hate, fear, or exclusion takes over a country or is threatened in its control by civil unrest (Anderson and Anderson 2012, 2017). The same is true of China. Every new dynasty consolidated control by exterminating rivals. Sometimes this process reached levels that approach the psychopathic (Qin, Ming), sometimes it was more reasonable, but it always ran well beyond the rational. Disruptions during dynasties were always met with mass killing. The difference from modern regimes are that China did not stand on an ideology of ethnic hate, so did not usually kill along ethnic or religious lines. Massacres were what is now called "politicide": anyone suspect of actively opposing the regime was eliminated. Usually, their families were also exterminated, making kinship the main variable, as opposed to the religion and ethnicity that structure mass killing in the modern world. Small intractable ethnic groups were, however, decimated or exterminated, the most conspicuous case being the Qing massacre of the Dzungars. It may have been inspired by Russian practice.

Genocide is the coward's way of waging war. A brave leader confident of his or her power will take on the enemy, internal or external. A weak, frightened one will go after a helpless minority. ("His or her" is not mere political correctness; recall Empress Wu's quickness to take on Korea, and the Trung sisters' resistance to Han.)

Final death blows were of three simple types. Qin, Han, Sui, Qing, and many of the minor and regional dynasties fell to their own leading generals when the emperor was ineffectual. Invasions by semiperipheral marcher states, always northwestern or northern neighbors, ended Xia, Shang, Zhou, Song, Liao, Jin, Ming, and several of the Northern and Southern Dynasties. The third type—spontaneous uprising in the most populous part of China—was seen solely in the case of Ming. All the early and most of the late dynasties fell to semiperipheral marklords, while all the "great" dynasties in between fell when they became so rotten that their own leading generals (or, in the Ming case, a plebeian rebel) stepped in and took over.

This is an odd but not unique pattern; it is like that seen in India and Persia, where most conquest dynasties were Central Asian, but some, especially in Persia, were set up by leading generals.

A typical dynasty began with an initial period of consolidation, usually troubled by coups or rebellions. This consolidation period is usually followed by a golden age of peace and prosperity, often all too brief (Tang's lasted only a few glorious decades) but sometimes very long, as in Ming and Qing. This gives way to increasing rebellions, palace intrigues, and corruption, and final collapse into chaos. These correspond perfectly with the r, K, omega, and alpha phases of Hollings' cycle.

Entrenched reactionary interests often reached a stage at which they were losing so much from change that they staked everything on fighting change. This was less obvious in China than it is in today's world, where fossil fuel corporations are driving a worldwide turn to fascism, but it certainly did happen in China. Highly traditionalist branches of the royal family provided one type of reactionary resistance. Landlords, especially giant families, provided another. Court blocs enriched by corruption provided another. Ibn Khaldun's observation that such problems increase over time, as population, wealth, and elite numbers increase, is confirmed.

A point of importance is that the massive shift of China's center of population and wealth from northwest to southeast, the change from aristocratic to partly meritocratic governance, the consolidation of autocracy in imperial hands, and other profound changes in China in the medieval period did not greatly affect the timing, nature, causes, or dynamics of dynastic cycles. The course of Qing was much like the course of Han. China changed enormously, but its cycles kept on their courses. This is not surprising to anyone using Ibn Khaldun's theories; they were, after all, developed through consideration of Islamic and European dynasties, and tested by scholars like Peter Turchin from modern European data. The farther from China we get, the more different the cycles, but still the structural similarities are astonishingly similar.

Susan Mann Jones and Philip Kuhn summarize Chinese and western ideas on dynastic decline: "Dynastic decline has traditionally implied a loss of moral and administrative vigour among the bureaucracy...an ebbing of centralized power and its accretion in the hands of regional satraps, a disruption of the balanced tension between state and society.... An image of dynastic decline also emerges...from the exploitation, careerism and inefficiency of local government" (Mann Jones and Kuhn 1978: 107).

However, a combination of high legitimacy and effective bureaucracy often allowed a rotten imperial government to last for decades. The extreme ranges were from Qin and Sui—which fell apart almost as soon as their powerful founders died—to Ming, which hung on for over two centuries of misrule. The long dynasties of Korea were also interesting in this regard. Chosŏn in particular survived some appalling misrule to survive over 500 years.

"Heaven is high and the emperor is far away," the proverb ran. But imperial China depended on highly complex and fine-tuned networks of trade, tribute, communication, and food distribution. Disrupting these by violent warfare over any significant time led to devastating excess mortality. The Taiping Rebellion of the 1840s was a particularly well-documented case. A few thousand died in battle, but the ultimate toll was in the tens of millions, because agriculture and commerce were disrupted; the vast majority of the deaths were from starvation or from local breakdowns of law and order systems (Spence 1997). The breakdown of Japan from 1573 to 1603 was similar.

10.2 Korea and Japan

The chronicles of Korea, Japan, Vietnam, and Central Asia show totally different patterns, not evidently related to climate change and not clearly predicted by Ibn Khaldun models. What does predict them is the situation in China, the huge core society in the East Asian world-system.

Korea's dynastic unification under Silla in 657 was directly caused by Tang interference. It was a result of Tang's own success and flourishing. Silla's fall in 938 is—as might be expected—a clear follow-up or knock-on effect of the fall of Tang. The changeover of Koryŏ to Chosŏn was the direct and clear result of the power of Ming. These were the only important dynastic transitions in Korea's amazingly stable premodern history.

In Japan, the single Yamato dynasty held ceremonial power throughout history, and still does so, but actual power was held by a succession of great families. The rise and fall of these regimes follows Ibn Khaldun's model and Peter Turchin's long-cycles model quite well. A family would last until it became so ramified, detached from the people, and overextended that it could not cope. The Fujiwara family in the Heian period, for instance, demonstrates Turchin's principle of elite overproduction; countless Fujiwara branches vied for power in an ever-shifting web of cooperation or competition. Then the Minamoto lineage suffered the same fate in the Kamakura period. Only the Tokugawa managed to keep their lineage together, falling in the end not to climate or family strife but to an odd mix of modernists and reactionaries, held together only by opportunism during a spell of Tokugawa weakness.

Japan's course was not so related to China as was Korea's, but the transition from Nara to Heian was related to affairs in Tang and is not related to anything climatic. It happened during a relatively pleasant climatic period. The fall of Heian and the rise of the Kamakura bakufu (the key year being 1185) reflects a very warm and wet period, which among other things caused devastating floods in Kyoto. The Kamakura bakufu gave way to the Ashikaga in 1336, during a cool, dry period. Ashikaga was stressed by bad weather through its entire troubled career, to its final collapse in 1573. The wars that followed, and the rise of the Tokugawa shogunate, are strongly correlated with some of the worst of the Little Ice Age. Finally, the early seventeenth century, arguably the coldest time in the Little Ice Age, was associated with the winding down of the wars and the takeover of the Tokugawas. Climate may have affected Japanese history, but changes occurred during bad, good, and ordinary climatic moments.

The bakufu changes in Japan show some typical Ibn Khaldun cyclic effects. Successful military commanders began them. They rose to great heights culturally and economically, then declined as rising population, increasing palace politics, increasing inequality and distance of elite and multitude, and military challenges conspired to stress them to the breaking point. The full panoply of Khaldunian decline is hard to find in the stories of the bakufu transitions. The stability of the Imperial dynasty and the instability of the military families kept the government from the full extent of slow deterioration into corruption and bureaucratic inertia that marked China's dynastic declines. The fall of a bakufu dynasty was typically associated with building rivalries between the Emperor's court and the bakufu government, this conflict being complicated by the shifting loyalties of the daimyos.

Japan profited by its insular position. Protected by sea and storms from invasion, let alone conquest, it never had to fight invaders, except for the brief Mongol invasions foiled by "divine winds."

10.3 Vietnam

Vietnam's fate is so tied to China that it had little opportunity to react to climate, though bad times were reflected in weakening dynasties. Strong dynasties (Han, Yuan, Ming) conquered it and held it with varying success. During weak periods in China, Vietnam became independent. The Medieval Warm Period, so advantageous elsewhere in East Asia, hurt Vietnam badly, causing regional droughts, probably most severe in the 900s and 1100–1200s.

Climate has influenced Vietnam's history, along with other factors, but nothing determinative stands out. The usual droughts and floods occurred, sometimes coinciding with dramatic events. Climate softened Vietnam for French takeover, but one doubts if it was decisive. The French were determined to have an empire.

In the East Asia was a world-system, Vietnam had to deal directly with the Chinese. Its history consisted of fighting off the Chinese Empire or trying to recover from the effort. China influenced Vietnam powerfully in language, religion, worldview, governance, and everything else cultural. Vietnam retained its language and a cultural core.

10.4 Central Asia

The states of Central Asia, from nomadic herding polities like the Mongols to oasis-based agricultural civilizations like the Uighurs, rose and fell in response to currents in China and western Asia. The rise of Buddhism and the later rise of Islam sent powerful currents of religious enthusiasm—coupled with military might in the latter case—washing over the region. Trade rose and fell with the power of the great states at the ends of the Silk Routes. Not only China and Persia, but India and Tibet, influenced trade and sometimes extended military power. Tibet remained a major player in southeast Central Asian politics and war until it was definitively brought into Beijing's orbit by the Manchus.

Early powers, notably the Xiongnu, Xianbei, Gök Turks, and Xixia, had their hours of glory, but faded, especially when caught between China to the east and other Central Asian powers to the north or west. They are gone so completely that we do not even know what language the Xiongnu spoke. The later Khitan and Jurchen succeeded better and are better known, but they too faded, though the Khitan survived as the Karakitai for a long time, and the Jurchen's heirs included the triumphant Manchus.

Mongolia rose and fell spectacularly. The Medieval Warm Period gave it an opportunity at the exact time when China was disunited and vulnerable. After the end of medieval warmth, Mongolia declined again.

The Uighur, Turkic, and Sogdian states were much more stable and long-lasting. The Sogdians eventually were religiously and linguistically converted to Islam and Farsi, but in that form they retain power in southwestern Central Asia. The Turkic states are resurgent today in Kazakhstan, Kirgizstan, and Uzbekistan, but are crushed in China. All these local states had their own dynastic dynamics, often reprising Ibn Khaldun's model very well. The Little Ice Age broke Central Asian power and wealth, but it is now slowly recovering, though inept environmental and agricultural policies may yet ruin it again.

10.5 Putting It Together: World-System Dynamics

The conclusion of all this effort is that climate change has influenced the East Asian world-system, but only as part of the back story. It was sometimes highly influential in Central Asian and Chinese history, but not very influential directly in Korea, Japan, or Vietnam. Their histories and dynastic cycles were so heavily influenced by China, or in the case of Japan by local political contingencies, that climate could not play a major role.

On the other hand, Ibn Khaldun's theory of dynastic cycles, especially as recently updated by Peter Turchin, is highly explanatory, though it fails to predict exact dates. Particularly notable are the confirmations of the importance of semiperipheral marcher states ("barbarians"), population increase including "elite overproduction," and the growth of corruption and amoral self-serving over time as elites become more distant from masses. The last of these links macro-level events to micro-level human decisions, which is clearly the most necessary and possibly the most difficult thing to do in explaining history.

People, especially rulers, often made decisions with one eye cocked on what was happening in neighboring countries. Mongols, Turks, and Serbi harried the borders, taking over whole countries when they could. Koreans and Vietnamese calculated how to maintain cultural and, if possible, political independence in the presence of the vast Chinese force next door. Japanese tried to balance international trade and raid. All depended heavily on international commerce to supply their needs and wants.

World-systems theory is nudged a bit toward an agent-based view of history, as opposed to one privileging back-story forces, and toward a view that looks at power and control as well as economics.

References

Anderson, E. N., & Anderson, B. A. (2012). *Warning signs of genocide*. Walnut Creek, CA: AltaMira.

References

Anderson, E. N., & Anderson, B. A. (2017). *Halting genocide in America*. Chesterfield, MO: Mira.
Brooke, J. L. (2014). *Climate change and the course of global history*. Cambridge: Cambridge University Press.
Chen, Q. (2015). Climate shocks, dynastic cycles and nomadic conquests: Evidence from historical China. *Oxford Economic Papers, 67*, 185–2024.
Fan, K.-w. (2010). Climatic change and dynastic cycles in Chinese history: A review essay. *Climatic Change, 101*, 565–573.
Giddens, A. (1984). *The constitution of society*. Berkeley: University of California Press.
Hinsch, B. (1988). Climatic change and history in China. *Journal of Asian History, 22*, 131–159.
Kang, D., Shaw, M., & Fu, R. T.-m. (2016). Measuring war in early modern East Asia, 1368-1841: Introducing Chinese and Korean language sources. *International Studies Quarterly, 60*, 766–777.
Khazanov, A. M. (1984). *Nomads and the outside world*. Cambridge: Cambridge University Press. Tr. Julia Crookenden (Russian original 1983).
Kidder, T. R., Haiwang, L., Storozum, M. J., & Zhen, Q. (2016). New perspectives on the collapse and regeneration of the Han dynasty. In R. K. Faulseit (Ed.), *Beyond collapse: Archaeological perspectives on resilience, revitalization, and transformation in complex societies* (pp. 70–98). Carbondale, IL: Southern Illinois University Press.
Latour, B. (2005). *Reassembling the social: An introduction to actor-network-theory*. Oxford: Oxford University Press.
Lee, H. F. (2018). Internal wars in history: Triggered by natural disasters or by socio-ecological catastrophes? *The Holocene, 28*, 1071–1081.
Lee, H. F., Pei, Q., Zhang, D. D., & Choi, K. P. K. (2015). Quantifying the intra-regional precipitation variability in Northwestern China over the past 1,400 years. *PLoS One*. https://doi.org/10.1371/journal.pone.0131693.
Lee, H. F., & Zhang, D. D. (2010). Natural disasters in Northwestern China, 1270-1949. *Climate Research, 41*, 245–257.
Lee, H. F., Zhang, D. D., Pei, Q., Jia, X., & Yue, R. (2016). Demographic impact of climate change on Northwestern China in the late imperial era. *Quaternary International*. https://doi.org/10.1016/j.quaint.2016.06.029.
Lee, H. F., Zhang, D. D., Pei, Q., & Yue, P. H. R. (2017). Quantitative analysis of the impact of droughts and floods on internal wars in China over the last 500 years. *Science China Earth Science, 60*, 2078–2088.
Li, J., Dodson, J., Yan, H., Zhang, D. D., Zhang, X., Xu, Q., Lee, H. F., Pei, Q., Cheng, B., Li, C., Ni, J., Sun, A., Lu, F., & Zong, Y. (2017). Quantifying climatic variability in monsoonal Northern China over the past 2200 years and its role in driving Chinese dynastic changes. *Quaternary Science Reviews, 159*, 35–46.
Mann Jones, S., & Kuhn, P. A. (1978). Dynastic decline and the roots of rebellion. In J. K. Fairbank (Ed.), *The Cambridge history of China. Vol. 10, late Ch'ing, 1800-1911, part I* (pp. 107–162). Cambridge: Cambridge University Press.
McDermott, J. P., & Yoshinobu, S. (2015). Economic change in China, 960-1279. In J. W. Chaffee & D. Twitchett (Eds.), *The Cambridge history of China. Vol. 5, Part two: Sung China, 960-1279* (pp. 321–436). Cambridge: Cambridge University Press.
Parker, G. (2013). *Global crisis: War, climate change and catastrophe in the seventeenth century*. New Haven, CT: Yale University Press.
Pei, Q., Li, G., Zhang, D. D., & Lee, H. F. (2016a). Temperature and precipitation effects on agrarian economy in late imperial China. *Environmental Research Letters, 11*, 064008.
Pei, Q., Zhang, D. D., & Lee, H. F. (2016b). Contextualizing human migration in different agro-ecological zones in ancient China. *Quaternary International, 426*, 65–74.
Spence, J. (1997). *God's chosen son: The Taiping heavenly Kingdom of Hong Xiuquan*. New York: W. W. Norton.
Su, Y., He, J., Fang, X., & Teng, J. (2018). Transmission pathways of China's historical climate change impacts based on a food security framework. *The Holocene, 28*, 1564–1573.

Turchin, P. (2016). *Ages of discord*. Chaplin, CT: Beresta Books.
Wei, Z., Rosen, A. M., Fang, X., Su, Y., & Zhang, X. (2015). Macro-economic cycles related to climate change in dynastic China. *Quaternary Research, 83*, 13–23.
Yin, J., Su, Y., & Fang, X. (2016). Climate change and social vicissitudes in China over the past two millennia. *Quaternary Research*. https://doi.org/10.1016/j.yqres.2016.07.003.
Zhang, D., Pei, Q., Lee, H. F., Zhang, J., Chang, C. Q., Li, B., Li, J., & Zhang, X. (2014). The pulse of imperial China: A quantitative analysis of long-term geopolitical and climatic cycles. *Global Ecology and Biogeography*. https://doi.org/10.1111/geb.12247.
Zhang, D. D., Zhang, J., Lee, H. F., & He, Y.-q. (2007). Climate change and war frequency in Eastern China over the last millennium. *Human Ecology, 35*, 403–414.
Zhang, L. (2016). *The river, the plain, and the state: An environmental drama in Northern Song China, 1048-1128*. New York: Cambridge University Press.
Zhang, Y., Meyers, P. A., Liu, X., Wang, G., Li, X., Yang, Y., & Wen, B. (2016). Holocene climate changes in the Central Asia Mountain region inferred from a peat sequence from the Altai Mountains, Xinjiang, Northwestern China. *Quaternary Science Reviews, 152*, 19–30.
Zhang, Y., Yang, P., Tong, C., Liu, X., Zhang, Z., Wang, G., & Meyers, P. A. (2018). Palynological record of Holocene vegetation and climate change in a high-resolution peat profile from the Xinjiang Altai Mountains, Northwestern China. *Quaternary Science Reviews, 201*, 111–123.

Comparisons: Cycles and Empires in Agrarian Worlds

11.1 Cycles Rising

The back story to historical change is summed up by Bruce Campbell (2016: 22) as "climate, ecosystems, microbes, humans, biology, society." One could also list them as climate, ecological factors, disease, demography, and secular economic change. Human-caused but *longue durée* matters such as soil erosion, deforestation, desertification, watercourse management, irrigation works, medical science, developing technology, and entrenched moral systems affect all these, nowhere more than in East Asia, where they came over time to be as much as part of the environment as mountains and rivers.

The mid story is current economic and political management: shifting patterns of trade, rising or falling corruption, governmentality in all senses, inexorable growth over time within regimes of bureaucracy and hierarchy, foreign pressures, and the like. Human emotions are marshalled, deployed, and managed at this level, especially by charismatic leaders who command 'asabiyah.

The front story—the immediate story on the ground—is one of child emperors, mad emperors, invading armies, charismatic rebels, key battles that could have gone either way, and what passed for "history" in earlier times—the old "kings and battles" stories. The pendulum has swung from that view to a view so focused on the *longue dureé* that the actual timing of events can no longer be explained; everything is the result of vast, impersonal forces. Grand forces can explain why Song fell, but cannot explain why it fell in 1279. Only such contingent factors as the personalities of Kublai Khan and Huizong, the demographic accident that gave the dying dynasty three young boy emperors in a row, and the specific decisions of high government officials, can explain that.

11.2 Models and Traditions

One of the most striking things in world history—and not often noted by comparative historians—is the contrast between East Asia's long-lived dynasties and the constant changes of dynasty that afflicted the western world. Korea's dynasties had long periods of rule, and Japan has had only one dynasty, though the real power was held by shoguns who lost power at long-cyclic intervals. By contrast, the Roman Empire lurched from coup to coup. Some emperors lasted only a few months, and rare was the dynasty that came anywhere close to a full Ibn Khaldun cycle. This was especially true in the decline centuries (200–500), but even the later Byzantine Empire, with its more stable and bureaucratic government, had frequent changes of dynasty. The usual scenario was for a general from the frontier regions to become rapidly powerful through local victories, build a power base in the army, and sweep down on Rome from his borderland position (Harper 2017). Sometimes two or three generals at once attacked, either fighting each other or dividing up the empire. Changes of ruling dynasty continued in England, France, and elsewhere, until the Hapsburg and other interlocked families established themselves on the thrones of Europe. The Tudors and Stuarts each lasted less than a Khaldunian century.

In China, this sort of revolving-door coup situation happened in the disunion period following the fall of Han and continuing until the consolidation of Tang. It happened yet again between Tang and Song. Japan, Vietnam and Korea had their periods of disunion. One reason in the east is the success of Qin and especially Han at developing a centralized, bureaucratic, empire that could rule no matter how incompetent the emperor was. There were other reasons, including the geographical fractionation of Europe, its chronic religious wars, and its dependence on shifting trade. One must also mention the successful development in Former Han of a ruling ideology that combined Confucian responsibility and individual duty with Legalist rule of law and detailed specification of bureaucratic and judicial responsibilities. The Confucian tendency to support family over state and the Legalist tendency toward totalitarianism were balanced out and relatively controlled in this combination (a point missed in the glib saying that Han policy was merely the iron hand of Legalism in the velvet Confucian glove). Japan, Korea, and Vietnam all adopted these measures.

Some of the credit for East Asia's stability must go to one of history's least appreciated geniuses: Emperor Wen of Han. He maintained the centralized, autocratic tradition of Qin, but stocked the bureaucracy with brilliant courtiers. He improved the meritocratic promotion system. He broke the power of the great families, notably including rival branches of his own Liu lineage. He instituted low taxes and many public works. His outreach through these to the mass of Chinese offset his necessarily merciless behavior toward rivals in his family. This was the key: he combined consolidation of power with genuinely popular and beneficial steps for the whole country. He had enough sense to publicize the latter more than the former, and thus got a reputation as a good, beneficial ruler that survived despite his summary treatment of his less trustworthy relatives. Under him and his successors, China developed a system of imperial-bureaucratic rule that lasted for two millennia (Pines 2009, 2012).

This combination of autocracy and morality carried on, and all emperors from then to 1911 had to pay lip service to it. Even Wen Di's far less benevolent successor, Wu Di (r. 140–87 BCE), saw the good sense of it.

Always, we can see Kant's classic principles of aggregation and differentiation at work (Kant 1978 [1798]). People see a major but vague, ill-defined, and shaded border, and they over sharpen it by drawing in it a firm (but artificial and thus constantly shifting) line separating "us" from "them." In China, "we" may be verging on nomadism, herding sheep, drinking milk; "they" may be growing millet and brewing beer, weaving cloth and building houses; but "we" are still the settled Chinese and "they" are still the nomad barbarians (Barfield 1989). David Bello (2016) sees this as true essentialization, comparable to racism in the west; this claim is somewhat exaggerated, but, especially for the conquest dynasties and the Qing Dynasty, Bello has a case.

Bicultural or "marginal" men had a particularly good shot at the brass ring, especially in the Middle Ages, when the nomads and central Asians were relatively powerful. The founders of the Sui and Tang Dynasties were frontier generals and apparently part Turkic. Similarly, a disproportionate percentage of the Roman emperors who seized power by military coup got their power being generals on barbarian frontiers. Very often they relied heavily on barbarian troops, and some were part barbarian by heredity. Ibn Khaldun's marshallers of 'asabiyah were semiperipheral marklords.

11.3 East Asian Agriculture as Relief

"Does anyone today seriously doubt that diseases, or climates, or plants make history as much as any empire? At the same time, is it possible to articulate the role of diseases, plants, or climate abstracted form accumulation, empire, or class?" So writes Jason Moore (2015: 48). Moore reminds us that people are not separate from nature, but a part of it, and human processes are part of the big world. He traces the implications of this for theories of value, metabolism, and much more, taking us far beyond the theme of the present work, but usefully informing it.

Within the Old World, there is a long-standing difference between dryland west and rice-based east. From central Asia and north-central India westward, wheat, barley, and livestock were the basis of civilization. This combination allows extensive as opposed to intensive farming. Over the western world until the twentieth century, low yields were the rule. Grain yielded 300–500 pounds per acre, and livestock were largely cattle and sheep raised in vast, extended range operations or by nomadic pastoralists. This could be done on a small-farm basis, but ancient Mesopotamia already had large temple and royal estates. By Roman times, there were vast plantations, worked by slaves or serfs. Production per acre and per worker was low, so large areas had to be used. This sort of agriculture was highly susceptible to drought, war, unrest, and social change. The decline and fall of Rome, the stagnation of Byzantium, the decline of the Near East after 1300, and other long dismal periods were related to this ecological problem.

Some areas got self-locked into dryland and livestock systems that caused degradation over time, but could not be changed because there simply was no capital or free time or space. People could not afford to change, because they depended year to year on the system that was slowly killing them. Northwest Africa was one such system (McNeill 1992). Extensive agriculture led to deforestation, overgrazing, soil loss, and other environmental problems, but no one could do anything about it, given the poverty and the constant wars and invasions that troubled the region. Ethiopia today is in a similar trap, with low-level dry-grain agriculture destroying the land, but no clear way to change.

On the other hand, wheat and barley are resistant to drought. Barley tolerates incredible levels of drought, heat, salinity, and soil impoverishment. Of North China's original staples, broomcorn or panic millet is also drought-tolerant, but foxtail millet is less drought-resistant than barley, preferring wetter conditions.

Latin America, especially before Columbus, was trapped by maize, and this problem has now spread to China, where maize has replaced millets. It is a highly demanding crop, feeding heavily and requiring considerable and regular water. It is grown in open rows, leading to massive erosion when raised on slopes. Its demanding nature and susceptibility to even slight drought or flood makes it a risky crop.

Rice was the main grain throughout east and south India, Southeast Asia, and east Asia north to central China and central Korea. Rice responds well to intensification. Yields of 2500 pounds per acre per crop were typical of wet-rice agriculture in later traditional times, and some areas got two or three crops per year. Extensive livestock raising was not practical, and animals were grown on small farms. The dynamics of rice agriculture makes small owner-operated farms more successful, and large slave-worked estates never caught on, though they were tried everywhere. They tended to disappear over time, as economic dynamism and population growth led to more rationalization and intensification of the systems. The result was a singular lack of long Dark Age periods. China's interdynastic periods, and even the periods of chaotic disunion in the 200s and 900s A.D., were not long and were not as extreme as the western European Dark Ages. Korea, Japan and Vietnam did not even have this much disunion. Korea and Japan have suffered amazingly little from cycles and dark ages.

Southeast Asia had some shakier periods, notably with the decline of Angkor, but again never had anything like the long cycles of decline that often characterized the west. One may suspect that this stable system encouraged the ideology of concern and compassion for all life that dominates religious life in the region (Hinduism, Buddhism, Jainism, Southeast Asian animism). Rice requires intensive human manipulation. It is thus buffered against natural fluctuations. It is relatively reliable except at dry, cold margins of its range.

China and the rest of East Asia endured essentially annual catastrophes, incident on a population that always pressed far too close on a steadily diminishing resource base (Anderson 1988; Elvin 2004; Li 2007). As Kenneth Pomeranz (2000, 2010) emphasizes, China did well—given the challenges faced—through the eighteenth century, and rebounded successfully in the late twentieth. China suffered from wars before and after the collapse of the Qing Dynasty and the subsequent eventual rise of Communism. But at no time was the civilization even remotely threatened.

Serious droughts 8200, 5200, and 4200 years ago in the Near East caused major disruptions and abandonments of settlements, but significantly failed to bring down the cultures or even halt their progressive trends. By contrast, shifting loci of power and trade brought the Mesopotamian civilizations down, in a perfectly fine climatic time, some 2500 years ago. The Greeks and Persians expanded, taking over the region. Much, including the ancient languages, was lost.

Chase-Dunn et al. (2007) have recently argued that the now well-known correlation of population growth, urbanization, and imperial cycles in the western and eastern ends of Eurasia can be related to climatic cycles. This challenging and important hypothesis awaits better resolution of the dating involved. Favorable climate around 3000 BCE, sharp changes for the worse around 2000, recovery after 500 BCE, deterioration again from 200 CE peaking around 600, increasing warmth and the Medieval Warm Period at 900–1200, and the Little Ice Age all show effects across Eurasia.

During the droughts associated with the Medieval Warm Period, some central regimes failed or weakened, localism increased, and outsiders could conquer the areas (Yoffee 2005, 2006). The Mon and Khmer, for instance, fell because of military pressure from neighbors that gained relative strength from MWP conditions. They failed to rise again because the trade naturally went to those stronger neighbors—Burmese, Thai, Vietnamese (Lieberman 2003). These seized the best ports and other trading locations. The vast city of Angkor, which at its peak was as large in area as Los Angeles and may have had a million people, was deserted. Recent claims that environmental damage—deforestation, siltation, and so on—caused the decline (BBC 2007) indicate that environmental deterioration added to the stress. Drought was certainly a factor.

Vietnam's major crisis of the late sixteenth century, eventually resolved by a restoration of the long-running Le dynasty (Lieberman 2003: 394ff; see Lieberman 2009), was part of the same regional phenomenon. Central Asia hit a low point in power, unity, and empire-building, and here the Little Ice Age was most certainly a factor (though devastation of the caravan trade by expanding sea trade was evidently much more important).

An interesting mediator between ecological problems and decline has been proposed by Janssen and Scheffer (2004). They propose that "sunk costs" led civilizations to continue practices long after these had become ecologically costly. A good idea is hard to abandon, and we all are aware that people would rather keep doing something that has worked in the past, even when it is failing now. People also are much more concerned with losing what they have than with getting more. A given strategy gets "locked in" by political interest groups, by the difficulty and expense of changing, and by the fact that retraining people to do something different would be almost impossibly hard to do. Civilizations seem to have continued irrigating, or building monuments, or burning forest long after these had become counterproductive.

11.4 Innovation and Progress: Golden Ages and Leaden Ages

"Chalices of wood, priests of gold
  Ireland had in the church of old;
  Things today are not so good—
  Chalices of gold, priests of wood."
Traditional Irish verse, my translation

We need to separate cycles from ongoing progress, which has been slow but clearly active throughout the human career.

A consistent feature of cycles is a period at the height of the K phase in which everything goes right. Such periods occasionally rise to the level of "Golden Ages," a term made analytically useful by A. L. Kroeber's detailed analysis in his book *Configurations of Culture Growth* (1944). Today, with the world caught in a dangerous downward spiral of pollution and overexploitation, we soberly focus on decline and fall. It seems worthwhile to look in some detail at the opposite cases.

High Tang—the period between about 650 and 755—is generally regarded by Chinese as the Golden Age of premodern China. Lesser golden ages are alleged for early Han from about 180 to 87 BCE (i.e., the reigns of the Wen, Jing, and Wu emperors) and Northern Song before 1100. The early 1400s were a golden age for Ming, especially the peaceful reign of the Xuande Emperor (r. 1424–1435; Chan 1988: 304). The troubled end of Ming was, amazingly, a Golden Age for literature (producing, among other things, Feng Menglong's stories, and the novel *Jin Ping Mei*) and painting (Dong Qichang, Huang Xiangjian, many others). The terrible events of the time prevented it from being a full Golden Age of politics, science, and the like, showing limits that constrain such ages.

Japan had its Golden Ages, including the Heian and mid-Tokugawa periods. Korea reached high points in thirteenth-century Koryo and fifteenth-century Chosŏn. Central Asia, though not the Chinese portion, had its Golden Age from 800 to 1200 (Beckwith 2013; Starr 2013).

The philosophical and intellectual golden age of Warring States China is noteworthy because at the same time Greece and India did about equally well, and the Middle East was not to be ignored. This seems to have been an international event, the "Axial Age." The climate was good in Axial times (cf. Chew 2001, 2006), but not good enough to explain much. Some have argued that the Axial Age followed the coming of iron weapons and large, bureaucratic states—a deadly combination that produced far more murderous wars, and thus led philosophers to seek decent, humane alternatives (Sanderson 2008). Rise of trade and competition between emerging states was clearly involved. It was a time of ferment and of small competing states, where having an advantage in knowing useful skills or even abstract science gave some advantage.

Conversely, China hit a relatively—but only relatively—leaden age in the eighteenth century, thanks in large part to the Qianlong Emperor's literary inquisition. This seems rather like the far longer and more dismal leaden ages in the west: the Dark Ages, the Byzantine Empire (1000 years of little accomplishment), and other times with astonishingly little great science or art. The common themes are reactionary economics (notably the Byzantine Empire's basis in slave trade) and a corresponpidingly reactionary government. Much more research is needed on this matter before one can say anything definite.

Randall Collins, a sociologist attending especially to the dynamics of direct interaction, analyzed philosophy—including its peak periods—in terms of lineages and social networks (Collins 1998). Fortunately for the future of such scholarship, Kroeber and Collins are in good agreement about when the Golden Ages were and who were the "great" (influential, highly regarded) philosophers. The main difference is that Collins has the advantage of two generations of research on Asian traditions, and could add more Asian data.

An interesting conclusion of Collins is that the entire worldwide history of philosophy includes *only two* loners, Wang Chong (in first-century China) and—of all people—Ibn Khaldun. Every other significant philosopher was involved in a lineage, and usually was in a major node of intellectual networks—a great university or equivalent center. Ibn Khaldun's loner status is not his fault. Ibn Khaldun had a wide circle of talkmates, and surely would have had students but for the horribly troubled times in which he lived. He had to stay on the move. His students died in petty local wars.

Wang Chong thus emerges as the only important thinker who was a genuine loner. By his own boasting, he reveled in being a contrarian, nonconformist, and freethinker. His brilliantly incisive, rigorous writings were typically directed against the sacred "truths" of his time (such as astrology; see Wang 1907). It cost him; his brilliant discourses survived and had some influence, but were too captious to have the influence exerted by inferior but "networked" contemporaries. Wang Chong would surely have had more effect on Chinese thought if he had been more sociable.

11.5 Cycles' End

Termination comes when, as W. B. Yeats wrote: "Things fall apart; the center cannot hold." Bureaucrats and generals serve themselves and their personal groups at the expense of the state. Ordinary people become bandits, then outright rebels. The state dissolves into little eddies and local currents.

As Chinese historians noted, truly evil people become visible in leadership late in an Ibn Khaldun cycle. They compete savagely for place, even in the best of times, but especially when the legitimate rewards of high place are few and the illegitimate ones many. The competition in turn makes the bad worse, as Lord Acton's Law holds: "Power tends to corrupt, and absolute power tends to corrupt absolutely." In a declining cycle, such people mobilize the discontents, fears, anxieties, and defensive tendencies of the ordinary people. In an upbound cycle, ordinary people find more reason to hope, and are less easily mobilized by appeals to fear and hate.

Unrest spreads from marginal or poor segments of the population (always somewhat disaffected, in any regime) to more powerful groups. Revolutions, from France to the socialist and communist revolutions of the twentieth century, succeeded only when they could get intellectuals and military leaders on board. Success comes only when local unrest, whether or not initiated by the poor, spreads to such local elites as are on the outs (for whatever reason) with local governmental powers. If this succeeds, other, more neutral elites get involved. Only when a huge percentage of national elite and military persons are on board would a dynasty fall. Even the Communist revolution was not a lower-class movement; the Communists included many elite individuals, and Mao Zedong himself was a relatively affluent farmer's son turned intellectual.

At this point it is necessary to introduce some comparative material from the important work of Norman Yoffee (2005, 2010, 2012; Yoffee and Cowgill 1988). He finds that early cities were fragile. This "fragility" showed itself in their usually declining, falling to enemies, or simply being deserted, at the end of roughly one or two centuries. (This would fit Ibn Khaldun's model—one or two cycles.) Mesopotamia's early city-states—the earliest being Uruk, urban by 3300—fell within a short time, and no empire formed till Sargon of Akkad put his together in 2350 BC, but it fell before 2000. Dynasties and empires were short-lived; the Assyrian Empire rose in 1900 BC and fell by 1760, rose and fell again 1350–1170 and again 850–650; during this time the Assyrian people kept right on going. (It was the last of these that in 722 sacked Israel and carried off the Ten Lost Tribes.) The Babylonians and ancient Egyptians had similar dynamics. Yoffee compares this with the early Chinese towns and cities, which usually declined in a century or two (though Zhengzhou continued to the present—a 3500-year run). Maya, Aztec and Andean cities did little better, though Teotihuacan ruled the Mexican world from 100 BC to 600 CE and even then remained a substantial town. Chaco Canyon with 2000 people was a lesser site and thus had a lesser run: 1050 to 1140 AD. The Indus Valley civilization lasted about 2600–1900 BC, but then fell, never to rise again.

Maya civilization grew and achieved greatness in the rather optimal climate between 500 BCE and 500 CE. It survived with a hitch—a noticeable pause—the cold, dry period from 550 to 650. It then collapsed in the Medieval Warm Period, which brought massive and long-lasting droughts to the area. These droughts not only devastated agriculture, they even removed drinking water; much of the Yucatan Peninsula and Maya Lowlands is without surface water. People had to store water, dig wells, or find caves with permanent sources. These all proved inadequate in drought times (Gill 2000). Also, hotter weather led to more plant diseases, and probably more human diseases as well.

There was more. Some areas had already been devastated by war, and collapsed before the droughts (Demarest 2004). Not all the Maya world collapsed, only the central portions; the northern Yucatan Peninsula and the southern highlands continued to be urbanized and civilized while cities, literacy, and high culture disappeared in the central lowlands. The claim that the Maya collapsed because of sheer ecological folly (Diamond 1997, 2005) and the counterclaim that they did not collapse at all (McAnany and Gallareta 2010) do not bear up under investigation.

Mayaland would have recovered with the return of cooler, moister weather in the 1300s, but by then the trade routes had shifted to the coast. This is certainly one reason, possibly the only important reason, why the central lowlands never recovered.

Trade, contact, and communication had focused around the geographical center of the lowland world. After that center collapsed, trade shifted to the coasts, and stayed there, carried by canoes.

In this case, we cannot see the micropolitics—we have no way of knowing what went on in the cities, or what people said and thought as agriculture became increasingly unsuccessful.

Many other New World societies collapsed or suffered sharp setbacks during the Medieval Warm Period, which seems to have been dry very widely. It devastated the Four Corners, hit the Mississippi Valley, ruined much of the Andes, and generally caused woe. One major reason was maize, vulnerable to even brief droughts.

Of course, in all these cases, even the famed Maya "collapse," the people themselves kept going. Yoffee tells that he was once involved in an excavation of a Hopi kiva dating to around 1300 AD, with teenage Hopi helping the excavation work. The archaeologists recovered a particular item, and the Hopi workers immediately said that its companion piece must be in the pit, because they are always together. The companion turned up ten minutes later. Purely oral transmission and ritual practice had preserved the association for 700 years (Yoffee 2012).

The end of a cycle occurs when a polity's commitments increase beyond its citizens' willingness to fund them. Military overextension is merely one part of this. Corruption, as noted in Chap. 2, is the most pervasive aspect of decline. The most consistent theme that plays across all Ibn Khaldun cycles, longer cycles, and histories of civilizations is that people will work for progress: economic betterment, greater knowledge, more personal freedom, and a better life generally. However, they will work at this only so long as they can see real personal opportunity in it.

William Thompson writes: "Then as now, governments were evaluated for their ability to fulfill their coordination and protection responsibilities. They were no doubt also blamed for climate-induced economic deterioration, even if there was not much that could be done about it.... In periods of severe economic deterioration, governmental legitimacy could be expected to suffer.... Unpaid armies were more likely to rebel. Provincial governors were more likely to act more autonomously from a weakening central government...." (Thompson 2007: 167).

Marx had some inkling of cycles, from Hegel's writings. Marx saw that the easiest way to unite people (the working masses, in his case) was to unite them against a common enemy. He saw that the workers outnumbered the bosses and the rich, and therefore could win, eventually, if they hung together. He saw that they could then set up a communist state. What he did not see was that even a communist state would inevitably have leaders, who would inevitably become disunited and corrupted once they had securely taken over and eliminated the old regimes.

Military overextension is notoriously associated with downfalls, but the surprising thing is how often that is not the case. Many of the great military overextensions of history—including the Ottoman Empire under Suleyman the Magnificent, Han China under Wu Di, the British in the early twentieth century, the Mongols under Genghis Khan's grandchildren (cf. De Rachewiltz 2004; Weatherford 2004)—did not lead to immediate collapse. The empires contracted, but survived, sometimes for centuries. Conversely, most of China's dynastic collapses did not follow from military overextension; it was absent from the last years of Han, Tang, and Ming. Tang had clearly overextended militarily in the 750s, and was trimmed back savagely by expanding Islam; Tang rallied and survived.

China continues. Zhengzhou, currently rounding out 3500 years as a major regional city, may be the oldest city in the world. It is certainly the oldest city that speaks the same language (much changed, but recognizable) that it did at its founding.

11.6 Fall and Collapse of Entire Regions or Civilizations

Jared Diamond, in his famous book *Collapse* (2005), surveys collapsed cultures in hopes of proving that environmental stupidity did them in. He found only a few cases of collapse, and most were very local: the fall of Pueblo III and the Hohokam in the Southwest of North America, Cahokia in the Mississippi Valley, a few medieval Norsemen in Greenland, a couple of tiny islands in otherwise-successful Polynesia. The only large-scale collapses he could find were Angkor, the Classic Maya, and the fall of Rome.

A large percentage of the above—the pueblos, the Maya, Angkor, Cahokia, and others—collapsed during the Medieval Warm Period, specifically during periods of major drought. Dry, hot conditions were clearly part of the problem. Drought is too simplistic an explanation, but it was a major part of the picture. Diamond's claim that sheer folly was the cause does not hold up.

Moreover, to Diamond's miserably short list, we can oppose the many instances when environmental catastrophe did not do a culture in. The extreme case is Iceland, virtually depopulated over centuries by the Little Ice Age (especially in the plague sieges of the fifteenth century), and by volcanoes, especially the eruption of Laki in 1783, which led to the death of about a quarter of the population. Iceland had its share of overgrazing and overfishing also. Iceland kept right on, culture unchanged, and is now one of the richest countries. Societies are resilient; plagues and famines, for instance, never seem to bring them down (see e.g. Lawler 2010; Tainter 1988, 2006).

All Diamond's cases are problematic. A major session at the American Anthropological Association convention in 2006 took on Diamond directly. A subsequent volume (McAnany and Yoffee 2010) brought the papers together and added several new ones. A large revisionist literature has flourished since (Lawler 2010). Some of it denies collapse even in cases where no reasonable person could agree with the denial, as in the fall of Rome or the collapse of the central Maya lowlands. Even in such cases, however, collapse is relative and usually temporary. Rome is greater than ever today, and the Maya are more numerous than in their Classic Period.

Some vaunted "collapses" were merely fluctuations in long-term dynamics. The Pueblo peoples moved, but are still numerous, in new homes. Collapse is usually gradual, in which case some would not call it "collapse." Rome took centuries to decline and fall. The Maya crash took 200 years and was largely confined to the tropical lowlands.

Still, Diamond has a point. Some of the "refutations" are no more than "it was more complicated than that," a line that could be said in response to literally every meaningful statement that could be made by a human being. Others said simply that "Diamond says the glass is half empty, but we think it is half full." Ignoring these, we find much interesting material to explain. The Greenland failure was clearly due to climate change: the Little Ice Age froze the trade routes and made farming impossible after 500 years of good success (Berglund 2010). Diamond would reply that the Greenlanders could simply have switched to the Inuit lifestyle and stayed on.

Diamond's poster-child story, his tale of Easter Island, now appears to be entirely wrong (Hunt and Lipo 2006, 2010; Hunt and Rapu-Haoa 2006). People did not reach the island till 1200 A.D. They accidentally imported Southeast Asian rats (*Rattus exulans*). These ate the nuts of the palms, causing deforestation, and the eggs of birds, causing loss of those species. The people continued to survive, however, by gardening, and were fairly numerous and well-off when contacted by whites. Whites brought disease and slaving, and this, not ecocide, wiped out the Easter Islanders and their culture. (Hunt's conclusions have been challenged, but he has the evidence and his challengers are reduced to questioning it on very tenuous grounds.)

The Maya collapsed locally, but survived. The Spanish conquest was indeed facilitated by "guns, germs and steel" (Diamond 1997), but also by the Spanish success at divide-and-conquer strategies (Cahill 2010), a technique they had honed through the 800 years of war with Muslims in Iberia. In both Mexico and Peru, conquest depended on setting Indians against Indians more than on Diamond's trinity. In Peru, a tiny, poorly armed band succeeded long before disease was established there, simply by backing the weaker side in a civil war, turning the tables on the stronger side by use of "guns and steel," and then turning on their proteges and murdering or imprisoning them through treachery.

On the other hand, the European conquest of the New World was unique in all history in terms of its sheer destructiveness, and Diamond's "guns, germs, and steel" (Diamond 1997) were indeed the worst instruments. Violence and disease reduced Native American populations, typically by 90–95%; many groups disappeared entirely. This was one case in which disease really changed history. Disease assisted the conquest, but neither wiped out states by itself nor destroyed the cultures in question. Armed conquest and direct violence at first, and powerful acculturative pressure later, were necessary for those to happen. The decline of population and economy in Central and South America after 1550 or 1600 was due almost entirely to disease.

Very rare is a collapse (as opposed to an invasion with disease) that reduces the population by 90% or more. Jared Diamond seems to have found just about all the cases in history. He may have missed a few ancient New World cases (see Millon 1988 for Teotihuacan, Mexico; Tiwanaku seems a possible case, according to information displays at the site, observed by this author Jan. 2005, though it is hard to imagine decline reaching 90%).

There are a few less dramatic collapses that Diamond missed. Central Asia's great medieval cultures have left very little legacy. The Mongols survive, a small group.

Sogdian and related East Iranian languages are virtually gone. The distantly related Tokharian languages are totally extinct. But the decline of these cultures took centuries.

Similarly, ancient Egyptian civilization died out almost without trace—but not before it had flourished, with its hieroglyphics and animal gods, hundreds of years into the Greco-Roman period. Christianity and then Islam finally ended it, not through "collapse" but through religious conversions that also led to linguistic, artistic, and demographic changes. Population did not crash and the economy kept flourishing.

The fall of Rome is everyone's favorite collapse to analyze (Heather 2006). Theories ranging from lead poisoning via pipes to malaria from the marshes formed by irrigation agriculture have been seriously entertained. A German historian listed 210 alleged reasons, ranging from accedia (loss of interest in life) to excessive bathing (Ward-Perkins 2005: 33–34). Loss of topsoil was extreme in some areas, but in general Roman cultivation was land-sparing. Recall, though, that Rome depended on a far less intensive and far more environmentally sensitive agriculture than China's. Edward Gibbon (1995 [1776-1788]) saw a mix of corruption, boorish military emperors seizing power in coups, barbarian attacks, and Christian fanaticism including wars based on trivial points of doctrine. He was scathing about the war fought over *homoousia* vs. *homoiousia*, noting the doctrinal differences were no greater than those in spelling (though he also noted there were real regional economic issues behind the war).

Peter Heather (2006), among others, argued for the rising power, sophistication, cultural and social complexity, and above all military organization of the barbarians, seeing the fall of Rome as occurring when the barbarians got good at Rome's own game of organized war. It is also noted by all scholars that trade and commerce broke down (Ward-Perkins 2005; cf. Spufford 2002), though perhaps not as much as usually thought (Hills 2007, but he clearly exaggerates in the opposite direction).

More recently, Kyle Harper (2017) has stressed the decline and end of the climatic optimum which we have met in regard to the decline and fall of Han. Its end marked the beginning of Rome's decline. One may note, however, that the "barbarians" who destroyed Rome and its western empire were much more vulnerable to cold and agricultural failure than Rome was, since they had simple agriculture and were attacking from the frozen north. Obviously something more than climate was operating.

Later, the savage cold period from 536 to around 650 ruined the truncated Byzantine Empire. The empire was very susceptible to cold and drought because of the rough lands it held and the low-yield agriculture practiced there. Worse, bubonic plague, already established in central Asia and probably locally in Europe (Damgaard et al. 2018), spread in a vast and unprecedented pandemic throughout the empire in 541. The empire never recovered, partly because the expanding Persian and Islamic states would not let it. Harper notes the corruption, military overextension, and power-jockeying, but oddly does not stress the importance of slavery, in spite of his own work on Roman slaving (Harper 2011). My conclusion from reading the relevant sources and comparing modern slavery is that an empire based on slave labor and the slave trade could not compete in any way with more vibrant civic cultures. More research is obviously necessary. Harper is careful not to claim cold and plague as *the* causes, but has certainly established them as major crises that the Byzantine state could not handle.

It appears that civilizations—in the cultural sense (Yoffee 2005)—end only when conquest is extreme and total, and foreign domination continues for centuries.

Central Asia and the Near East declined from about 1300 until relatively modern times. At first this was largely because of the incredible orgies of murder that spread over them; the Mongol conquests and Tamerlane's incredible destruction (Marozzi 2004) were only the most spectacular cases. The Black Death then swept an already weakened world (Dols 1977). Trade then changed, with sea trade taking over from caravans, and the Europeans taking over the sea trade. With all this, the reactionary, anti-intellectual, anti-progress philosophy of Ash'arism (a puritanical form of Islam) appeared. It continued to increase at the expense of more enlightened and liberal Islamic traditions. (The extreme of broadly Ash'arite thought is seen today in Wahhabism—though the latter would probably have horrified the early Ash'arites.) Finally, the Little Ice Age impacted Central Asia for centuries. Truly, this was a multicausal decline. One assumes that most declines, and even most wars and meltdowns, are multicausal, with conquests, trade shifts, disease, culture, and even climate all involved.

Not only in China, but also in Japan in the sixteenth century, Korea in the eighteenth, Europe in the fourteenth, and elsewhere, "Malthusian" collapses have more to do with governmental failure than with population outrunning food production potential. In the fourteenth century, Andalucia had an agriculture modern enough and diffusible enough to save all Europe. But Europe did not copy or adapt it, partly because it was being done by the despised Muslims. So Europe starved. (For a more nuanced look at the fourteenth century crisis, see Tuchman 1978.) China and Japan rallied in the seventeenth century and fed their masses easily.

The Four Horsemen of the Apocalypse—war, famine, plague, and death (or poverty)—limit populations in premodern states. Average life expectancies in those states were 30–40 years (Bengtsson et al. 2004). Epidemics routinely ravaged Europe and China in historic times, and never brought down a single country. Claims that they "contributed" to dynastic falls in China, or transitions in Japan, because a plague occurred at the same time as a dynastic fall, are rendered silly when one recalls that East Asia had epidemics of some sort almost every year, and all remained rather local. East Asia never had anything comparable to the plague that hit Europe in 1346–48. Even that appalling plague failed to bring down a single European state or culture. Harper traces the decline of Byzantium partly to plague (Harper 2017), but Byzantium's decline was contemporary with the rise of Baghdad, Antioch, Sevilla, and other cities, plague or no. Byzantium had other problems as well as plague.

The Black Death—bubonic plague—that lowered Europe's population drastically in the 1340s (Benedictow 2016; Campbell 2016) was not just a matter of disease. It followed years of population increase without agricultural modernization. This in turn was not blind Malthusian forces at work; it was the product of a stagnating feudal system that created maximal incentives to fight, minimal incentives to develop (Tuchman 1978). Scholars like Benedictow may write as if the bubonic plague struck an innocent Europe with bomb-like force and killed millions. What really happened was that Europe's dense, poorly fed population, living in crowded and unsanitary conditions and already wracked with diseases, inevitably succumbed when a new disease triggered general collapse. The weather was extremely cold and uncertain. It was wet in the plague epicenters of Central Asia, allowing rodents to reproduce. It was devastatingly cold, locally wet, and uneven in Europe, causing harvest failures and mass starvation even before the plague struck this weakened population (Campbell 2016).

We have known for decades that the bubonic plague did not kill all the victims of the 1345–1353 epidemics (though DNA analysis has now proved that it did kill most of them). Many diseases erupted in the breakdown of what little health care and sanitation had existed as of 1346. Starvation and neglect killed many sick who might have recovered, and probably many healthy people too, especially children.

East Asia's environmental problems track bad government. Good government did what it could to fix the mess (see studies in Elvin and Liu 1998; also Von Glahn 2015). The causes of declines and collapses remain controversial, and are certainly various (Tainter 1988, 2006).

Worldwide, the causes of the fall of civilizations remain obscure and controversial (see Yoffee and Cowgill 1988 for disparate theories and accounts). All manner of mystical ideas and racist theories have been floated, only to fail when subjected to even the most minimal level of testing (Tainter 1988). Scholars have noticed that theories of collapse of earlier civilizations—notably Rome and the Maya—always seem to reflect the theorists' personal concerns about their own times. The causes for the fall of the Maya, in the literature if not in reality, neatly track the headlines of the twentieth century: revolution, ecological exhaustion, big government, war, climate change, and so on.

Several different types of collapse can be identified.

Type 0. Normal governmental change (democratic, successional, or by revolving-door coup as in China's fractured states between 906 and 960); not an actual decline, fall, or significant change.

Type 1. "Regime change" with major policy and governance change, often some degree of chaos or violence and with imposition of conquerors' language, but ultimately no radical or long-lasting effect: This sort of change was common in ancient Mesopotamia; Diamond called many cases "collapses." Yoffee (2010) properly pointed out that they were no such thing. The cycles of East Asia could be classified here. They certainly did not involve collapse of the civilization.

Type 2. Major, relatively permanent redirection, involving change in policy and governance, but without major violence or social breakdown. This occurred in China with the rise of Han Wu Di in 140 BCE, the end of the Yongle period of the Ming Dynasty and the reversal of Yongle policies, and several other times.

Type 3. Civil war and massive redirection of government, but no major population loss or Dark Age period. This never happened in China.

Type 4: Genuine fall or collapse leading to long Dark Ages but not to the end of cultures or civilizations. This occurred in China 220–581 and 906–960, and in much of Central Asia after 1400. It also occurred in the central Maya lowlands after 900.

Type 5: Total obliteration of an entire civilization. The few notable examples are the endings of the early Mesopotamian, Hittite, and ancient Egyptian civilizations. This never occurred in most of East Asia, but it did happen in southwest China, where the Dian culture and subsequent Nanzhao state (with its unique civilization) were conquered by the Mongols and disappeared.

References

Anderson, E. N. (1988). *The food of China*. New Haven, CT: Yale University Press.
Barfield, T. J. (1989). *The perilous frontier: Nomadic empires and China, 221 BC to AD 1757*. Cambridge, MA: Blackwell.
BBC News. (2007, August 14). Map reveals ancient urban sprawl. *BBC News Online*.
Beckwith, C. (2013). *Warriors of the cloisters: The central Asian origins of science in the medieval world*. Princeton: Princeton University Press.
Bello, D. (2016). *Across forest, steppe, and mountain: Environment, identity and empire in Qing China's borderlands*. Cambridge: Cambridge University Press.
Benedictow, O. (2016). *The black death and subsequent plague epidemics in Scandinavian countries: Perspectives and controversies*. Berlin: De Gruyter.
Bengtsson, T., Campbell, C., & Lee, J. A. (Eds.). (2004). *Life under pressure: Mortality and living standards in Europe and Asia, 1700–1900*. Cambridge, MA: MIT Press.
Berglund, J. (2010). Did the Medieval Norse Society in Greenland really fail? In P. A. McAnany & N. Yoffee (Eds.), *Questioning collapse: Human resilience, ecological vulnerability, and the aftermath of empire* (pp. 45–70). Cambridge: Cambridge University Press.
Cahill, D. (2010). Advanced Andeans and backward Europeans: Structure and agency in the collapse of the Inca empire. In P. A. McAnany & N. Yoffee (Eds.), *Questioning collapse: Human resilience, ecological vulnerability, and the aftermath of empire* (pp. 207–238). Cambridge: Cambridge University Press.
Campbell, B. M. S. (2016). *The great transition: Climate, disease and society in the Late-Medieval World*. Cambridge: Cambridge University Press.
Chan, H.-L. (1988). The Chien-Wen, Yung-Lo, Hung-His, and Hsüan-Te Reigns, 1399-1435. In F. Mote & D. Twitchett (Eds.), *Cambridge history of China, Vol. 7, The Ming Dynasty, 1368–1644, Part 1* (pp. 182–304). Cambridge: Cambridge University Press.
Chase-Dunn, C., Hall, T. D., & Turchin, P. (2007). World-systems in the biogeosphere: Urbanization, state formation and climate change since the iron age. In A. Hornborg & C. Crumley (Eds.), *The world system and the earth system: Global socioenvironmental change and sustainability since the Neolithic* (pp. 132–148). Walnut Creek, CA: Left Coast Press.
Chew, S. (2001). *World ecological degradation: Accumulation, urbanization, and deforestation*. Lanham, MD: Alta Mira Press, Rowman and Littlefield Publishers.
Chew, S. (2006). Dark ages. In B. Gills & W. Thompson (Eds.), *Globalization and global history* (pp. 163–202). London: Routledge.
Collins, R. (1998). *The sociology of philosophies*. Cambridge, MA: Harvard University Press.
Damgaard, P. d. B., Marchi, N., Rasmussen, S., Peyrot, M., Renaud, G., Kradin, N., & Willerslev, E. (2018). 137 ancient human genomes from across the Eurasian steppes. *Nature, 557*, 369–374.
Demarest, A. (2004). *Ancient Maya: The rise and fall of a rainforest civilization*. Cambridge: Cambridge University Press.

De Rachewiltz, I. (2004). *The secret history of the Mongols: A Mongolian epic chronicle of the thirteenth century* (Vol. 2, 1347 pp). Leiden: Brill.
Diamond, J. (1997). *Guns, germs and steel: The fates of human societies*. New York: W.W. Norton.
Diamond, J. (2005). *Collapse: How societies choose to fail or succeed*. New York: Viking.
Dols, M. (1977). *The black death in the Middle East*. Princeton: Princeton University Press.
Elvin, M. (2004). *The retreat of the elephants: An environmental history of China*. New Haven: Yale University Press.
Elvin, M., & Liu, T.-J. (Eds.). (1998). *Sediments of time: Environment and society in Chinese history*. Cambridge: Cambridge University Press.
Gibbon, E. (1995[1776–1788]). *The decline and fall of the Roman empire*. New York: Penguin.
Gill, R. (2000). *The great Maya droughts*. Albuquerque: University of New Mexico Press.
Harper, K. (2011). *Slavery in the late Roman world, AD 275–425*. Cambridge: Cambridge University Press.
Harper, K. (2017). *The fate of Rome: Climate, disease, and the end of an empire*. Princeton: Princeton University Press.
Heather, P. (2006). *The fall of the Roman empire: A new history of Rome and the barbarians*. London: Oxford University Press.
Hunt, T. L., & Lipo, C. P. (2006). Late colonization of Easter Island. *Science, 311*, 1603–1606.
Hunt, T. L., & Lipo, C. P. (2010). Ecological catastrophe, collapse, and the myth of 'ecocide' on Rapa Nui (Easter Island). In P. A. McAnany & N. Yoffee (Eds.), *Questioning collapse: Human resilience, ecological vulnerability, and the aftermath of empire* (pp. 21–44). Cambridge: Cambridge University Press.
Hunt, T. L., & Rapu-Haoa, S. (2006). *What (really) happened on Rapa Nui? Ecological Catastrophe and cultural collapse*. In Paper, American Anthropological Association, Annual Conference, San Jose, CA.
Janssen, M. A., & Scheffer, M. (2004). Overexploitation of renewable resources by ancient societies and the role of sunk-cost effects. *Ecology and Society, 9*(1), 6.
Kant, I. (1978). *Anthropology from a pragmatic point of view* (V. L. Dowdell, Trans.) (Ger. Orig. 1798). Carbondale: Southern Illinois University Press.
Kroeber, A. L. (1944). *Configurations of culture growth*. Berkeley: University of California Press.
Lawler, A. (2010). Collapse? What collapse? Societal change revisited. *Science, 300*, 907–908.
Li, L. (2007). *Fighting famine in North China: State, market, and environmental decline, 1690s–1990s*. Stanford: Stanford University Press.
Lieberman, V. (2003). *Strange parallels: Southeast Asia in global context, c. 800–1830. Vol. 1: Integration on the mainland*. Cambridge: Cambridge University Press.
Lieberman, V. (2009). *Strange parallels: Southeast Asia in global context, c. 800–1830. Vol. 2: Mainland mirrors: Europe, Japan, China, South Asia, and the islands*. Cambridge: Cambridge University Press.
Marozzi, J. (2004). *Tamerlane: Sword of Islam, conqueror of the world*. London: Da Capo.
McAnany, P., & T. Gallareta N. (2010). Bellicose rulers and climatological peril? Retrofitting twenty-first-century woes on eighth-century Maya Society. In P. McAnany & N. Yoffee (Eds.), *Questioning collapse: Human resilience, ecological vulnerability, and the aftermath of empire* (pp. 142–175). Cambridge: Cambridge University Press.
McAnany, P. A., & Yoffee, N. (Eds.). (2010). *Questioning collapse: Human resilience, ecological vulnerability, and the aftermath of empire*. Cambridge: Cambridge University Press.
McNeill, J. R. (1992). *The mountains of the Mediterranean world*. Cambridge: Cambridge University Press.
Millon, R. (1988). The last years of Teotihuacan dominance. In N. Yoffee & G. L. Cowgill (Eds.), *The collapse of ancient states and civilizations* (pp. 102–164). Tucson: University of Arizona Press.
Moore, J. (2015). *Capitalism in the web of life: Ecology and the accumulation of capital*. New York: Verso.

Pines, Y. (2009). *Envisioning eternal empire: Chinese political thought of the warring states era*. Honolulu: University of Hawai'i Press.
Pines, Y. (2012). *The everlasting empire: The political culture of ancient China and its imperial legacy*. Princeton: Princeton University Press.
Pomeranz, K. (2000). *The great divergence: China, Europe, and the making of the modern world economy*. Princeton, NJ: Princeton University Press.
Pomeranz, K. (2010). Calamities without collapse: Environment, economy, and society in China, ca. 1800–1949. In P. A. McAnany & N. Yoffee (Eds.), *Questioning collapse: Human resilience, ecological vulnerability, and the aftermath of empire* (pp. 71–110). Cambridge: Cambridge University Press.
Sanderson, S. (2008). Adaptation, evolution, and religion. *Religion, 38*, 141–156.
Starr, S. F. (2013). *Lost enlightenment: Central Asia's Golden age from the Arab conquest to Tamerlane*. Princeton: Princeton University Press.
Tainter, J. A. (1988). *The collapse of complex societies*. Cambridge: Cambridge University Press.
Tainter, J. A. (2006). Archaeology of overshoot and collapse. *Annual Reviews in Anthropology, 35*, 59–74.
Thompson, W. (2007). Climate, water, and political-economic crises in ancient Mesopotamia and Egypt. In A. Hornborg & C. Crumley (Eds.), *The world system and the earth system: Global socioenvironmental change and sustainability since the Neolithic* (pp. 163–179). Walnut Creek, CA: Left Coast Press.
Tuchman, B. (1978). *A distant mirror: The calamitous 14th century*. New York: Knopf.
Von Glahn, R. (2015). *The economic history of China*. Cambridge: Cambridge University Press.
Wang, C. (1907). *Lun-Heng* (F. Alfred, Trans.). Leipzig: Otto Harrassowitz.
Weatherford, J. (2004). *Genghis Khan and the making of the modern world*. New York: Three Rivers Press.
Yoffee, N. (2005). *Myths of the archaic state*. Cambridge: Cambridge University Press.
Yoffee, N. (2010). Collapse in ancient Mesopotamia: What happened, what didn't. In P. A. McAnany & N. Yoffee (Eds.), *Questioning collapse: Human resilience, ecological vulnerability, and the aftermath of empire* (pp. 176–203). Cambridge: Cambridge University Press.
Yoffee, N. (2012, January 6). *The evolution of fragility*. Talk, UCLA (Fowler Museum).
Yoffee, N., & Cowgill, G. (Eds.). (1988). *The collapse of ancient states and civilizations*. Tucson: University of Arizona Press.

What East Asia's Dynamics Teach Us about Climate, Society, and Change in the Modern and Future World

12

> "The only thing anyone ever learned from the study of history is that no one ever learned anything from the study of history." G.W. Hegel (as quoted by Hayden White 1987: 82)

Today, the world is facing climate change on an unprecedented scale. The basic change is global warming, due partly to solar radiation increase (still rebounding from the Little Ice Age) but largely to the effects of greenhouse gases, especially CO_2 but also methane, water vapor, and industrial pollutant gases. The world is expected to be about 2 °C warmer by 2100. This will not be evenly distributed. Colder areas—high latitudes and altitudes—will warm more. World rainfall will increase, but will be concentrated in already rainy areas, and probably in monsoon lands like East Asia. The drier parts of the world will get notably drier and hotter. It is highly likely that vast tracts of the Middle East, North Africa, and the American Southwest will be virtually uninhabitable by 2100, because of extreme heat and lack of water. Recent studies show that if CO_2 doubles, world temperature will rise about 2.8 °C (Cox et al. 2018).

Rising sea levels will deepen the world oceans by perhaps as much as a meter by 2100, but opinions differ considerably about actual sea-level changes, because of difficulties in predicting thermal expansion, heat radiation, ice melting, and other variables.

An editorial in *Nature* (2018) warns us not to ascribe conflict to climate without hard evidence. "Many studies that link global warming to civil unrest are biased and exacerbate stigma" (p. 275). They note that Syria's civil war has been blamed on climate (drought), but neighboring Jordan had no such war. Sudan's conflicts have similarly been blamed on climate, but neighboring countries had no such trouble and Sudan has a long history of violent repression.

A study by Wendy Barnaby (2009) showed that wars do not normally start over water; people would rather negotiate. Whether this will be true in future, as resources truly run out and negotiation offers no hope, is unclear. At least we know that, so far, the links between climate change, resource shortage, and conflict are indirect and unclear.

Fresh water development is utterly inadequate to deal with the predicted needs of the 2040s. The Colorado, Rio Grande, Nile, and Yellow Rivers rarely or never reach the sea. So much of the Amu Darya is taken for irrigation that the Aral Sea has almost entirely dried up. The same fate is overtaking the formerly huge Lake Urmia in Iran. None of the drier parts of the world is dealing with its future water problems. Seawater desalination is the only serious hope, but it remains expensive, and research on it is surprisingly thin. Meanwhile, increasingly desperate local shortages are managed by using groundwater, overdrafting resources that are often irreplaceable "fossil" water from the Pleistocene, and at best require many years of recharge for every year of current downdraft.

The best projections of availability of game, fish, and other wild resources is that they will be commercially extinct by 2050. Africa's wildlife is disappearing as fast as America's did in the late nineteenth century. Sea fish will be effectively extinct after 2050; almost all fish will come from aquaculture (Worm 2016). Forest resources will be depleted by that time, except for the boreal forests of Russia, Scandinavia, Canada, and limited parts of the United States. These boreal forests will soon follow other forests into oblivion, however, because of global warming. Tropical forests are disappearing at the rate of many square kilometers a day. Reserves are rarely protected once there is any pressure on them.

Medical research is not being adequately targeted toward new diseases of humans, crops, or domestic animals, or the rapid evolution of antibiotic-resistant bacteria. Without major increases in research in those areas, the world will certainly be afflicted with enormous epidemics affecting humans, major crop plants, and livestock. As traditional varieties of crops are abandoned and a rapidly shrinking number of agribusiness firms controls world seed markets, monocrop cultivation of only a very few varieties of staple crops becomes the rule. This is frankly suicidal; it guarantees epidemics.

Giant corporations become ever more proficient at "externalizing" their costs of production, i.e., passing the costs on to the public. Subsidies, lax enforcement of rules, special tax breaks, special exemption from pollution control laws, and other governmental favors are now critical to the survival of the fossil-fuel corporations and similar firms. They plow much of the subsidy money back into lobbying and campaign donations, and control whole governments around the world. They thus succeed in passing the vast majority of their real costs onto the general public. This not only costs the public; it enables the giant firms to stay highly profitable even though their real cost/benefit ratios are fantastically unfavorable.

All these problems are soluble, but solving them is prevented by the lock-ins characteristic of the omega phase of a cycle. The world economy is dominated by giant primary-production interests—private or governmental oil corporations, state enterprises, mining multinationals, agribusiness conglomerates, and the like. These deploy antiquated policies. They control politics by teaming up with ideologies of hatred, usually either Communism (as in China and North Korea) or extremist right-wing religions: Salafi and Wahhabi Islam, right-wing Christianity, ultra-orthodox

Judaism, Burmese Buddhism, and so on. If this continues, there is no hope of adequate attention to needed changes. Overfishing, unsustainable logging, wasteful water use, increasing pollution, decreasing medical research, and decline of investment in human capital will all continue. The modern world's commitments to oil, coal, groundwater pumping, water-intensive crops in desert areas, and ever heavier agrochemical use are examples of economic lock-in. We know we are destroying ourselves in the long run, but have too much invested to stop. They will drive collapse and perhaps open space for rebirth.

Ibn Khaldun cycles are still occurring. Industrialization, capitalism, the Enlightenment, and other modern phenomena have not changed them. Clearest are the cycles of the United States. We have also observed chaos alternating with stability in modern China and in many other modern states. Chaotic times—Holling's alpha periods—reset the clock. The Great Depression, WWII, and decolonization reset clocks for most of the world. The United States had its great change of rule in 1932. Mexico also reset its clock about that time with the leadership of Lazaro Cardenas. Most of Europe, and Japan, had their great changes of regimes in 1945. India followed in 1948, China in 1949, and other countries in due course as they threw off colonial chains or modernized their governance.

This allows us to predict serious stress points. The United States suffered a dramatic regime change in 2016–2017, with the collapse of the 240-year-old Enlightenment tradition of freedom and justice, and the rise of a fascist regime dedicated to eliminating those benefits. The fantastic levels of corruption in the modern United States guarantee collapse. Most of the corruption is quite legal. Firms donate all they wish to politicians who vote on the issues that are critical to those firms, up to and including actual government subsidies. It should be obvious that a legislator should be forced to recuse himself from voting for subsidies for a firm that funds his campaign. No such rules exist. The inevitable result is that the most evil interests—the ones that depend heavily on subsidies and that most harm the general good—are the biggest donors: fossil fuels, mining, corporate large-scale monoculture, arms and weaponry, and shady enterprises like gambling and the riskier forms of high finance.

Ibn Khaldun's theory predicts serious problems involving regime change in most of the world between 2020 and 2050. There is much evidence that this will indeed be the case. Governments today are turning away from agricultural research and innovation. More and more of the world's agriculture is controlled by giant agribusiness firms. These firms become dinosaurs: too big to conquer but too reactionary to survive. The world they create is the world of the plantations of old: enslaved or starvation-waged labor, political tyranny and corruption, massive lying by the government, and massive manipulation of racist, ethnic, religious, and political hatred to divide the public against itself and leave the giant reactionary firms in place.

We are also facing the classic feature of the omega phase of a cycle: more and more people seeing their best interest lies in working for themselves at the expense of the common good. We have lost Smithian competition—competition to see who can better serve the wants and needs of consumers—and entered a stage of Randian competition: competition to see who can most effectively cheat the consumers by developing oligopolies, externalizing costs, corrupting politicians, and, in general, making their money by acting at the expense of the collectivity.

The future of most countries is already visible in the current situations of North Korea, China, Iran, Saudi Arabia, Sudan, and several other nations: a dinosauric oligarchy (oil interests in three of those five cases) ruling through appeal to extremist ideologies. Turkey, Poland, Hungary, Venezuela, the United States, Russia, and many other countries are moving rapidly in the same direction. Only ethnic or religious extremism can produce faithful and motivated bureaucrats to serve such a corrupt regime. In the United States, for instance, the stupidity and unreliability of ordinary white supremacists and racists forces the government to rely on right-wing religious interests when there are not enough trained servants available from oil and other primary production sectors. Elite overproduction is evident in most of the countries, for example in the huge and internally fractured Saudi royal family and in the higher levels of American administration. World hunger problems are due to failure of governments to invest in agriculture and other subsistence production, not to population density per se. Many governments, including those mentioned above, but especially the United States, show the military overstretch of a polity in its omega phase. Climate change, population growth, resource depletion, unsustainable agriculture that is beginning to collapse, and insidious increase of diseases are the far-back story, the long-term and wide-flung stresses that lie behind increasing conflict, fear, stress, hatred, and violence.

Contra current intellectual fads, this is not a matter of "capitalism" (still less of "neoliberalism," a term so corrupted by misuse that it can no longer be said to have any agreed-on meaning). Capitalist countries do indeed show all the problems, but so do communist and socialist ones. The nearest to adequate coping is found in the mixed systems of Scandinavia, but the Scandinavian countries too are overfishing, extracting oil, and otherwise acting against their long-term interests.

The result, unless there is a total reversal of the course of events worldwide, will be increasing autocracy, leading inevitably to loss of accountability, increasing selfishness among elites, corruption, reactionary policies, and consequent worsening of the resource crisis. Progressive fraying of the production sectors will lead to more and more unrest, with coups, rebellions, and meltdowns occurring. These in turn will produce more and more genocides. Genocide has been the increasingly common response of challenged autocracies in this overpopulated world, where need to save one's labor force has been replaced by need to thin out an unmanageable excess of mouths to feed. The incredible callousness of the Republican Party in the United States over failures of health care and immigration management, and resulting unnecessary deaths, shows that this is not restricted to actual murder by police states. Similar "bureaupathy" leads to massive displacement by big dams and other projects, and to mass famines due to inept or evil food policies (Howard-Hassman 2016).

The Enlightenment gave the world the ideals of democracy, freedom, and rapid discovery science, and these spread to East Asia and have taken root there to varying degrees. We now face the hope of a second Enlightenment, with falling birth rates, more efficient use of resources, and sharp limitations on waste, pollution, and destruction of natural resources. This has also come to East Asia, but has only begun to take effect. Instead of moving in that direction, most countries today are consolidating power in the hands of fewer and fewer people—autocratic governments and giant firms.

East Asia in dynastic times showed the classic pattern of agrarian empires: Autocratic rule through bureaucracy over a vast mass of relatively independent farmers and other workers. The Enlightenment then came with the rise of the trade-commerce-communication economy. Enlightenment governance now succeeds in the outward-looking, trade-based margins of East Asia: South Korea and Taiwan. The counter-Enlightenment of tyranny, corruption, surveillance, and genocide has dominated China and North Korea for most of the last several decades. China's experiment with opening its economy and politics has been substantially reversed, despite successes.

Governments are increasingly resorting to genocide, and above all to calculated and targeted mass starvation and displacement, to manage their populations. In a world where international war is increasingly dangerous and deadly, and where population has expanded beyond any possible accommodation, governments deal with their resource problems by thinning their own population rather than by the time-honored method of trying to take from the neighbors. Mass starvation has become common as a method of governance (Howard-Hassman 2016), and can be expected to become more so.

The recent successes of China, North Korea, Iran, Cuba, Myanmar, and many other countries at maintaining brutal dictatorships over many decades have shown that current techniques of power maintenance—surveillance, intimidation, oppression, imprisonment, genocide, and more—are exceedingly effective. The age of revolutions is over. Current violent civil unrest in Afghanistan, Syria, Sudan, South Sudan, Somalia, and other nations are leading nowhere except to give more and more power to the most extreme and anti-democratic forces. Tunisia is the only country in the twenty-first century to have had a revolutionary change for the better.

The ecological and environmental stresses that will occur after 2020 will not be met by flexible open regimes, but, in most countries, by dictatorships governed—de facto—by fossil fuel, mining, and agribusiness corporations. However, regime change after a cycle is inevitable, and the effect of successful control will be to make disaffected elites even more violent and powerful when they eventually arise and stage coups or revolutions.

This book has demonstrated that the problems of economic and technological lock-in, governmental sclerosis, corruption by productive interests, Malthusian squeeze, increasing conservatism in the face of increasing need for change, and failure of leadership are all part of one highly predictable cyclic pattern. That pattern has revealed itself in every civilization in history, and is so close to Holling's general cycle of natural populations that one must conclude it is part of a general pattern.

The coming resource squeeze, worldwide, will most likely return the world and all its larger nations to the agrarian pattern: economic stagnation, income stagnation, and Malthusian squeeze. With rapid advance of wealth checked by progressive exhaustion of resources, the natural response will be a return to the agrarian imperial pattern: a tiny ruling class with unimaginable wealth; a small bureaucracy, well off but somewhat insecure, below it; a small middle class; a vast mass—about two-thirds of the population—at the margin of subsistence; and an underclass of 10–20% of the population who are at the edge of starvation, and who usually succumb to starvation, malnutrition, and disease at an early age. This is what East Asia's states and every other agrarian state endured for thousands of years, and it is by far the most probable future for a world where resources are depleted and frontiers closed. The major difference from agrarian empires is that today's successful surveillance and repression, combined with the routine use of extremist ethnic and religious fanaticism to marshal support, will lead to more effective control and much more killing.

In the end, the kernel of an Ibn Khaldun collapse is that the interests of the rich and powerful become increasingly incompatible with the interests of the rest, and the government remains too dependent on the rich and powerful to deal with that conundrum.

This does *not* mean that it is necessary. It could be fixed easily and quickly if people could break out of the lock-ins created by giant primary-production interests. That will take new leaders who can deploy 'asabiyah in an environmental and social revolution. Marxism, inseparably wedded to heavy industry and the fossil-fuel economy, is inadequate. Current "green" ideologies have been visionary and idealistic, but singularly lacking in plans for getting from "here" to "there." We need a generation of Genghis Khans: leaders who can organize to create a new world. They must be peaceful ecological innovators, not wild warlords, but they must have Genghis' ability to combine grand vision with pragmatic organization.

References

Barnaby, W. (2009). Do nations go to war over water? *Nature, 458*, 282–283.
Cox, P. M., Huntingford, C., & Williamson, M. S. (2018). Emergent constraint on equilibrium climate sensitivity from global temperature variability. *Nature, 553*, 319–322.
Howard-Hassman, R. E. (2016). *State food crimes*. Cambridge: Cambridge University Press.
Nature Editorial. (2018). Climate conflict. *Nature, 554*, 275–276.
White, H. (1987). *The content of the form*. Baltimore: Johns Hopkins University Press.
Worm, B. (2016). Averting a global fisheries disaster. *Proceedings of the National Academy of Sciences, 113*, 4895–4897.